Graduate Texts in Mathematics

(continued after index)

Mark R. Sepanski

Compact Lie Groups

Springer

Mark R. Sepanski
Department of Mathematics
Baylor University
Waco, TX 76798-7328
USA
mark_sepanski@baylor.edu

Mathematics Subject Classification (2000): 22E15, 17B20

ISBN-13: 978-1-4419-2138-3
e-ISBN-10: 0-387-49158-9
e-ISBN-13: 978-0-387-49158-5

Printed on acid-free paper.

springer.com   (TXQ/SB)

To Laura, Sarah, Ben, and Shannon

Contents

Preface

As an undergraduate, I was offered a reading course on the representation theory of finite groups. When I learned this basically meant studying homomorphisms from groups into matrices, I was not impressed. In its place I opted for a reading course on the much more glamorous sounding topic of multilinear algebra. Ironically, when I finally took a course on representation theory from B. Kostant in graduate school, I was immediately captivated.

In broad terms, representation theory is simply the study of symmetry. In practice, the theory often begins by classifying all the ways in which a group acts on vector spaces and then moves into questions of decomposition, unitarity, geometric realizations, and special structures. In general, each of these problems is extremely difficult. However in the case of compact Lie groups, answers to most of these questions are well understood. As a result, the theory of compact Lie groups is used extensively as a stepping stone in the study of noncompact Lie groups.

Regarding prerequisites for this text, the reader must first be familiar with the definition of a group and basic topology. Secondly, elementary knowledge of differential geometry is assumed. Students lacking a formal course in manifold theory will be able to follow most of this book if they are willing to take a few facts on faith. This mostly consists of accepting the existence of an invariant integral in §1.4.1. In a bit more detail, the notion of a submanifold is used in §1.1.3, the theory of covering spaces is used in §1.2, §1.3, §4.2.3, and §7.3.6, integral curves are used in §4.1.2, and Frobenius' theorem on integral submanifolds is used in the proof of Theorem 4.14. A third prerequisite is elementary functional analysis. Again, students lacking formal course work in this area can follow most of the text if they are willing to assume a few facts. In particular, the Spectral Theorem for normal bounded operators is used in the proof of Theorem 3.12, vector-valued integration is introduced in §3.2.2, and the Spectral Theorem for compact self-adjoint operators is used in the proof of Lemma 3.13.

The text assumes no prior knowledge of Lie groups or Lie algebras and so all the necessary theory is developed here. Students already familiar with Lie groups can quickly skim most of Chapters 1 and 4. Similarly, students familiar with Lie algebras can quickly skim most of Chapter 6.

The book is organized as follows. Chapter 1 lays out the basic definitions, examples, and theory of compact Lie groups. Though the construction of the spin groups in §1.3 is very important to later representation theory and mathematical physics, this material can be easily omitted on a first reading. Doing so allows for a more rapid transition to the harmonic analysis in Chapter 3. A similar remark holds for the construction of the spin representations in §2.1.2.4. Chapter 2 introduces the concept of a finite-dimensional representation. Examples, Schur's Lemma, unitarity, and the canonical decomposition are developed here. Chapter 3 begins with matrix coefficients and character theory. It culminates in the celebrated Peter–Weyl Theorem and its corresponding Fourier theory.

Up through Chapter 3, the notion of a Lie algebra is unnecessary. In order to progress further, Chapter 4 takes up their study. Since this book works with compact Lie groups, it suffices to consider linear Lie groups which allows for a fair amount of differential geometry to be bypassed. Chapter 5 examines maximal tori and Cartan subalgebras. The Maximal Torus Theorem, Dynkin's Formula, the Commutator Theorem, and basic structural results are given. Chapter 6 introduces weights, roots, the Cartan involution, the Killing form, the standard $\mathfrak{sl}(2, \mathbb{C})$, various lattices, and the Weyl group. Chapter 7 uses all this technology to prove the Weyl Integration Formula, the Weyl Character Formula, the Highest Weight Theorem, and the Borel–Weil Theorem.

Since this work is intended as a textbook, most references are given only in the bibliography. The interested reader may consult [61] or [34] for brief historical outlines of the theory. With that said, there are a number of resources that had a powerful impact on this work and to which I am greatly indebted. First, the excellent lectures of B. Kostant and D. Vogan shaped my view of the subject. Notes from those lectures were used extensively in certain sections of this text. Second, any book written by A. Knapp on Lie theory is a tremendous asset to all students in the field. In particular, [61] was an extremely valuable resource. Third, although many other works deserve recommendation, there are four outstanding texts that were especially influential: [34] by Duistermaat and Kolk, [72] by Rossmann, [70] by Onishchik and Vinberg, and [52] by Hoffmann and Morris. Many thanks also go to C. Conley who took up the onerous burden of reading certain parts of the text and making helpful suggestions. Finally, the author is grateful to the Baylor Sabbatical Committee for its support during parts of the preparation of this text.

Mark Sepanski

March 2006

Compact Lie Groups

1

Compact Lie Groups

1.1 Basic Notions

1.1.1 Manifolds

Lie theory is the study of symmetry springing from the intersection of algebra, analysis, and geometry. Less poetically, Lie groups are simultaneously groups and manifolds. In this section, we recall the definition of a manifold (see [8] or [88] for more detail). Let $n \in \mathbb{N}$.

Definition 1.1. An n-dimensional *topological manifold* is a second countable (i.e., possessing a countable basis for the topology) Hausdorff topological space M that is locally homeomorphic to an open subset of $\mathbb{R}^n$.

This means that for all $m \in M$ there exists a homeomorphism $\varphi : U \to V$ for some open neighborhood U of m and an open neighborhood V of $\mathbb{R}^n$. Such a homeomorphism φ is called a *chart*.

Definition 1.2. An n-dimensional smooth *manifold* is a topological manifold M along with a collection of charts, $\{\varphi_\alpha : U_\alpha \to V_\alpha\}$, called an *atlas*, so that
(1) $M = \cup_\alpha U_\alpha$ and
(2) For all α, β with $U_\alpha \cap U_\beta \neq \emptyset$, the *transition map* $\varphi_{\alpha,\beta} = \varphi_\beta \circ \varphi_\alpha^{-1} : \varphi_\alpha(U_\alpha \cap U_\beta) \to \varphi_\beta(U_\alpha \cap U_\beta)$ is a smooth map on $\mathbb{R}^n$.

It is an elementary fact that each atlas can be completed to a unique maximal atlas containing the original. By common convention, a manifold's atlas will always be extended to this completion.

Besides $\mathbb{R}^n$, common examples of manifolds include the *n-sphere*,

$$S^n = \{x \in \mathbb{R}^{n+1} \mid \|x\| = 1\},$$

where $\|\cdot\|$ denotes the standard Euclidean norm, and the *n-torus*,

$$T^n = \underbrace{S^1 \times S^1 \times \cdots \times S^1}_{n \text{ copies}}.$$

Another important manifold is real *projective space*, $\mathbb{P}(\mathbb{R}^n)$, which is the n-dimensional compact manifold of all lines in $\mathbb{R}^{n+1}$. It may be alternately realized as $\mathbb{R}^{n+1}\backslash\{0\}$ modulo the equivalence relation $x \sim \lambda x$ for $x \in \mathbb{R}^{n+1}\backslash\{0\}$ and $\lambda \in \mathbb{R}\backslash\{0\}$, or as S^n modulo the equivalence relation $x \sim \pm x$ for $x \in S^n$. More generally, the *Grassmannian*, $\mathrm{Gr}_k(\mathbb{R}^n)$, consists of all k-planes in $\mathbb{R}^n$. It is a compact manifold of dimension $k(n-k)$ and reduces to $\mathbb{P}(\mathbb{R}^{n-1})$ when $k = 1$.

Write $M_{n,m}(\mathbb{F})$ for the set of $n \times m$ matrices over $\mathbb{F}$ where $\mathbb{F}$ is either $\mathbb{R}$ or $\mathbb{C}$. By looking at each coordinate, $M_{n,m}(\mathbb{R})$ may be identified with $\mathbb{R}^{nm}$ and $M_{n,m}(\mathbb{C})$ with $\mathbb{R}^{2nm}$. Since the determinant is continuous on $M_{n,n}(\mathbb{F})$, we see $\det^{-1}\{0\}$ is a closed subset. Thus the *general linear group*

$$(1.3) \qquad GL(n, \mathbb{F}) = \{g \in M_{n,n}(\mathbb{F}) \mid g \text{ is invertible}\}$$

is an open subset of $M_{n,n}(\mathbb{F})$ and therefore a manifold. In a similar spirit, for any finite-dimensional vector space V over $\mathbb{F}$, we write $GL(V)$ for the set of invertible linear transformations on V.

1.1.2 Lie Groups

Definition 1.4. A *Lie group* G is a group and a manifold so that
(1) the *multiplication* map $\mu : G \times G \to G$ given by $\mu(g, g') = gg'$ is smooth and
(2) the *inverse* map $\iota : G \to G$ by $\iota(g) = g^{-1}$ is smooth.

A trivial example of a Lie group is furnished by $\mathbb{R}^n$ with its additive group structure. A slightly fancier example of a Lie group is given by S^1. In this case, the group structure is inherited from multiplication in $\mathbb{C}\backslash\{0\}$ via the identification

$$S^1 \cong \{z \in \mathbb{C} \mid |z| = 1\}.$$

However, the most interesting example of a Lie group so far is $GL(n, \mathbb{F})$. To verify $GL(n, \mathbb{F})$ is a Lie group, first observe that multiplication is smooth since it is a polynomial map in the coordinates. Checking that the inverse map is smooth requires the standard linear algebra formula $g^{-1} = \mathrm{adj}(g)/\det g$, where the $\mathrm{adj}(g)$ is the transpose of the matrix of cofactors. In particular, the coordinates of $\mathrm{adj}(g)$ are polynomial functions in the coordinates of g and $\det g$ is a nonvanishing polynomial on $GL(n, \mathbb{F})$ so the inverse is a smooth map.

Writing down further examples of Lie groups requires a bit more machinery. In fact, most of our future examples of Lie groups arise naturally as subgroups of $GL(n, \mathbb{F})$. To this end, we next develop the notion of a Lie subgroup.

1.1.3 Lie Subgroups and Homomorphisms

Recall that an (immersed) *submanifold* N of M is the image of a manifold N' under an injective immersion $\varphi : N' \to M$ (i.e., a one-to-one smooth map whose differential has full rank at each point of N') together with the manifold structure on N

making $\varphi : N' \to N$ a diffeomorphism. It is a familiar fact from differential geometry that the resulting topology on N may not coincide with the relative topology on N as a subset of M. A submanifold N whose topology agrees with the relative topology is called a *regular* (or *imbedded*) submanifold.

Defining the notion of a Lie subgroup is very similar. Essentially the word homomorphism needs to be thrown in.

Definition 1.5. A *Lie subgroup* H of a Lie group G is the image in G of a Lie group H' under an injective immersive homomorphism $\varphi : H' \to G$ together with the Lie group structure on H making $\varphi : H' \to H$ a diffeomorphism.

The map φ in the above definition is required to be smooth. However, we will see in Exercise 4.13 that it actually suffices to verify that φ is continuous.

As with manifolds, a Lie subgroup is *not* required to be a regular submanifold. A typical example of this phenomenon is constructed by wrapping a line around the torus at an irrational angle (Exercise 1.5). However, regular Lie subgroups play a special role and there happens to be a remarkably simple criterion for determining when Lie subgroups are regular.

Theorem 1.6. *Let G be a Lie group and $H \subseteq G$ a subgroup (with no manifold assumption). Then H is a regular Lie subgroup if and only if H is closed.*

The proof of this theorem requires a fair amount of effort. Although some of the necessary machinery is developed in §4.1.2, the proof lies almost entirely within the purview of a course on differential geometry. For the sake of clarity of exposition and since the result is only used to efficiently construct examples of Lie groups in §1.1.4 and §1.3.2, the proof of this theorem is relegated to Exercise 4.28. While we are busy putting off work, we record another useful theorem whose proof, for similar reasons, can also be left to a course on differential geometry (e.g., [8] or [88]). We note, however, that a proof of this result follows almost immediately once Theorem 4.6 is established.

Theorem 1.7. *Let H be a closed subgroup of a Lie group G. Then there is a unique manifold structure on the quotient space G/H so the projection map $\pi : G \to G/H$ is smooth, and so there exist local smooth sections of G/H into G.*

Pressing on, an immediate corollary of Theorem 1.6 provides an extremely useful method of constructing new Lie groups. The corollary requires the well-known fact that when $f : H \to M$ is a smooth map of manifolds with $f(H) \subseteq N$, N a *regular* submanifold of M, then $f : H \to N$ is also a smooth map (see [8] or [88]).

Corollary 1.8. *A closed subgroup of a Lie group is a Lie group in its own right with respect to the relative topology.*

Another common method of constructing Lie groups depends on the Rank Theorem from differential geometry.

Definition 1.9. A *homomorphism of Lie groups* is a smooth homomorphism between two Lie groups.

Theorem 1.10. *If G and G' are Lie groups and $\varphi : G \to G'$ is a homomorphism of Lie groups, then φ has constant rank and $\ker \varphi$ is a (closed) regular Lie subgroup of G of dimension $\dim G - \mathrm{rk}\,\varphi$ where $\mathrm{rk}\,\varphi$ is the rank of the differential of φ.*

Proof. It is well known (see [8]) that if a smooth map φ has constant rank, then $\varphi^{-1}\{e\}$ is a closed regular submanifold of G of dimension $\dim G - \mathrm{rk}\,\varphi$. Since $\ker \varphi$ is a subgroup, it suffices to show that φ has constant rank. Write l_g for left translation by g. Because φ is a homomorphism, $\varphi \circ l_g = l_{\varphi(g)} \circ \varphi$, and since l_g is a diffeomorphism, the rank result follows by taking differentials. □

1.1.4 Compact Classical Lie Groups

With the help of Corollary 1.8, it is easy to write down new Lie groups. The first is the *special linear group*

$$SL(n, \mathbb{F}) = \{g \in GL(n, \mathbb{F}) \mid \det g = 1\}.$$

As $SL(n, \mathbb{F})$ is a closed subgroup of $GL(n, \mathbb{F})$, it follows that it is a Lie group.

Using similar techniques, we next write down four infinite families of compact Lie groups collectively known as the *classical compact Lie groups*: $SO(2n + 1)$, $SO(2n)$, $SU(n)$, and $Sp(n)$.

1.1.4.1 $SO(n)$ The *orthogonal group* is defined as

$$O(n) = \{g \in GL(n, \mathbb{R}) \mid g^t g = I\},$$

where g^t denotes the transpose of g. The orthogonal group is a closed subgroup of $GL(n, \mathbb{R})$, so Corollary 1.8 implies that $O(n)$ is a Lie group. Since each column of an orthogonal matrix is a unit vector, we see that topologically $O(n)$ may be thought of as a closed subset of $S^{n-1} \times S^{n-1} \times \cdots \times S^{n-1} \subseteq \mathbb{R}^{n^2}$ (n copies). In particular, $O(n)$ is a compact Lie group.

The *special orthogonal group* (or *rotation group*) is defined as

$$SO(n) = \{g \in O(n) \mid \det g = 1\}.$$

This is a closed subgroup of $O(n)$, and so $SO(n)$ is also a compact Lie group.

Although not obvious at the moment, the behavior of $SO(n)$ depends heavily on the parity of n. This will become pronounced starting in §6.1.4. For this reason, the special orthogonal groups are considered to embody two separate infinite families: $SO(2n + 1)$ and $SO(2n)$.

1.1.4.2 $SU(n)$ The *unitary group* is defined as

$$U(n) = \{g \in GL(n, \mathbb{C}) \mid g^*g = I\},$$

where g^* denotes the complex conjugate transpose of g. The unitary group is a closed subgroup of $GL(n, \mathbb{C})$, and so $U(n)$ is a Lie group. As each column of a unitary matrix is a unit vector, we see that $U(n)$ may be thought of, topologically, as a closed subset of $S^{2n-1} \times S^{2n-1} \times \cdots \times^{2n-1} \subseteq \mathbb{R}^{2n^2}$ (n copies). In particular, $U(n)$ is a compact Lie group.

Likewise, the *special unitary group* is defined as

$$SU(n) = \{g \in U(n) \mid \det g = 1\}.$$

As usual, this is a closed subgroup of $U(n)$, and so $SU(n)$ is also a compact Lie group. The special case of $n = 2$ will play an especially important future role. It is straightforward to check (Exercise 1.8) that

$$(1.11) \qquad SU(2) = \left\{ \begin{pmatrix} a & -\overline{b} \\ b & \overline{a} \end{pmatrix} \mid a, b \in \mathbb{C} \text{ and } |a|^2 + |b|^2 = 1 \right\}$$

so that topologically $SU(2) \cong S^3$.

1.1.4.3 $Sp(n)$ The final compact classical Lie group, the *symplectic group*, ought to be defined as

$$(1.12) \qquad Sp(n) = \{g \in GL(n, \mathbb{H}) \mid g^*g = I\},$$

where $\mathbb{H} = \{a + ib + jc + kd \mid a, b, c, d \in \mathbb{R}\}$ denotes the *quaternions* and g^* denotes the quaternionic conjugate transpose of g. However, $\mathbb{H}$ is a noncommutative division algebra, so understanding the meaning of $GL(n, \mathbb{H})$ takes a bit more work. Once this is done, Equation 1.12 will become the honest definition of $Sp(n)$.

To begin, view $\mathbb{H}^n$ as a *right* vector space with respect to scalar multiplication and let $M_{n,n}(\mathbb{H})$ denote the set of $n \times n$ matrices over $\mathbb{H}$. By using matrix multiplication on the *left*, $M_{n,n}(\mathbb{H})$ may therefore be identified with the set of $\mathbb{H}$-linear transformations of $\mathbb{H}^n$. Thus the old definition of $GL(n, \mathbb{F})$ in Equation 1.3 can be carried over to define $GL(n, \mathbb{H}) = \{g \in M_{n,n}(\mathbb{H}) \mid g \text{ is an invertible transformation of } \mathbb{H}^n\}$.

Verifying that $GL(n, \mathbb{H})$ is a Lie group, unfortunately, requires more work. In the case of $GL(n, \mathbb{F})$ in §1.1.2, that work was done by the determinant function which is no longer readily available for $GL(n, \mathbb{H})$. Instead, we embed $GL(n, \mathbb{H})$ into $GL(2n, \mathbb{C})$ as follows.

Observe that any $v \in \mathbb{H}$ can be uniquely written as $v = a + jb$ for $a, b \in \mathbb{C}$. Thus there is a well-defined $\mathbb{C}$-linear isomorphism $\vartheta : \mathbb{H}^n \to \mathbb{C}^{2n}$ given by $\vartheta(v_1, \ldots, v_n) = (a_1, \ldots, a_n, b_1, \ldots, b_n)$ where $v_p = a_p + jb_p$, $a_p, b_p \in \mathbb{C}$. Use this to define a $\mathbb{C}$-linear injection of algebras $\widetilde{\vartheta} : M_{n,n}(\mathbb{H}) \to M_{n,n}(\mathbb{C})$ by $\widetilde{\vartheta}X = \vartheta \circ X \circ \vartheta^{-1}$ for $X \in M_{n,n}(\mathbb{H})$ with respect to the usual identification of matrices as linear maps. It is straightforward to verify (Exercise 1.12) that when X is uniquely written as $X = A + jB$ for $A, B \in M_{n,n}(\mathbb{C})$, then

$$(1.13) \qquad \widetilde{\vartheta}(A + jB) = \begin{pmatrix} A & -\overline{B} \\ B & \overline{A} \end{pmatrix},$$

where $\overline{A}$ denotes complex conjugation of A. Thus $\widetilde{\vartheta}$ is a $\mathbb{C}$-linear algebra isomorphism from $M_{n,n}(\mathbb{H})$ to

$$M_{2n,2n}(\mathbb{C})_{\mathbb{H}} \equiv \{ \begin{pmatrix} A & -\overline{B} \\ B & \overline{A} \end{pmatrix} \mid A, B \in M_{n,n}(\mathbb{C}) \}.$$

An alternate way of checking this is to first let r_j denote scalar multiplication by j on $\mathbb{H}^n$, i.e., *right* multiplication by j. It is easy to verify (Exercise 1.12) that $\vartheta r_j \vartheta^{-1} z = J\overline{z}$ for $z \in \mathbb{C}^{2n}$ where

$$J = \begin{pmatrix} 0 & -I_n \\ I_n & 0 \end{pmatrix}.$$

Since ϑ is a $\mathbb{C}$-linear isomorphism, the image of $\widetilde{\vartheta}$ consists of all $Y \in M_{2n,2n}(\mathbb{C})$ commuting with $\vartheta r_j \vartheta^{-1}$ so that $M_{2n,2n}(\mathbb{C})_{\mathbb{H}} = \{ Y \in M_{2n}(\mathbb{C}) \mid YJ = J\overline{Y} \}$.

Finally, observe that X is invertible if and only if $\widetilde{\vartheta}X$ is invertible. In particular, $M_{n,n}(\mathbb{H})$ may be thought of as $\mathbb{R}^{4n^2}$ and, since $\det \circ \widetilde{\vartheta}$ is continuous, $GL(n, \mathbb{H})$ is the open set in $M_{n,n}(\mathbb{H})$ defined by the complement of $(\det \circ \widetilde{\vartheta})^{-1}\{0\}$. Since $GL(n, \mathbb{H})$ is now clearly a Lie group, Equation 1.12 shows that $Sp(n)$ is a Lie group by Corollary 1.8. As with the previous examples, $Sp(n)$ is compact since each column vector is a unit vector in $\mathbb{H}^n \cong \mathbb{R}^{4n}$.

As an aside, Dieudonné developed the notion of determinant suitable for $M_{n,n}(\mathbb{H})$ (see [2], 151–158). This quaternionic determinant has most of the nice properties of the usual determinant and it turns out that elements of $Sp(n)$ always have determinant 1.

There is another useful realization for $Sp(n)$ besides the one given in Equation 1.12. The isomorphism is given by $\widetilde{\vartheta}$ and it remains only to describe the image of $Sp(n)$ under $\widetilde{\vartheta}$. First, it is easy to verify (Exercise 1.12) that $\widetilde{\vartheta}(X^*) = (\widetilde{\vartheta}X)^*$ for $X \in M_{n,n}(\mathbb{H})$, and thus $\widetilde{\vartheta}Sp(n) = U(2n) \cap M_{2n,2n}(\mathbb{C})_{\mathbb{H}}$. This answer can be reshaped further. Define

$$Sp(n, \mathbb{C}) = \{ g \in GL(2n, \mathbb{C}) \mid g^t Jg = J \}$$

so that $U(2n) \cap M_{2n,2n}(\mathbb{C})_{\mathbb{H}} = U(2n) \cap Sp(n, \mathbb{C})$. Hence $\widetilde{\vartheta}$ realizes the isomorphism:

$$(1.14) \qquad Sp(n) \cong U(2n) \cap M_{2n,2n}(\mathbb{C})_{\mathbb{H}}$$
$$= U(2n) \cap Sp(n, \mathbb{C}).$$

1.1.5 Exercises

Exercise 1.1 Show that S^n is a manifold that can be equipped with an atlas consisting of only two charts.

Exercise 1.2 **(a)** Show that $\mathrm{Gr}_k(\mathbb{R}^n)$ may be realized as the rank k elements of $M_{n,k}(\mathbb{R})$ modulo the equivalence relation $X \sim Xg$ for $X \in M_{n,k}(\mathbb{R})$ of rank k and $g \in GL(k, \mathbb{R})$. Find another realization showing that $\mathrm{Gr}_k(\mathbb{R}^n)$ is compact.
(b) For $S \subseteq \{1, 2, \ldots, n\}$ with $|S| = k$ and $X \in M_{n,k}(\mathbb{R})$, let $X|_S$ be the $k \times k$ matrix obtained from X by keeping only those rows indexed by an element of S, let $U_S = \{X \in M_{n,k}(\mathbb{R}) \mid X|_S \text{ is invertible}\}$, and let $\varphi_S : U_S \to M_{(n-k),k}(\mathbb{R})$ by $\varphi_S(X) = [X(X|_S)^{-1}]|_{S^c}$. Use these definitions to show that $\mathrm{Gr}_k(\mathbb{R}^n)$ is a $k(n - k)$ dimensional manifold.

Exercise 1.3 **(a)** Show that conditions (1) and (2) in Definition 1.4 may be replaced by the single condition that the map $(g_1, g_2) \to g_1 g_2^{-1}$ is smooth.
(b) In fact, show that condition (1) in Definition 1.4 implies condition (2).

Exercise 1.4 If U is an open set containing e in a Lie group G, show there exists an open set $V \subseteq U$ containing e, so $VV^{-1} \subseteq U$, where VV^{-1} is $\{vw^{-1} \mid v, w \in V\}$.

Exercise 1.5 Fix $a, b \in \mathbb{R}\backslash\{0\}$ and consider the subgroup of T^2 defined by $R_{a,b} = \{(e^{2\pi iat}, e^{2\pi ibt}) \mid t \in \mathbb{R}\}$.
(a) Suppose $\frac{a}{b} \in \mathbb{Q}$ and $\frac{a}{b} = \frac{p}{q}$ for relatively prime $p, q \in \mathbb{Z}$. As t varies, show that the first component of $R_{a,b}$ wraps around S^1 exactly p-times, while the second component wraps around q-times. Conclude that $R_{a,b}$ is closed and therefore a regular Lie subgroup diffeomorphic to S^1.
(b) Suppose $\frac{a}{b} \notin \mathbb{Q}$. Show that $R_{a,b}$ wraps around infinitely often without repeating. Conclude that $R_{a,b}$ is a Lie subgroup diffeomorphic to $\mathbb{R}$, but *not* a regular Lie subgroup (c.f. Exercise 5.*).
(c) What happens if a or b is 0?

Exercise 1.6 **(a)** Use Theorem 1.10 and the map $\det : GL(n, \mathbb{R}) \to \mathbb{R}$ to give an alternate proof that $SL(n, \mathbb{R})$ is a Lie group and has dimension $n^2 - 1$.
(b) Show the map $X \to XX^t$ from $GL(n, \mathbb{R})$ to $\{X \in M_{n,n}(\mathbb{R}) \mid X^t = X\}$ has constant rank $\frac{n(n+1)}{2}$. Use the proof of Theorem 1.10 to give an alternate proof that $O(n)$ is a Lie group and has dimension $\frac{n(n-1)}{2}$.
(c) Use the map $X \to XX^*$ on $GL(n, \mathbb{C})$ to give an alternate proof that $U(n)$ is a Lie group and has dimension n^2.
(d) Use the map $X \to XX^*$ on $GL(n, \mathbb{H})$ to give an alternate proof that $Sp(n)$ is a Lie group and has dimension $2n^2 + n$.

Exercise 1.7 For a Lie group G, write $Z(G) = \{z \in G \mid zg = gz, \text{ all } g \in G\}$ for the *center* of G. Show
(a) $Z(U(n)) \cong S^1$ and $Z(SU(n)) \cong \mathbb{Z}/n\mathbb{Z}$ for $n \geq 2$,
(b) $Z(O(2n)) \cong \mathbb{Z}/2\mathbb{Z}$, $Z(SO(2n)) \cong \mathbb{Z}/2\mathbb{Z}$ for $n \geq 2$, and $Z(SO(2)) = SO(2)$,
(c) $Z(O(2n + 1)) \cong \mathbb{Z}/2\mathbb{Z}$ for $n \geq 1$, and $Z(SO(2n + 1)) = \{I\}$ for $n \geq 1$,
(d) $Z(Sp(n)) \cong \mathbb{Z}/2\mathbb{Z}$.

Exercise 1.8 Verify directly Equation 1.11.

Exercise 1.9 (a) Let $A \subseteq GL(n, \mathbb{R})$ be the subgroup of diagonal matrices with positive elements on the diagonal and let $N \subseteq GL(n, \mathbb{R})$ be the subgroup of upper triangular matrices with 1's on the diagonal. Using Gram-Schmidt orthogonalization, show multiplication induces a diffeomorphism of $O(n) \times A \times N$ onto $GL(n, \mathbb{R})$. This is called the *Iwasawa* or *KAN* decomposition for $GL(n, \mathbb{R})$. As topological spaces, show that $GL(n, \mathbb{R}) \cong O(n) \times \mathbb{R}^{\frac{n(n+1)}{2}}$. Similarly, as topological spaces, show that $SL(n, \mathbb{R}) \cong SO(n) \times \mathbb{R}^{\frac{(n+2)(n-1)}{2}}$.
(b) Let $A \subseteq GL(n, \mathbb{C})$ be the subgroup of diagonal matrices with positive real elements on the diagonal and let $N \subseteq GL(n, \mathbb{C})$ be the subgroup of upper triangular matrices with 1's on the diagonal. Show that multiplication induces a diffeomorphism of $U(n) \times A \times N$ onto $GL(n, \mathbb{C})$. As topological spaces, show $GL(n, \mathbb{C}) \cong U(n) \times \mathbb{R}^{n^2}$. Similarly, as topological spaces, show that $SL(n, \mathbb{C}) \cong SU(n) \times \mathbb{R}^{n^2-1}$.

Exercise 1.10 Let $N \subseteq GL(n, \mathbb{C})$ be the subgroup of upper triangular matrices with 1's on the diagonal, let $\overline{N} \subseteq GL(n, \mathbb{C})$ be the subgroup of lower triangular matrices with 1's on the diagonal, and let W be the subgroup of permutation matrices (i.e., matrices with a single one in each row and each column and zeros elsewhere). Use Gaussian elimination to show $GL(n, \mathbb{C}) = \bigsqcup_{w \in W} \overline{N} w N$. This is called the *Bruhat* decomposition for $GL(n, \mathbb{C})$.

Exercise 1.11 (a) Let $P \subseteq GL(n, \mathbb{R})$ be the set of positive definite symmetric matrices. Show that multiplication gives a bijection from $P \times O(n)$ to $GL(n, \mathbb{R})$.
(b) Let $H \subseteq GL(n, \mathbb{C})$ be the set of positive definite Hermitian matrices. Show that multiplication gives a bijection from $H \times U(n)$ to $GL(n, \mathbb{C})$.

Exercise 1.12 (a) Show that $\widetilde{\vartheta}$ is given by the formula in Equation 1.13.
(b) Show $\vartheta r_j \vartheta^{-1} z = J \overline{z}$ for $z \in \mathbb{C}^{2n}$.
(c) Show that $\widetilde{\vartheta}(X^*) = (\widetilde{\vartheta} X)^*$ for $X \in M_{n,n}(\mathbb{H})$.

Exercise 1.13 For $v, u \in \mathbb{H}^n$, let $(v, u) = \sum_{p=1}^n v_p \overline{u_p}$.
(a) Show that $(Xv, u) = (v, X^* u)$ for $X \in M_{n,n}(\mathbb{H})$.
(b) Show that $Sp(n) = \{g \in M_n(\mathbb{H}) \mid (gv, gu) = (v, u), \text{ all } v, u \in \mathbb{H}^n\}$.

1.2 Basic Topology

1.2.1 Connectedness

Recall that a topological space is *connected* if it is not the disjoint union of two nonempty open sets. A space is *path connected* if any two points can be joined by a continuous path. While in general these two notions are distinct, they are equivalent for manifolds. In fact, it is even possible to replace continuous paths with smooth paths.

The first theorem is a technical tool that will be used often.

Theorem 1.15. *Let G be a connected Lie group and U a neighborhood of e. Then U generates G, i.e., $G = \bigcup_{n=1}^{\infty} U^n$ where U^n consists of all n-fold products of elements of U.*

Proof. We may assume U is open without loss of generality. Let $V = U \cap U^{-1} \subseteq U$ where U^{-1} is the set of all inverses of elements in U. This is an open set since the inverse map is continuous. Let $H = \cup_{n=1}^{\infty} V^n$. By construction, H is an open subgroup containing e. For $g \in G$, write $gH = \{gh \mid h \in H\}$. The set gH contains g and is open since left multiplication by g^{-1} is continuous. Thus G is the union of all the open sets gH. If we pick a representative $g_\alpha H$ for each coset in G/H, then $G = \amalg_\alpha (g_\alpha H)$. Hence the connectedness of G implies that G/H contains exactly one coset, i.e., $eH = G$, which is sufficient to finish the proof. $\qquad\square$

We still lack general methods for determining when a Lie group G is connected. This shortcoming is remedied next.

Definition 1.16. If G is a Lie group, write G^0 for the connected component of G containing e.

Lemma 1.17. *Let G be a Lie group. The connected component G^0 is a regular Lie subgroup of G. If G^1 is any connected component of G with $g_1 \in G^1$, then $G^1 = g_1 G^0$.*

Proof. We prove the second statement of the lemma first. Since left multiplication by g_1 is a homeomorphism, it follows easily that $g_1 G^0$ is a connected component of G. But since $e \in G^0$, this means that $g_1 \in g_1 G^0$ so $g_1 G^0 \cap G^1 \neq \emptyset$. Since both are connected components, $G^1 = g_1 G^0$ and the second statement is finished.

Returning to the first statement of the lemma, it clearly suffices to show that G^0 is a subgroup. The inverse map is a homeomorphism, so $(G^0)^{-1}$ is a connected component of G. As above, $(G^0)^{-1} = G^0$ since both components contain e. Finally, if $g_1 \in G^0$, then the components $g_1 G^0$ and G^0 both contain g_1 since $e, g_1^{-1} \in G^0$. Thus $g_1 G^0 = G^0$, and so G^0 is a subgroup, as desired. $\qquad\square$

Theorem 1.18. *If G is a Lie group and H a connected Lie subgroup so that G/H is also connected, then G is connected.*

Proof. Since H is connected and contains e, $H \subseteq G^0$, so there is a continuous map $\pi : G/H \to G/G^0$ defined by $\pi(gH) = gG^0$. It is trivial that G/G^0 has the discrete topology with respect to the quotient topology. The assumption that G/H is connected forces $\pi(G/H)$ to be connected, and so $\pi(G/H) = eG^0$. However, π is a surjective map so $G/G^0 = eG^0$, which means $G = G^0$. $\qquad\square$

Definition 1.19. Let be G a Lie group and M a manifold.
(1) An *action* of G on M is a smooth map from $G \times M \to M$, denoted by $(g, m) \to g \cdot m$ for $g \in G$ and $m \in M$, so that:
    (i) $e \cdot m = m$, all $m \in M$ and
    (ii) $g_1 \cdot (g_2 \cdot m) = (g_1 g_2) \cdot m$ for all $g_1, g_2 \in G$ and $m \in M$.
(2) The action is called *transitive* if for each $m, n \in M$, there is a $g \in G$, so $g \cdot m = n$.
(3) The *stabilizer* of $m \in M$ is $G^m = \{g \in G \mid g \cdot m = m\}$.

If G has a transitive action on M and $m_0 \in M$, then it is clear (Theorem 1.7) that the action of G on m_0 induces a diffeomorphism from G/G^{m_0} onto M.

Theorem 1.20. *The compact classical groups, $SO(n)$, $SU(n)$, and $Sp(n)$, are connected.*

Proof. Start with $SO(n)$ and proceed by induction on n. As $SO(1) = \{1\}$, the case $n = 1$ is trivial. Next, observe that $SO(n)$ has a transitive action on S^{n-1} in $\mathbb{R}^n$ by matrix multiplication. For $n \geq 2$, the stabilizer of the north pole, $N = (1, 0, \ldots, 0)$, is easily seen to be isomorphic to $SO(n - 1)$ which is connected by the induction hypothesis. From the transitive action, it follows that $SO(n)/SO(n)^N \cong S^{n-1}$ which is also connected. Thus Theorem 1.18 finishes the proof.

For $SU(n)$, repeat the above argument with $\mathbb{R}^n$ replaced by $\mathbb{C}^n$ and start the induction with the fact that $SU(1) \cong S^1$. For $Sp(n)$, repeat the same argument with $\mathbb{R}^n$ replaced by $\mathbb{H}^n$ and start the induction with $Sp(1) \cong \{v \in \mathbb{H} \mid |v| = 1\} \cong S^3$.   □

1.2.2 Simply Connected Cover

For a connected Lie group G, recall that the *fundamental group*, $\pi_1(G)$, is the homotopy class of all loops at a fixed base point. The Lie group G is called *simply connected* if $\pi_1(G)$ is trivial.

Standard covering theory from topology and differential geometry (see [69] and [8] or [88] for more detail) says that there exists a unique (up to isomorphism) *simply connected cover* $\widetilde{G}$ of G, i.e., a connected, simply connected manifold $\widetilde{G}$ with a *covering* (or *projection*) map $\pi : \widetilde{G} \to G$. Recall that being a covering map means π is a smooth surjective map with the property that each $g \in G$ has a connected neighborhood U of g in G so that the restriction of π to each connected component of $\pi^{-1}(U)$ is a diffeomorphism onto U.

Lemma 1.21. *If H is a discrete normal subgroup of a connected Lie group G, then H is contained in the center of G.*

Proof. For each $h \in H$, consider $C_h = \{ghg^{-1} \mid g \in G\}$. Since C_h is the continuous image of the connected set G, C_h is connected. Normality of H implies $C_h \subseteq H$. Discreteness of H and connectedness of C_h imply that C_h is a single point. As h is clearly in C_h, this shows that $C_h = \{h\}$, and so h is central.   □

Theorem 1.22. *Let G be a connected Lie group.*
(1) The connected simply connected cover $\widetilde{G}$ is a Lie group.
(2) If π is the covering map and $\widetilde{Z} = \ker \pi$, then $\widetilde{Z}$ is a discrete central subgroup of $\widetilde{G}$.
(3) π induces a diffeomorphic isomorphism $G \cong \widetilde{G}/\widetilde{Z}$.
(4) $\pi_1(G) \cong \widetilde{Z}$.

Proof. Because coverings satisfy the *lifting property* (e.g., for any smooth map f of a connected simply connected manifold M to G with $m_0 \in M$ and $g_0 \in \pi^{-1}(f(m_0))$, there exists a unique smooth map $\widetilde{f} : M \to \widetilde{G}$ satisfying $\pi \circ \widetilde{f} = f$ and $\widetilde{f}(m_0) = g_0$), the Lie group structure on G lifts to a Lie group structure on $\widetilde{G}$, making π a homomorphism. To see this, consider the map $s : \widetilde{G} \times \widetilde{G} \to G$ by $f(\widetilde{g}, \widetilde{h}) =$

$\pi(\widetilde{g})\pi(\widetilde{h})^{-1}$ and fix some $\widetilde{e} \in \pi^{-1}(e)$. Then there is a unique lift $\widetilde{s} : \widetilde{G} \times \widetilde{G} \to \widetilde{G}$ so that $\pi \circ \widetilde{s} = s$. To define the group structure $\widetilde{G}$, let $\widetilde{h}^{-1} = \widetilde{s}(\widetilde{e}, \widetilde{h})$ and $\widetilde{g}\widetilde{h} = \widetilde{s}(\widetilde{g}, \widetilde{h}^{-1})$. It is straightforward to verify that this structure makes $\widetilde{G}$ into a Lie group and π into a homomorphism (Exercise 1.21).

Hence we have constructed a connected simply connected Lie group $\widetilde{G}$ and a covering homomorphism $\pi : \widetilde{G} \to G$. Since π is a covering and a homomorphism, $\widetilde{Z} = \ker \pi$ is a discrete normal subgroup of $\widetilde{G}$ and so central by Lemma 1.21. Hence π induces a diffeomorphic isomorphism from $\widetilde{G}/\widetilde{Z}$ to G. The statement regarding $\pi_1(G)$ is a standard result from the covering theory of deck transformations (see [8]). □

Lemma 1.23. $Sp(1)$ and $SU(2)$ are simply connected and isomorphic to each other. Either group is the simply connected cover of $SO(3)$, i.e., $SO(3)$ is isomorphic to $Sp(1)/\{\pm 1\}$ or $SU(2)/\{\pm I\}$.

Proof. The isomorphism from $Sp(1)$ to $SU(2)$ is given by $\widetilde{\vartheta}$ in §1.1.4.3. Since either group is topologically S^3, the first statement follows.

For the second statement, write $(\cdot, \cdot)$ for the real inner product on $\mathbb{H}$ given by $(u, v) = \mathrm{Re}(u\bar{v})$ for $u, v \in \mathbb{H}$. By choosing an orthonormal basis $\{1, i, j, k\}$, we may identify $\mathbb{H}$ with $\mathbb{R}^4$ and $(\cdot, \cdot)$ with the standard Euclidean dot product on $\mathbb{R}^4$. Then $1^\perp = \{v \in \mathbb{H} \mid (1, v) = 0\}$ is the set of *imaginary* (or *pure*) *quaternions*, $\mathrm{Im}(\mathbb{H})$, spanned over $\mathbb{R}$ by $\{i, j, k\}$. In particular, we may identify $O(3)$ with $O(\mathrm{Im}(\mathbb{H})) \equiv \{\mathbb{R}\text{-linear maps } T : \mathrm{Im}(\mathbb{H}) \to \mathrm{Im}(\mathbb{H}) \mid (Tu, Tv) = (u, v) \text{ all } u, v \in \mathrm{Im}(\mathbb{H})\}$ and the connected component $O(\mathrm{Im}(\mathbb{H}))^0$ with $SO(3)$.

Define a smooth homomorphism $\mathrm{Ad} : Sp(1) \to O(\mathrm{Im}(\mathbb{H}))^0$ by $(\mathrm{Ad}(g))(u) = gu\bar{g}$ for $g \in Sp(1)$ and $u \in \mathrm{Im}(\mathbb{H})$. To see this is well defined, first view $\mathrm{Ad}(g)$ as an $\mathbb{R}$-linear transformation on $\mathbb{H}$. Using the fact that $g\bar{g} = 1$ for $g \in Sp(1)$, it follows immediately that $\mathrm{Ad}(g)$ leaves $(\cdot, \cdot)$ invariant. As $\mathrm{Ad}(g)$ fixes 1, $\mathrm{Ad}(g)$ preserves $\mathrm{Im}(\mathbb{H})$. Thus $\mathrm{Ad}(g) \in O(\mathrm{Im}(\mathbb{H}))^0$ since $Sp(1)$ is connected.

It is well known that $SO(3)$ consists of all rotations (Exercise 1.22). To show Ad is surjective, it therefore suffices to show that each rotation lies in the image of Ad. Let $v \in \mathrm{Im}(\mathbb{H})$ be a unit vector. Then v can be completed to a basis $\{v, u, w\}$ of $\mathrm{Im}(\mathbb{H})$ sharing the same properties as the $\{i, j, k\}$ basis. It is a simple calculation to show that $\mathrm{Ad}(\cos\theta + v \sin\theta)$ fixes v and is a rotation through an angel of 2θ in the uw plane (Exercise 1.23). Hence Ad is surjective. The same calculation also shows that $\ker \mathrm{Ad} = \{\pm 1\}$. Since the simply connected cover is unique, the proof is finished. □

In §6.3.3 we develop a direct method for calculating $\pi_1(G)$. For now we compute the fundamental group for the classical compact Lie groups by use of a higher homotopy exact sequence.

Theorem 1.24. (1) $\pi_1(SO(2)) \cong \mathbb{Z}$ and $\pi_1(SO(n)) \cong \mathbb{Z}/2\mathbb{Z}$ for $n \geq 3$.
(2) $SU(n)$ is simply connected for $n \geq 2$.
(3) $Sp(n)$ is simply connected for $n \geq 1$.

Proof. Start with $SO(n)$. As $SO(2) \cong S^1$, $\pi_1(SO(2)) \cong \mathbb{Z}$. Recall from the proof of Theorem 1.20 that $SO(n)$ has a transitive action on S^{n-1} with stabilizer isomorphic to $SO(n-1)$. From the resulting exact sequence, $\{1\} \to SO(n-1) \to SO(n) \to S^{n-1} \to \{1\}$, there is a long exact sequence of higher homotopy groups (e.g., see [51] p. 296)

$$\cdots \to \pi_2(S^{n-1}) \to \pi_1(SO(n-1)) \to \pi_1(SO(n)) \to \pi_1(S^{n-1}) \to \cdots.$$

For $n \geq 3$, $\pi_1(S^{n-1})$ is trivial, so there is an exact sequence

$$\pi_2(S^{n-1}) \to \pi_1(SO(n-1)) \to \pi_1(SO(n)) \to \{1\}.$$

Since $\pi_2(S^{n-1})$ is trivial for $n \geq 4$, induction on the exact sequence implies $\pi_1(SO(n)) \cong \pi_1(SO(3))$ for $n \geq 4$. It only remains to show that $\pi_1(SO(3)) \cong \mathbb{Z}/2\mathbb{Z}$, but this follows from Lemma 1.23 and Theorem 1.22.

For $SU(n)$, as in the proof of Theorem 1.20, there is an exact sequence $\{1\} \to SU(n-1) \to SU(n) \to S^{2n-1} \to \{1\}$. Since $\pi_1(S^{2n-1})$ and $\pi_2(S^{2n-1})$ are trivial for $n \geq 3$ (actually for $n = 2$ as well, though not useful here), the long exact sequence of higher homotopy groups implies that $\pi_1(SU(n)) \cong \pi_1(SU(2))$ for $n \geq 2$. By Lemma 1.23, $\pi_1(SU(2))$ is trivial.

For $Sp(n)$, the corresponding exact sequence is $\{1\} \to Sp(n-1) \to Sp(n) \to S^{4n-1} \to \{1\}$. Since $\pi_1(S^{4n-1})$ and $\pi_2(S^{4n-1})$ are trivial for $n \geq 2$ (actually for $n = 1$ as well), the resulting long exact sequence implies $\pi_1(Sp(n)) \cong \pi_1(Sp(1))$ for $n \geq 1$. By Lemma 1.23, $\pi_1(Sp(1))$ is trivial. $\quad\square$

As an immediate corollary of Theorems 1.22 and 1.24, there is a connected simply connected double cover of $SO(n)$, $n \geq 3$. That simply connected Lie group is called $\mathrm{Spin}_n(\mathbb{R})$ and it fits in the following exact sequence:

$$(1.25) \qquad \{1\} \to \mathbb{Z}/2\mathbb{Z} \to \mathrm{Spin}_n(\mathbb{R}) \to SO(n) \to \{I\}.$$

Lemma 1.23 shows $\mathrm{Spin}_3(\mathbb{R}) \cong SU(2) \cong Sp(1)$. For larger n, an explicit construction of $\mathrm{Spin}_n(\mathbb{R})$ is given in §1.3.2.

1.2.3 Exercises

Exercise 1.14 For a connected Lie group G, show that even if the second countable hypothesis is omitted from the definition of manifold, G is still second countable.

Exercise 1.15 Show that an open subgroup of a Lie group is closed.

Exercise 1.16 Show that $GL(n, \mathbb{C})$ and $SL(n, \mathbb{C})$ are connected.

Exercise 1.17 Show that $GL(n, \mathbb{R})$ has two connected components: $GL(n, \mathbb{R})^0 = \{g \in GL(n, \mathbb{R}) \mid \det g > 0\}$ and $\{g \in GL(n, \mathbb{R}) \mid \det g < 0\}$. Prove $SL(n, \mathbb{R})$ is connected.

Exercise 1.18 Show $O(2n + 1) \cong SO(2n + 1) \times (\mathbb{Z}/2\mathbb{Z})$ as both a manifold and a group. In particular, $O(2n + 1)$ has two connected components with $O(2n + 1)^0 = SO(2n + 1)$.

Exercise 1.19 (a) Show $O(2n) \cong SO(2n) \times (\mathbb{Z}/2\mathbb{Z})$ as a manifold. In particular, $O(2n)$ has two connected components with $O(2n)^0 = SO(2n)$.
(b) Show that $O(2n)$ is *not* isomorphic to $SO(2n) \times (\mathbb{Z}/2\mathbb{Z})$ as a group. Instead show that $O(2n)$ is isomorphic to a semidirect product $SO(2n) \rtimes (\mathbb{Z}/2\mathbb{Z})$. Describe explicitly the multiplication structure on $SO(2n) \rtimes (\mathbb{Z}/2\mathbb{Z})$ under its isomorphism with $O(2n)$.

Exercise 1.20 Show $U(n) \cong (SU(n) \times S^1)/(\mathbb{Z}/n\mathbb{Z})$ as both a manifold and a group. In particular, $U(n)$ is connected.

Exercise 1.21 Check the details in the proof of Theorem 1.22 to carefully show that the Lie group structure on G lifts to a Lie group structure on $\widetilde{G}$, making the covering map $\pi : \widetilde{G} \to G$ a homomorphism.

Exercise 1.22 Let $\mathcal{R}_3 \subseteq GL(3, \mathbb{R})$ be the set of rotations in $\mathbb{R}^3$ about the origin. Show that $\mathcal{R}_3 = SO(3)$.

Exercise 1.23 (a) Let $v \in \text{Im}(\mathbb{H})$ be a unit vector. Show that v can be completed to a basis $\{v, u, w\}$ of $\text{Im}(\mathbb{H})$, sharing the same properties as the $\{i, j, k\}$ basis.
(b) Show $\text{Ad}(\cos\theta + v\sin\theta)$ from the proof of Lemma 1.23 fixes v and acts by a rotation through an angle 2θ on the $\mathbb{R}$-span of $\{u, w\}$.

Exercise 1.24 Let $\mathfrak{su}(2) = \left\{ \begin{pmatrix} ix & -\overline{b} \\ b & -ix \end{pmatrix} \mid b \in \mathbb{C}, \ x \in \mathbb{R} \right\}$ and $(X, Y) = \frac{1}{2}\text{tr}(XY^*)$ for $X, Y \in \mathfrak{su}(2)$. Define $(\text{Ad}\,g)X = gXg^{-1}$ for $g \in SU(2)$ and $X \in \mathfrak{su}(2)$. Modify the proof of Lemma 1.23 to directly show that the map $\text{Ad} : SU(2) \to SO(3)$ is well defined and realizes the simply connected cover of $SO(3)$ as $SU(2)$.

1.3 The Double Cover of $SO(n)$

At the end of §1.2.2 we saw that $SO(n)$, $n \geq 3$, has a simply connected double cover called $\text{Spin}_n(\mathbb{R})$. The proof of Lemma 1.23 gave an explicit construction of $\text{Spin}_3(\mathbb{R})$ as $Sp(1)$ or $SU(2)$. The key idea was to first view $SO(3)$ as the set of rotations in $\mathbb{R}^3$ and then use the structure of the quaternion algebra, $\mathbb{H}$, along with a conjugation action to realize each rotation uniquely up to a $\pm$-sign.

This section gives a general construction of $\text{Spin}_n(\mathbb{R})$. The algebra that takes the place of $\mathbb{H}$ is called the Clifford algebra, $C_n(\mathbb{R})$, and instead of simply constructing rotations, it is more advantageous to use a conjugation action that constructs all reflections.

1.3.1 Clifford Algebras

Alhough the entire theory of Clifford algebras easily generalizes (Exercise 1.30), it is sufficient for our purposes here to work over $\mathbb{R}^n$ equipped with the standard Euclidean dot product $(\cdot, \cdot)$. Recall that the *tensor algebra* over $\mathbb{R}^n$ is $\mathcal{T}_n(\mathbb{R}) = \bigoplus_{k=0}^{\infty} \mathbb{R}^n \otimes \mathbb{R}^n \otimes \cdots \otimes \mathbb{R}^n$ (k copies) with a basis $\{1\} \cup \{x_{i_1} \otimes x_{i_2} \otimes \cdots \otimes x_{i_k} \mid 1 \leq i_k \leq n\}$, where $\{x_1, x_2, \ldots, x_n\}$ is a basis of $\mathbb{R}^n$.

Definition 1.26. The *Clifford algebra* is

$$C_n(\mathbb{R}) = \mathcal{T}_n(\mathbb{R})/\mathcal{I}$$

where $\mathcal{I}$ is the ideal of $\mathcal{T}_n(\mathbb{R})$ generated by

$$\{(x \otimes x + |x|^2) \mid x \in \mathbb{R}^n\}.$$

By way of notation for Clifford multiplication, write

$$x_1 x_2 \cdots x_k$$

for the element $x_1 \otimes x_2 \otimes \cdots \otimes x_k + \mathcal{I} \in C_n(\mathbb{R})$, where $x_1, x_2, \ldots, x_n \in \mathbb{R}^n$.

In particular,

$$(1.27) \qquad\qquad x^2 = -|x|^2$$

in $C_n(\mathbb{R})$ for $x \in \mathbb{R}^n$. Starting with the equality $xy + yx = (x + y)^2 - x^2 - y^2$ for $x, y \in \mathbb{R}^n$, it follows that Equation 1.27 is equivalent to

$$(1.28) \qquad\qquad xy + yx = -2(x, y)$$

in $C_n(\mathbb{R})$ for $x, y \in \mathbb{R}^n$.

It is a straightforward exercise (Exercise 1.25) to show that

$$C_0(\mathbb{R}) \cong \mathbb{R}, C_1(\mathbb{R}) \cong \mathbb{C}, \text{ and } C_2(\mathbb{R}) \cong \mathbb{H}.$$

More generally, define the *standard basis* for $\mathbb{R}^n$ to be $\{e_1, e_2, \ldots, e_n\}$, where $e_k = (0, \ldots, 0, 1, 0, \ldots, 0)$ with the 1 appearing in the kth entry. Clearly $\{1\} \cup \{e_{i_1} e_{i_2} \cdots e_{i_k} \mid k > 0, 1 \leq i_k \leq n\}$ spans $C_n(\mathbb{R})$, but this is overkill. First, observe that $C_n(\mathbb{R})$ inherits a filtration from $\mathcal{T}_n(\mathbb{R})$ by degree. Up to lower degree terms, Equation 1.28 can be used to commute adjacent e_{i_j} and Equation 1.27 can be used to remove multiple copies of e_{i_j} within a product $e_{i_1} e_{i_2} \cdots e_{i_k}$. An inductive argument on filtration degree therefore shows that

$$(1.29) \qquad\qquad \{1\} \cup \{e_{i_1} e_{i_2} \cdots e_{i_k} \mid 1 \leq i_1 < i_2 < \ldots i_k \leq n\}$$

spans $C_n(\mathbb{R})$, so $\dim C_n(\mathbb{R}) \leq 2^n$. In fact, we will shortly see Equation 1.29 provides a *basis* for $C_n(\mathbb{R})$ and so $\dim C_n(\mathbb{R}) = 2^n$. This will be done by constructing a *linear* isomorphism $\Psi : C_n(\mathbb{R}) \to \bigwedge \mathbb{R}^n$, where $\bigwedge \mathbb{R}^n = \bigoplus_{k=0}^{n} \bigwedge^k \mathbb{R}^n$ is the *exterior algebra* of $\mathbb{R}^n$.

To begin, we recall some multilinear algebra.

Definition 1.30. (1) For $x \in \mathbb{R}^n$, let *exterior multiplication* be the map $\epsilon(x)$: $\bigwedge^k \mathbb{R}^n \to \bigwedge^{k+1} \mathbb{R}^n$ given by

$$(\epsilon(x))(y) = x \wedge y$$

for $y \in \bigwedge^k \mathbb{R}^n$.

(2) For $x \in \mathbb{R}^n$, let *interior multiplication* be the map $\iota(x)$: $\bigwedge^k \mathbb{R}^n \to \bigwedge^{k-1} \mathbb{R}^n$ given by

$$(\iota(x))(y_1 \wedge y_2 \wedge \cdots \wedge y_k) = \sum_{i=1}^{k} (-1)^{i+1} (x, y_i)\, y_1 \wedge y_2 \wedge \cdots \wedge \widehat{y_i} \wedge \cdots \wedge y_k$$

for $y_i \in \mathbb{R}^n$, where $\widehat{y_i}$ means to omit the term.

It is straightforward (Exercise 1.26) from multilinear algebra that $\iota(x)$ is the adjoint of $\epsilon(x)$ with respect to the natural form on $\bigwedge \mathbb{R}^n$. In particular, $\epsilon(x)^2 = \iota(x)^2 = 0$ for $x \in \mathbb{R}^n$. It is also straightforward (Exercise 1.26) that

(1.31)
$$\epsilon(x)\iota(x) + \iota(x)\epsilon(x) = m_{|x|^2},$$

where $m_{|x|^2}$ is the operator that multiplies by $|x|^2$.

Definition 1.32. (1) For $x \in \mathbb{R}^n$, let $L_x : \bigwedge \mathbb{R}^n \to \bigwedge \mathbb{R}^n$ be given by $L_x = \epsilon(x) - \iota(x)$.

(2) Let $\Phi : T_n(\mathbb{R}) \to \mathrm{End}(\bigwedge \mathbb{R}^n)$ be the natural map of algebras determined by setting $\Phi(x) = L_x$ for $x \in \mathbb{R}^n$.

Observe that Equation 1.31 implies that $L_x^2 + m_{|x|^2} = 0$ so that $\Phi(\mathcal{I}) = 0$. In particular, Φ descends to $C_n(\mathbb{R})$.

Definition 1.33. (1) Abusing notation, let $\Phi : C_n(\mathbb{R}) \to \mathrm{End}(\bigwedge \mathbb{R}^n)$ be the map induced on $C_n(\mathbb{R})$ by the original map $\Phi : T_n(\mathbb{R}) \to \mathrm{End}(\bigwedge \mathbb{R}^n)$.

(2) Let $\Psi : C_n(\mathbb{R}) \to \bigwedge \mathbb{R}^n$ by $\Psi(v) = (\Phi(v))(1)$.

Explicitly, for $x_i \in \mathbb{R}^n$, $\Psi(x_1) = (\epsilon(x_1) - \iota(x_1))\, 1 = x_1$, and

$$\Psi(x_1 x_2) = (\epsilon(x_1) - \iota(x_1))\, (\epsilon(x_2) - \iota(x_2))\, 1$$
$$= (\epsilon(x_1) - \iota(x_1))\, x_2 = x_1 \wedge x_2 - (x_1, x_2).$$

In general,

(1.34)
$$\Psi(x_1 x_2 \cdots x_k) = x_1 \wedge x_2 \wedge \cdots \wedge x_k + \text{terms in } \bigoplus_{i \geq 1} \bigwedge^{k-2i} \mathbb{R}^n.$$

Equation 1.34 is easily established (Exercise 1.27) by induction on k. Also by induction on degree, it is an immediate corollary of Equation 1.34 that Ψ is surjective. A dimension count therefore shows that Ψ is a linear isomorphism. In summary:

Theorem 1.35. *The map* $\Psi : C_n(\mathbb{R}) \to \bigwedge \mathbb{R}^n$ *is a linear isomorphism of vector spaces, and so* $\dim C_n(\mathbb{R}) = 2^n$ *and Equation 1.29 provides a basis for* $C_n(\mathbb{R})$.

Thus, with respect to the standard basis (or any orthonormal basis for that matter), $C_n(\mathbb{R})$ has a particularly simple algebra structure. Namely, $C_n(\mathbb{R})$ is the $\mathbb{R}$-span of the basis $\{1\} \cup \{e_{i_1} e_{i_2} \cdots e_{i_k} \mid 1 \leq i_1 < i_2 < \cdots < i_k \leq n\}$ with the algebraic relations generated by $e_i^2 = -1$ and $e_i e_j = -e_j e_i$ when $i \neq j$.

1.3.2 Spin$_n(\mathbb{R})$ and Pin$_n(\mathbb{R})$

For the next definition, observe that $\mathcal{T}_n(\mathbb{R})$ breaks into a direct sum of the subalgebra generated by the tensor product of any even number of elements of $\mathbb{R}^n$ and the subspace generated the tensor product of any odd number of elements of $\mathbb{R}^n$. Since $\mathcal{I}$ is generated by elements of even degree, it follows that this decomposition descends to $\mathcal{C}_n(\mathbb{R})$.

Definition 1.36. (1) Let $\mathcal{C}_n^+(\mathbb{R})$ be the subalgebra of $\mathcal{C}_n(\mathbb{R})$ spanned by all products of an *even* number of elements of $\mathbb{R}^n$.
(2) Let $\mathcal{C}_n^-(\mathbb{R})$ be the subspace of $\mathcal{C}_n(\mathbb{R})$ spanned by all products of an *odd* number of elements of $\mathbb{R}^n$ so $\mathcal{C}_n(\mathbb{R}) = \mathcal{C}_n^+(\mathbb{R}) \oplus \mathcal{C}_n^-(\mathbb{R})$ as a vector space.
(3) Let the automorphism α, called the *main involution*, of $\mathcal{C}_n(\mathbb{R})$ act as multiplication by ± 1 on $\mathcal{C}_n^{\pm}(\mathbb{R})$.
(4) *Conjugation*, an anti-involution on $\mathcal{C}_n(\mathbb{R})$, is defined by

$$(x_1 x_2 \cdots x_k)^* = (-1)^k x_k \cdots x_2 x_1$$

for $x_i \in \mathbb{R}^n$.

The next definition makes sense for $n \geq 1$. However, because of Equation 1.25, we are really only interested in the case of $n \geq 3$ (see Exercise 1.34 for details when $n = 1, 2$).

Definition 1.37. (1) Let Spin$_n(\mathbb{R}) = \{g \in \mathcal{C}_n^+(\mathbb{R}) \mid gg^* = 1 \text{ and } gxg^* \in \mathbb{R}^n \text{ for all } x \in \mathbb{R}^n\}$.
(2) Let Pin$_n(\mathbb{R}) = \{g \in \mathcal{C}_n(\mathbb{R}) \mid gg^* = 1 \text{ and } \alpha(g)xg^* \in \mathbb{R}^n \text{ for all } x \in \mathbb{R}^n\}$. Note Spin$_n(\mathbb{R}) \subseteq$ Pin$_n(\mathbb{R})$.
(3) For $g \in$ Pin$_n(\mathbb{R})$ and $x \in \mathbb{R}^n$, define the homomorphism $\mathcal{A} :$ Pin$_n(\mathbb{R}) \to GL(n, \mathbb{R})$ by $(\mathcal{A}g)x = \alpha(g)xg^*$. Note $(\mathcal{A}g)x = gxg^*$ when $g \in$ Spin$_n(\mathbb{R})$.

Viewing left multiplication by $v \in \mathcal{C}_n(\mathbb{R})$ as an element of End$(\mathcal{C}_n(\mathbb{R}))$, use of the determinant shows that the set of invertible elements of $\mathcal{C}_n(\mathbb{R})$ is an open subgroup of $\mathcal{C}_n(\mathbb{R})$. It follows fairly easily that the set of invertible elements is a Lie group. As both Spin$_n(\mathbb{R})$ and Pin$_n(\mathbb{R})$ are closed subgroups of this Lie group, Corollary 1.8 implies that Spin$_n(\mathbb{R})$ and Pin$_n(\mathbb{R})$ are Lie groups as well.

Lemma 1.38. *$\mathcal{A}$ is a covering map of* Pin$_n(\mathbb{R})$ *onto* $O(n)$ *with* ker $\mathcal{A} = \{\pm 1\}$, *so there is an exact sequence*

$$\{1\} \to \{\pm 1\} \to \text{Pin}_n(\mathbb{R}) \overset{\mathcal{A}}{\to} O(n) \to \{I\}.$$

Proof. $\mathcal{A}$ *maps* Pin$_n(\mathbb{R})$ *into* $O(n)$: Let $g \in$ Pin$_n(\mathbb{R})$ and $x \in \mathbb{R}^n$. Using Equation 1.27 and the fact that conjugation on $\mathbb{R}^n$ is multiplication by -1, we calculate

$$|(\mathcal{A}g)x|^2 = -\left(\alpha(g)xg^*\right)^2 = -\left(\alpha(g)xg^*\right)\left(\alpha(g)xg^*\right) = \alpha(g)xg^*\left(\alpha(g)xg^*\right)^*$$
$$= \alpha(g)xg^*gx^*\alpha(g)^* = \alpha(g)xx^*\alpha(g)^* = -\alpha(g)x^2\alpha(g)^* = |x|^2 \, \alpha(gg^*)$$
$$= |x|^2.$$

Thus $\mathcal{A}g \in O(n)$.

$\mathcal{A}$ *maps* $\text{Pin}_n(\mathbb{R})$ *onto* $O(n)$: It is well known (Exercise 1.32) that each orthogonal matrix is a product of reflections. Thus it suffices to show that each reflection lies in the image of $\mathcal{A}$. Let $x \in S^{n-1}$ be any unit vector in $\mathbb{R}^n$ and write r_x for the reflection across the plane perpendicular to x. Observe $xx^* = -x^2 = |x|^2 = 1$. Thus $\alpha(x)xx^* = -xxx^* = -x$. If $y \in \mathbb{R}^n$ and $(x, y) = 0$, then Equation 1.28 says $xy = -yx$ so that $\alpha(x)yx^* = xyx = -x^2y = y$. Hence $x \in \text{Pin}_n(\mathbb{R})$ and $\mathcal{A}x = r_x$.

$\ker\mathcal{A} = \{\pm 1\}$: Since $\mathbb{R} \cap \text{Pin}_n(\mathbb{R}) = \{\pm 1\}$ and both elements are clearly in $\ker\mathcal{A}$, it suffices to show that $\ker\mathcal{A} \subseteq \mathbb{R}$. So suppose $g \in \text{Pin}_n(\mathbb{R})$ with $\mathcal{A}g = I$. As $g^* = g^{-1}$, $\alpha(g)x = xg$ for all $x \in \mathbb{R}^n$. Expanding g with respect to the standard basis from Equation 1.29, we may uniquely write $g = e_1 a + b$, where a, b are linear combinations of 1 and monomials in $e_2, e_3, \dots, e_n$. Looking at the special case of $x = e_1$, we have $\alpha(e_1 a + b)e_1 = e_1(e_1 a + b)$ so that $-e_1\alpha(a)e_1 + \alpha(b)e_1 = -a + e_1 b$. Since a and b contain no e_1's, $\alpha(a)e_1 = e_1 a$ and $\alpha(b)e_1 = e_1 b$. Thus $a + e_1 b_1 = -a + e_1 b$ which implies that $a = 0$ so that g contains no e_1. Induction similarly shows that g contains no e_k, $1 \le k \le n$, and so $g \in \mathbb{R}$.

$\mathcal{A}$ *is a covering map*: From Theorem 1.10, π has constant rank with $N = \text{rk}\,\pi = \dim \text{Pin}_n(\mathbb{R})$ since $\ker\pi = \{\pm 1\}$. For any $g \in \text{Pin}_n(\mathbb{R})$, the Rank Theorem from differential geometry ([8]) says there exists cubical charts (U, φ) of g and (V, ψ) of $\pi(g)$ so that $\psi \circ \pi \circ \varphi^{-1}(x_1, \dots, x_N) = (x_1, \dots, x_N, 0, \dots, 0)$ with $\dim O(n) - N$ zeros. Using the second countability of $\text{Pin}_n(\mathbb{R})$ and the Baire category theorem, surjectivity of π implies $\dim O(n) = N$. In particular, π restricted to U is a diffeomorphism onto V. Since $\ker\pi = \{\pm 1\}$, π is also a diffeomorphism of $-U$ onto V. Finally, injectivity of π on U implies that $(-U) \cap U = \emptyset$ so that the connected components of $\pi^{-1}(V)$ are U and $-U$. $\square$

Lemma 1.39. $\text{Pin}_n(\mathbb{R})$ *and* $\text{Spin}_n(\mathbb{R})$ *are compact Lie groups with*

$$\text{Pin}_n(\mathbb{R}) = \{x_1 \cdots x_k \mid x_i \in S^{n-1} \text{ for } 1 \le k \le 2n\}$$
$$\text{Spin}_n(\mathbb{R}) = \{x_1 x_2 \cdots x_{2k} \mid x_i \in S^{n-1} \text{ for } 2 \le 2k \le 2n\}$$

and $\text{Spin}_n(\mathbb{R}) = \mathcal{A}^{-1}(SO(n))$.

Proof. We know from the proof of Lemma 1.38 that $\mathcal{A}x = r_x$ for each $x \in S^{n-1} \subseteq \text{Pin}_n(\mathbb{R})$. Since elements of $O(n)$ are products of at most $2n$ reflections and $\mathcal{A}$ is surjective with kernel $\{\pm 1\}$, this implies that $\text{Pin}_n(\mathbb{R}) = \{x_1 \cdots x_k \mid x_i \in S^{n-1} \text{ for } 1 \le k \le 2n\}$. The equality $\text{Spin}_n(\mathbb{R}) = \text{Pin}_n(\mathbb{R}) \cap C_n^+(\mathbb{R})$ then implies $\text{Spin}_n(\mathbb{R}) = \{x_1 x_2 \cdots x_{2k} \mid x_i \in S^{n-1} \text{ for } 2 \le 2k \le 2n\}$. In particular, $\text{Pin}_n(\mathbb{R})$ and $\text{Spin}_n(\mathbb{R})$ are compact. Moreover because $\det r_x = -1$, the last equality is equivalent to the equality $\text{Spin}_n(\mathbb{R}) = \mathcal{A}^{-1}(SO(n))$. $\square$

Theorem 1.40. *(1)* $\text{Pin}_n(\mathbb{R})$ *has two connected ($n \ge 2$) components with* $\text{Spin}_n(\mathbb{R}) = \text{Pin}_n(\mathbb{R})^0$.

(2) $\text{Spin}_n(\mathbb{R})$ *is the connected ($n \ge 2$) simply connected ($n \ge 3$) two-fold cover of* $SO(n)$. *The covering homomorphism is given by* $\mathcal{A}$ *with* $\ker\mathcal{A} = \{\pm 1\}$, *i.e., there is an exact sequence*

$$\{1\} \to \{\pm 1\} \to \mathrm{Spin}_n(\mathbb{R}) \overset{\mathcal{A}}{\to} SO(n) \to \{I\}.$$

Proof. For $n \geq 2$, consider the path $t \to \gamma(t) = \cos t + e_1 e_2 \sin t$. Since $\gamma(t) = e_1(-e_1 \cos t + e_2 \sin t)$, it follows that $\gamma(t) \in \mathrm{Spin}_n(\mathbb{R})$ and so $\{\pm 1\}$ are path connected in $\mathrm{Spin}_n(\mathbb{R})$. From Lemmas 1.38 and 1.39, we know that $\mathrm{Spin}_n(\mathbb{R})$ is a double cover of $SO(n)$ and so $\mathrm{Spin}_n(\mathbb{R})$ is connected. Thus, for $n \geq 3$, Theorem 1.24 and the uniqueness of connected simply connected coverings implies $\mathrm{Spin}_n(\mathbb{R})$ is the connected simply connected cover of $SO(n)$.

Finally, let $x_0 \in S^{n-1}$. Clearly $\mathrm{Pin}_n(\mathbb{R}) = x_0 \, \mathrm{Spin}_n(\mathbb{R}) \amalg \mathrm{Spin}_n(\mathbb{R})$. We know that $\mathcal{A}$ is a continuous map of $\mathrm{Pin}_n(\mathbb{R})$ onto $O(n)$. Since $O(n)$ is not connected but $x_0 \, \mathrm{Spin}_n(\mathbb{R})$ and $\mathrm{Spin}_n(\mathbb{R})$ are connected, $x_0 \, \mathrm{Spin}_n(\mathbb{R}) \amalg \mathrm{Spin}_n(\mathbb{R})$ cannot be connected. Thus $x_0 \, \mathrm{Spin}_n(\mathbb{R})$ and $\mathrm{Spin}_n(\mathbb{R})$ are the connected components of $\mathrm{Pin}_n(\mathbb{R})$. $\qquad\square$

1.3.3 Exercises

Exercise 1.25 Show $\mathcal{C}_0(\mathbb{R}) \cong \mathbb{R}$, $\mathcal{C}_1(\mathbb{R}) \cong \mathbb{C}$, and $\mathcal{C}_2(\mathbb{R}) \cong \mathbb{H}$.

Exercise 1.26 (a) Show $\iota(x) = \epsilon(x)^*$ with respect to the inner product on $\bigwedge \mathbb{R}^n$ induced by defining $(x_1 \wedge x_2 \wedge \cdots \wedge x_k, \; y_1 \wedge y_2 \wedge \cdots \wedge y_l)$ to be 0 when $k \neq l$ and to be $\det(x_i, y_j)$ when $k = l$.
(b) Show $\epsilon(x)\iota(x) + \iota(x)\epsilon(x) = m_{|x|^2}$ for any $x \in \mathbb{R}^n$.

Exercise 1.27 (a) Prove Equation 1.34.
(b) Prove Theorem 1.35.

Exercise 1.28 For $u, v \in \mathcal{C}_n(\mathbb{R})$, show that $uv = 1$ if and only if $vu = 1$.

Exercise 1.29 For $n \geq 3$, show that the polynomial $x_1^2 + \cdots + x_n^2$ is irreducible over $\mathbb{C}$. However, show that $x_1^2 + \cdots + x_n^2$ is a product of linear factors over $\mathcal{C}_n(\mathbb{R})$.

Exercise 1.30 Let $(\cdot, \cdot)$ be *any* symmetric bilinear form on $\mathbb{R}^n$ or $\mathbb{C}^n$. Generalize the notion of Clifford algebra in Definition 1.26 by replacing $x \otimes x + |x|^2$ by $x \otimes x - (x, x)$ in the definition of $\mathcal{I}$. Prove the analogue of Theorem 1.35 still holds. If $(\cdot, \cdot)$ has signature p, q on $\mathbb{R}^n$, the resulting Clifford algebra is denoted $\mathcal{C}_{p,q}(\mathbb{R})$ (so $\mathcal{C}_n(\mathbb{R}) = \mathcal{C}_{0,n}(\mathbb{R})$) and if $(\cdot, \cdot)$ is the negative dot product on $\mathbb{C}^n$, the resulting Clifford algebra is denoted by $\mathcal{C}_n(\mathbb{C})$.

Exercise 1.31 Show that there is an algebra isomorphism $\mathcal{C}_{n-1}(\mathbb{R}) \cong \mathcal{C}_n^+(\mathbb{R})$ induced by mapping $a + b$, $a \in \mathcal{C}_{n-1}^+(\mathbb{R})$ and $b \in \mathcal{C}_{n-1}^-(\mathbb{R})$ to $a + be_n$. Conclude that $\dim \mathcal{C}_n^+(\mathbb{R}) = 2^{n-1}$.

Exercise 1.32 Use induction on n to show that any $g \in O(n)$ may be written as a product of at most $2n$ reflections. *Hint:* If $g \in O(n)$ and $ge_1 \neq e_1$, show that there is a reflection r_1 so that $r_1 ge_1 = e_1$. Now use orthogonality and induction.

Exercise 1.33 Show that $\mathcal{A}(\cos t + e_1 e_2 \sin t) = \begin{pmatrix} \cos 2t & -\sin 2t & 0 \\ \sin 2t & \cos 2t & 0 \\ 0 & 0 & I_{n-2} \end{pmatrix}$.

Exercise 1.34 (a) Under the isomorphism $C_1(\mathbb{R}) \cong \mathbb{C}$ induced by $e_1 \to i$, show that $\text{Pin}(1) = \{\pm 1, \pm i\}$ and $\text{Spin}(1) = \{\pm 1\}$ with $\mathcal{A}(\pm 1) = I$ and $\mathcal{A}(\pm i) = -I$ on $\mathbb{R}i$.
(b) Under the isomorphism $C_2(\mathbb{R}) \cong \mathbb{H}$ induced by $e_1 \to i$, $e_2 \to j$, and $e_1 e_2 \to k$, show that $\text{Pin}(2) = \{\cos\theta + k\sin\theta, i\sin\theta + j\cos\theta\}$ and $\text{Spin}(2) = \{\cos\theta + k\sin\theta\}$ with $\mathcal{A}(\cos\theta + k\cos\theta)$ acting as rotation by 2θ in the ij-plane.

Exercise 1.35 (a) For n odd, show that the center of $\text{Spin}_n(\mathbb{R})$ is $\{\pm 1\}$.
(b) For n even, show that the center of $\text{Spin}_n(\mathbb{R})$ is $\{\pm 1, \pm e_1 e_2 \cdots e_n\}$.

Exercise 1.36 (a) Replace $\mathbb{R}$ by $\mathbb{C}$ in Definitions 1.36 and 1.37 to define $\text{Spin}_n(\mathbb{C})$ (c.f. Exercise 1.30). Modify the proof of Theorem 1.40 to show $\mathcal{A}$ realizes $\text{Spin}_n(\mathbb{C})$ as a connected double cover of $SO(n, \mathbb{C}) = \{g \in SL(n, \mathbb{C}) \mid (gx, gy) = (x, y) \text{ for all } x, y \in \mathbb{C}^n\}$, where $(\cdot, \cdot)$ is the negative dot product on $\mathbb{C}^n$.
(b) Replace $C_n(\mathbb{R})$ by $C_{p,q}(\mathbb{R})$ (Exercise 1.30) in Definitions 1.36 and 1.37 to define $\text{Spin}_{p,q}(\mathbb{R})$. Modify the proof of Theorem 1.40 to show that $\mathcal{A}$ realizes $\text{Spin}_{p,q}(\mathbb{R})$ as a double cover of $SO(p, q)^0$, where $SO(p, q) = \{g \in SL(n, \mathbb{R}) \mid (gx, gy) = (x, y) \text{ for all } x, y \in \mathbb{C}^n\}$ and $(\cdot, \cdot)$ has signature p, q on $\mathbb{R}^n$.
(c) For $p, q > 0$ but not both 1, show that $\text{Spin}_{p,q}(\mathbb{R})$ is connected. For $p = q = 1$, show that $\text{Spin}_{1,1}(\mathbb{R})$ has two connected components.

Exercise 1.37 (a) Let $\mathfrak{so}(n) = \{X \in M_{n,n}(\mathbb{R}) \mid X^t = -X\}$ and $\mathfrak{q} = \sum_{i \neq j} \mathbb{R}e_i e_j \subseteq C_n(\mathbb{R})$. Show that $\mathfrak{so}(n)$ and $\mathfrak{q}$ are closed under the bracket (Lie) algebra structure given by $[x, y] = xy - yx$.
(b) Show that there is a (Lie) bracket algebra isomorphism from $\mathfrak{so}(n)$ to $\mathfrak{q}$ induced by the map $E_{i,j} - E_{j,i} \to \frac{1}{2} e_i e_j$ where $\{E_{i,j}\}$ is the set of standard basis elements for $M_{n,n}(\mathbb{R})$.

1.4 Integration

1.4.1 Volume Forms

If $\Phi : M \to N$ is a smooth map of manifolds, write $d\Phi : T_p(M) \to T_{\Phi(p)}(N)$ for the *differential* of Φ where $T_p(M)$ is the *tangent space* of M at p. Write $\Phi^* : T^*_{\Phi(p)}(N) \to T^*_p(M)$ for the *pullback* of Φ where $T^*_p(M)$ is the *cotangent space* of M at p. As usual, extend the definition of the pullback to the exterior algebra, $\Phi^* : \bigwedge T^*_{\Phi(p)}(N) \to \bigwedge T^*_p(M)$, as a map of algebras.

If M is an n-dimensional manifold, M is said to be *orientable* if there exists a nonvanishing element $\omega_M \in \bigwedge^*_n(M)$ where $\bigwedge^*_n(M)$ is the *exterior n-bundle* of the cotangent bundle of M. When this happens, ω_M determines an *orientation* on M that permits integration of n-forms on M.

Suppose ω_M (ω_N) is a nonvanishing n-form providing an orientation on M (N). If Φ is a diffeomorphism, $\Phi : M \to N$, then $\Phi^* \omega_N = c\omega_M$ where c is a nonvanishing function on M. When $c > 0$, Φ is called *orientation preserving* and when $c < 0$, Φ is called *orientation reversing*. Similarly, a chart (U, φ) of M is said to be an *orientation preserving chart* if U is open and if ψ is orientation preserving with respect to the orientations provided by $\omega|_U$, i.e., ω restricted to U, and by the standard volume form on $\varphi(U) \subseteq \mathbb{R}^n$, i.e., $dx_1 \wedge dx_2 \wedge \cdots \wedge dx_n|_{\varphi(U)}$.

If ω is a continuous n-form compactly supported in U where (U, φ) is an orientation preserving chart, recall the *integral of* ω with respect to the orientation on M induced by ω_M is defined as

$$\int_M \omega = \int_{\varphi(U)} (\varphi^{-1})^* \omega.$$

As usual (see [8] or [88] for more detail), the requirement that ω be supported in U is removed by covering M with orientation preserving charts, multiplying ω by a partition of unity subordinate to that cover, and summing over the partition using the above definition on each chart.

The *change of variables* formula from differential geometry is well known. If $\Phi : M \to N$ is a diffeomorphism of oriented manifolds and ω' is any continuous compactly supported n-form on N, then

$$(1.41) \qquad \int_N \omega' = \pm \int_M \Phi^* \omega'$$

with the sign being a $+$ when Φ is orientation preserving and a $-$ when Φ is orientation reversing. A simple generalization of Equation 1.41 applicable to covering maps is also useful. Namely, if $\Psi : M \to N$ is an m-fold covering map of oriented manifolds and ω' is any continuous compactly supported n-form on N, then

$$(1.42) \qquad m \int_N \omega' = \pm \int_M \Psi^* \omega'$$

with the sign determined by whether Ψ is orientation preserving or orientation reversing. The proof follows immediately from Equation 1.41 by using a partition of unity argument and the definition of a covering (Exercise 1.39).

Finally, functions on M can be integrated by fixing a *volume form* on M. A volume form is simply a fixed choice of a nonvanishing n-form, ω_M, defining the orientation on M. If f is a continuous compactly supported function on M, integration is defined with respect to this volume form by

$$\int_M f = \int_M f \omega_M.$$

It is easy to see (Exercise 1.40) that switching the volume form ω_M to $c\omega_M$, for some $c \in \mathbb{R}\backslash\{0\}$, multiplies the value of $\int_M f$ by $|c|$ (for negative c, the orientation is switched as well as the form against which f is integrated). In particular, the value of $\int_M f$ depends only on the choice of volume form modulo $\pm\omega_M$.

1.4.2 Invariant Integration

Let G be a Lie group of dimension n.

Definition 1.43. **(1)** Write l_g and r_g for *left* and *right translation* by $g \in G$, i.e., $l_g(h) = gh$ and $r_g(h) = hg$ for $h \in G$.
(2) A volume form, ω_G, on G is called *left invariant* if $l_g^* \omega_G = \omega_G$ and *right invariant* if $r_g^* \omega_G = \omega_G$ for all $g \in G$.

Lemma 1.44. *(1) Up to multiplication by a nonzero scalar, there is a unique left invariant volume form on G.*
(2) If G is compact, up to multiplication by ± 1, there is a unique left invariant volume form, ω_G, on G, so $\int_G 1 = 1$ with respect to ω_G.

Proof. Since $\dim \bigwedge_n^*(G)_e = 1$, up to multiplication by a nonzero scalar, there is a unique choice of $\omega_e \in \bigwedge_n^*(G)_e$. This choice uniquely extends to a left invariant n-form, ω, by defining $\omega_g = l_{g^{-1}}^* \omega_e$. For part (2), recall that replacing the volume form ω by $c\omega$ multiplies the value of the resulting integral by $|c|$. Because G is compact, $\int_G 1$ is finite with respect to the volume form ω. Thus there is a unique c, up to multiplication by ± 1, so that $\int_G 1 = 1$ with respect to the volume form $c\omega$. $\qquad \square$

Definition 1.45. For compact G, let ω_G be a left invariant volume form on G normalized so $\int_G 1 = 1$ with respect to ω_G. For any $f \in C(G)$, define

$$\int_G f(g)\, dg = \int_G f = \int_G f \omega_G$$

with respect to the orientation given by ω_G. By using the Riesz Representation Theorem, dg is also used to denote its completion to a Borel measure on G called *Haar measure* (see [37] or [73] for details).

If G has a suitably nice parametrization, it is possible to use the relation $\omega_g = l_{g^{-1}}^* \omega_e$ to pull the volume form back to an explicit integral over Euclidean space (see Exercise 1.44).

Theorem 1.46. *Let G be compact. The measure dg is left invariant, right invariant, and invariant under inversion, i.e.,*

$$\int_G f(hg)\, dg = \int_G f(gh)\, dg = \int_G f(g^{-1})\, dg = \int_G f(g)\, dg$$

for $h \in G$ and f a Borel integrable function on G.

Proof. It suffices to work with continuous f. Left invariance follows from the left invariance of ω_G and the change of variables formula in Equation 1.41 (l_h is clearly orientation preserving):

$$\int_G f(hg)\, dg = \int_G (f \circ l_h) \omega_G = \int_G (f \circ l_h)(l_h^* \omega_G)$$
$$= \int_G l_h^*(f \omega_G) = \int_G f \omega_G = \int_G f(g)\, dg.$$

To address right invariance, first observe that l_g and r_g commute. Thus the n-form $r_g^* \omega_G$ is still left invariant. By Lemma 1.44, this means $r_g^* \omega_G = c(g)^{-1} \omega_G$ for some $c(g) \in \mathbb{R} \backslash \{0\}$. Because $r_g \circ r_h = r_{hg}$, it follows that the *modular function* $c : G \to \mathbb{R} \backslash \{0\}$ is a homomorphism. The compactness of G clearly forces $|c(g)| = 1$ (Exercise 1.41).

Since r_g is orientation preserving if and only if $c(g) > 0$, the definitions and Equation 1.41 imply that

$$\int_G f(gh)\, dg = \int_G (f \circ r_h) \omega_G = c(h) \int_G (f \circ r_h)(r_h^* \omega_G)$$

$$= c(h) \int_G r_h^*(f \omega_G) = c(h) \operatorname{sgn}(c(h)) \int_G f \omega_G = \int_G f(g)\, dg.$$

Invariance of the measure under the transformation $g \to g^{-1}$ is handled similarly (Exercise 1.42).   □

We already know that ω_G is the unique (up to ± 1) left invariant normalized volume form on G. More generally, the corresponding measure dg is the unique left invariant normalized Borel measure on G.

Theorem 1.47. *For compact G, the measure dg is the unique left invariant Borel measure on G normalized so G has measure 1.*

Proof. Suppose dh is a left invariant Borel measure on G normalized so G has measure 1. Then for nonnegative measurable f, definitions and the Fubini–Tonelli Theorem show that

$$\int_G f(g)\, dg = \int_G \int_G f(g)\, dg\, dh = \int_G \int_G f(gh)\, dg\, dh$$

$$= \int_G \int_G f(gh)\, dh\, dg = \int_G \int_G f(h)\, dh\, dg = \int_G f(h)\, dh,$$

which is sufficient to establish $dg = dh$.   □

1.4.3 Fubini's Theorem

Part of the point of Fubini's Theorem is to reduce integration in multiple variables to more simple iterated integrals. Here we examine a variant that is appropriate for compact Lie groups. In the special case where the H_i are compact Lie groups and $G = H_1 \times H_2$, Fubini's Theorem will simply say that

$$\int_{H_1 \times H_2} f(g)\, dg = \int_{H_1} \left(\int_{H_2} f(h_1 h_2)\, dh_2 \right) dh_1$$

for integrable f on G.

More generally, let G be a Lie group and H a closed subgroup of G, so (Theorem 1.7) G/H is a manifold. In general, G/H may not be orientable (Exercise 1.38). The next theorem tells us when G/H *is* orientable and how its corresponding measure relates to dg and dh, the invariant measures on G and H. Abusing notation, continue to write l_g for left translation by $g \in G$ on G/H.

Theorem 1.48. *Let G be a compact Lie group and H a closed subgroup of G. If l_h^* is the identity map on $\bigwedge_{\text{top}}^*(G/H)_{eH}$ for all $h \in H$ (which is always true when H is connected), then, up to scalar, G/H possesses a unique left G-invariant volume form, $\omega_{G/H}$, and a corresponding left invariant Borel measure, $d(gH)$. Up to ± 1, $\omega_{G/H}$ can be uniquely normalized, so*

$$\int_{G/H} F = \int_G F \circ \pi,$$

where $\pi : G \to G/H$ is the canonical projection and F is an integrable function on G/H. In this case,

$$\int_G f(g)\, dg = \int_{G/H} \left(\int_H f(gh)\, dh \right) d(gH),$$

where f is an integrable function on G.

Proof. Consider first the question of the existence of a left invariant volume form on G/H. As in the proof of Lemma 1.44, let $\omega_{eH} \in \bigwedge_{\text{top}}^*(G/H)_{eH}$. If it makes sense to define the form ω by setting $\omega_{gH} = l_{g^{-1}}^* \omega_{eH}$, then ω is clearly left invariant and unique up to scalar multiplication. However, this process is well defined if and only if $l_{g^{-1}}^* = l_{(gh)^{-1}}^*$ on $\bigwedge_{\text{top}}^*(G/H)_{eH}$ for all $h \in H$ and $g \in G$. Since $l_{gh} = l_g \circ l_h$, it follows that $\omega_{G/H}$ exists if and only if l_h^* is the identity map on $\bigwedge_{\text{top}}^*(G/H)_{eH}$ for all $h \in H$.

Since $\bigwedge_{\text{top}}^*(G/H)_{eH}$ is one-dimensional, $l_h^* \omega_{eH} = c(h) \omega_{eH}$ for $h \in H$ and some $c(h) \in \mathbb{R} \backslash \{0\}$. The equality $l_{hh'} = l_h \circ l_{h'}$ shows that $c : H \to \mathbb{R} \backslash \{0\}$ is a homomorphism. The compactness of G shows that $c(h) \in \{\pm 1\}$. If H is connected, the image of H under c must be connected and so $c(h) = 1$, which shows that $\omega_{G/H}$ exists.

Suppose that $\omega_{G/H}$ exists. Since dh is invariant, the function $g \to \int_H f(gh)\, dh$ may be viewed as a function on G/H. Working with characteristic functions, the assignment $f \to \int_{G/H} \left(\int_H f(gh)\, dh \right) d(gH)$ defines a normalized left invariant Borel measure on G. By Theorem 1.47, this measure must be dg and so the second displayed formula of this theorem is established. To see that the first displayed equation holds, let $f = F \circ \pi$. □

1.4.4 Exercises

Exercise 1.38 (a) Show that the antipode map, $x \to -x$, on S^{2n} is orientation reversing.

(b) Show $\mathbb{P}(\mathbb{R}^{2n})$ is not orientable.

(c) Find a compact Lie group G with a closed subgroup H, so $G/H \cong \mathbb{P}(\mathbb{R}^{2n})$.

Exercise 1.39 If $\Psi : M \to N$ is an m-fold covering map of oriented manifolds and ω' is any continuous compactly supported n-form on N, show that

$$m \int_N \omega' = \pm \int_M \Psi^* \omega'$$

with the sign determined by whether Ψ is orientation preserving or orientation reversing.

Exercise 1.40 If f is a continuous compactly supported function on an orientable manifold M, show that switching the volume form from ω_M to $c\omega_M$, for some $c \in \mathbb{R}\setminus\{0\}$, multiplies the value of $\int_M f$ by $|c|$.

Exercise 1.41 If G is a compact Lie group and $c : G \to \mathbb{R}\setminus\{0\}$ is a homomorphism, show that $c(g) \in \{\pm 1\}$ for all $g \in G$ and that $c(g) = 1$ if G is connected.

Exercise 1.42 (a) For f a continuous function on a compact Lie group G, show that $\int_G f(g^{-1})\,dg = \int_G f(g)\,dg$.
(b) If φ is a smooth automorphism of G, show that $\int_G f \circ \varphi = \int_G f$.

Exercise 1.43 Let G be the Lie group $\{\begin{pmatrix} x & y \\ 0 & 1 \end{pmatrix} \mid x, y \in \mathbb{R}$ and $x > 0\}$. Show that the left invariant measure is $x^{-2}dxdy$ but the right invariant measure is $x^{-1}dxdy$.

Exercise 1.44 Let G be a Lie group and $\varphi : U \to V \subseteq \mathbb{R}^n$ a chart of G with $e \in U$, $0 \in V$, and $\varphi(e) = 0$. Suppose f is any integrable function on G supported in U.
(a) For $x \in V$, write $g = g(x) = \varphi^{-1}(x) \in U$. Show the function $l_x = \varphi \circ l_{g^{-1}} \circ \varphi^{-1}$ is well defined on a neighborhood of x.
(b) Write $\left|\frac{\partial l_x}{\partial x}\big|_x\right|$ for the absolute value of the determinant of the Jacobian matrix of l_x evaluated at x, i.e., $\left|\frac{\partial l_x}{\partial x}\big|_x\right| = |\det J|$, where the Jacobian matrix J is given by $J_{i,j} = \frac{\partial (l_x)_j}{\partial x_i}\big|_x$. Pull back the relation $\omega_g = l_{g^{-1}}^*\omega_e$ to show that the left invariant measure dg can be *scaled* so that

$$\int_G f\,dg = \int_V (f \circ \varphi^{-1})(x) \left|\frac{\partial l_x}{\partial x}\Big|_x\right| dx_1 \ldots dx_n.$$

(c) Show that changing l_x to $r_x = \varphi \circ r_{g^{-1}} \circ \varphi^{-1}$ in part (b) gives an expression for the right invariant measure.
(d) Write $\{(\frac{\partial}{\partial x_i}\big|_y)\}_{i=1}^n$ for the standard basis of of $T_y(\mathbb{R}^n)$. Show that the Jacobian matrix J is the change of basis matrix for the bases $\{d(l_{g^{-1}} \circ \varphi^{-1})(\frac{\partial}{\partial x_i}\big|_x)\}_{i=1}^n$ and $\{d\varphi^{-1}(\frac{\partial}{\partial x_i}\big|_0)\}_{i=1}^n$ of $T_e(G)$, i.e., $d(l_{g^{-1}} \circ \varphi^{-1})(\frac{\partial}{\partial x_i}\big|_x) = \sum_j J_{i,j}d\varphi^{-1}(\frac{\partial}{\partial x_j}\big|_0)$.
(e) Fix a basis $\{v_i\}_{i=1}^n$ of $T_e(G)$. Let C be the change of basis matrix for the bases $\{d(l_{g^{-1}} \circ \varphi^{-1})(\frac{\partial}{\partial x_i}\big|_x)\}_{i=1}^n$ and $\{v\}_{i=1}^n$, i.e., $d(l_{g^{-1}} \circ \varphi^{-1})(\frac{\partial}{\partial x_i}\big|_x) = \sum_j C_{i,j}v_j$. After *rescaling* dg, conclude that

$$\int_G f\,dg = \int_V (f \circ \varphi^{-1})(x)\,|\det C|\,dx_1 \cdots dx_n.$$

(f) Let H be a closed subgroup of a compact Lie group G and now suppose $\varphi : U \to V \subseteq \mathbb{R}^n$ a chart of G/H with $eH \in U$, $0 \in V$, and $\varphi(e) = 0$. Suppose l_h^* is the identity map on $\bigwedge_{\text{top}}^*(G/H)_{eH}$ for all $h \in H$ (which is always true when H is connected) and F is any integrable function on G/H supported in U. Fix a basis

$\{v_i\}_{i=1}^n$ of $T_{eH}(G/H)$ and define C as in part (e). Show that that $d(gH)$ can be *scaled* so that

$$\int_{G/H} F \, d(gH) = \int_V (F \circ \varphi^{-1})(x) \, |\det C| \, dx_1 \cdots dx_n.$$

Exercise 1.45 (a) View $GL(n, \mathbb{R})$ as an open dense set in $M_{n,n}(\mathbb{R})$ and identify functions on $GL(n, \mathbb{R})$ with functions on $M_{n,n}(\mathbb{R})$ that vanish on the complement of $GL(n, \mathbb{R})$. Show that the left and right invariant measure on $GL(n, \mathbb{R})$ is given by

$$\int_{GL(n,\mathbb{R})} f(g) \, dg = \int_{M_{n,n}(\mathbb{R})} f(X) \, |\det X|^{-n} \, dX,$$

where dX is the standard Euclidean measure on $M_{n,n}(\mathbb{R}) \cong \mathbb{R}^{n^2}$. In particular, the invariant measure for the multiplicative group $\mathbb{R}^\times = \mathbb{R} \backslash \{0\}$ is $\frac{dx}{|x|}$.
(b) Show that the invariant measure for the multiplicative group $\mathbb{C}^\times$ is $\frac{dx\,dy}{x^2+y^2}$ with respect to the usual embedding of $\mathbb{C}^\times$ into $\mathbb{C} \cong \mathbb{R}^2$.
(c) Show that the invariant measure for the multiplicative group $\mathbb{H}^\times$ is $\frac{dx\,dy\,du\,dv}{(x^2+y^2+u^2+v^2)^2}$ with respect to the usual embedding of $\mathbb{H}^\times$ into $\mathbb{H} \cong \mathbb{R}^4$.

Exercise 1.46 (a) On S^2, show that the $SO(3)$ normalized invariant measure is given by the integral $\frac{1}{4\pi} \int_0^\pi \int_0^{2\pi} F(\cos\theta \sin\phi, \sin\theta \sin\phi, \cos\phi) \sin\phi \, d\theta d\phi$.
(b) Let f be the function on $SO(3)$ that maps a matrix to the determinant of the lower right 2×2 submatrix. Evaluate $\int_{SO(3)} f$.

Exercise 1.47 Let

$$\alpha(\theta) = \begin{pmatrix} \cos\theta & -\sin\theta & 0 \\ \sin\theta & \cos\theta & 0 \\ 0 & 0 & 1 \end{pmatrix}, \beta(\theta) = \begin{pmatrix} 1 & 0 & 0 \\ 0 & \cos\theta & -\sin\theta \\ 0 & \sin\theta & \cos\theta \end{pmatrix},$$

$$\text{and } \gamma(\theta) = \begin{pmatrix} \cos\theta & 0 & -\sin\theta \\ 0 & 1 & 0 \\ \sin\theta & 0 & \cos\theta \end{pmatrix}.$$

(a) Verify that $(\cos\theta \sin\phi, \sin\theta \sin\phi, \cos\phi) = \alpha(\theta)\beta(\phi)e_3$ where $e_3 = (0, 0, 1)$. Use the isomorphism $S^2 \cong SO(3)/SO(2)$ to show that each element $g \in SO(3)$ can be written as $g = \alpha(\theta)\beta(\phi)\alpha(\psi)$ for $0 \leq \theta, \psi < 2\pi$ and $0 \leq \phi \leq \pi$ and that (θ, ϕ, ψ) is unique when $\phi \neq 0, \pi$. The coordinates (θ, ϕ, ψ) for $SO(3)$ are called the *Euler angles*.
(b) Viewing the map $(\theta, \phi, \psi) \to g = \alpha(\theta)\beta(\phi)\alpha(\psi)$ as a map into $M_{3,3}(\mathbb{R}) \cong \mathbb{R}^9$, show that

$$g^{-1}\frac{\partial g}{\partial \theta} = \beta'(0) \sin\phi \cos\psi + \gamma'(0) \sin\phi \sin\psi + \alpha'(0) \cos\phi$$

$$g^{-1}\frac{\partial g}{\partial \phi} = \beta'(0) \sin\psi - \gamma'(0) \cos\psi$$

$$g^{-1}\frac{\partial g}{\partial \psi} = \alpha'(0).$$

For $0 < \theta, \psi < 2\pi$ and $0 < \phi < \pi$, conclude that the inverse of the map $(\theta, \phi, \psi) \rightarrow \alpha(\theta)\beta(\phi)\alpha(\psi)$ is a chart for an open dense subset of $SO(3)$.

(c) Use Exercise 1.44 to show that the invariant integral on $SO(3)$ is given by

$$\int_{SO(3)} f(g)\,dg = \frac{1}{8\pi^2} \int_0^{2\pi} \int_0^\pi \int_0^{2\pi} f(\alpha(\theta)\beta(\phi)\alpha(\psi)) \sin \phi\, d\theta d\phi d\psi$$

for integrable f on $SO(3)$.

Exercise 1.48 Let

$$\alpha(\theta) = \begin{pmatrix} e^{i\frac{\theta}{2}} & 0 \\ 0 & e^{-i\frac{\theta}{2}} \end{pmatrix} \text{ and } \beta(\theta) = \begin{pmatrix} \cos\frac{\theta}{2} & -\sin\frac{\theta}{2} \\ \sin\frac{\theta}{2} & \cos\frac{\theta}{2} \end{pmatrix}.$$

As in Exercise 1.47, show that the invariant integral on $SU(2)$ is given by

$$\int_{SU(2)} f(g)\,dg = \frac{1}{8\pi^2} \int_0^{2\pi} \int_0^\pi \int_0^{2\pi} f(\alpha(\theta)\beta(\phi)\alpha(\psi)) \sin \phi\, d\theta d\phi d\psi$$

for integrable f on $SU(2)$.

2

Representations

Lie groups are often the abstract embodiment of symmetry. However, most frequently they manifest themselves through an action on a vector space which will be called a representation. In this chapter we confine ourselves to the study of finite-dimensional representations.

2.1 Basic Notions

2.1.1 Definitions

Definition 2.1. A *representation* of a Lie group G on a finite-dimensional complex vector space V is a homomorphism of Lie groups $\pi : G \to GL(V)$. The *dimension* of a representation is $\dim V$.

Technically, a representation should be denoted by the pair (π, V). When no ambiguity exists, it is customary to relax this requirement by referring to a representation (π, V) as simply π or as V. Some synonyms for expressing the fact that (π, V) is a representation of G include the phrases V is a *G-module* or G *acts on* V. As evidence of further laziness, when a representation π is clearly understood it is common to write

$$gv \text{ or } g \cdot v \text{ in place of } (\pi(g))(v)$$

for $g \in G$ and $v \in V$.

Although smoothness is part of the definition of a homomorphism (Definition 1.9), in fact we will see that continuity of π is sufficient to imply smoothness (Exercise 4.13). We will also eventually need to deal with infinite-dimensional vector spaces. The additional complexity of infinite-dimensional spaces will require a slight tweaking of our definition (Definition 3.11), although the changes will not affect the finite-dimensional case.

Two representations will be called equivalent if they are the same up to, basically, a change of basis. Recall that $\text{Hom}(V, V')$ is the set of all linear maps from V to V'.

Definition 2.2. Let (π, V) and (π', V') be finite-dimensional representations of a Lie group G.
(1) $T \in \mathrm{Hom}(V, V')$ is called an *intertwining operator* or *G-map* if $T \circ \pi = \pi' \circ T$.
(2) The set of all G-maps is denoted by $\mathrm{Hom}_G(V, V')$.
(3) The representations V and V' are *equivalent*, $V \cong V'$, if there exists a bijective G-map from V to V'.

2.1.2 Examples

Let G be a Lie group. A representation of G on a finite-dimensional vector space V smoothly assigns to each $g \in G$ an invertible linear transformation of V satisfying

$$\pi(g)\pi(g') = \pi(gg')$$

for all $g, g' \in G$. Although surprisingly important at times, the most boring example of a representation is furnished by the map $\pi : G \to GL(1, \mathbb{C}) = \mathbb{C}\backslash\{0\}$ given by $\pi(g) = 1$. This one-dimensional representation is called the *trivial representation*. More generally, the action of G on a vector space is called *trivial* if each $g \in G$ acts as the identity operator.

2.1.2.1 Standard Representations
Let G be $GL(n, \mathbb{F})$, $SL(n, \mathbb{F})$, $U(n)$, $SU(n)$, $O(n)$, or $SO(n)$. The *standard representation* of G is the representation on $\mathbb{C}^n$ where $\pi(g)$ is given by matrix multiplication on the left by the matrix $g \in G$. It is clear that this defines a representation.

2.1.2.2 SU(2)
This example illustrates a general strategy for constructing new representations. Namely, if a group G acts on a space M, then G can be made to act on the space of functions on M (or various generalizations of functions).

Begin with the standard two-dimensional representation of $SU(2)$ on $\mathbb{C}^2$ where $g\eta$ is simply left multiplication of matrices for $g \in SU(2)$ and $\eta \in \mathbb{C}^2$. Let

$$V_n(\mathbb{C}^2)$$

be the vector space of holomorphic polynomials on $\mathbb{C}^2$ that are homogeneous of degree n. A basis for $V_n(\mathbb{C}^2)$ is given by $\{z_1^k z_2^{n-k} \mid 0 \leq k \leq n\}$, so $\dim V_n(\mathbb{C}^2) = n + 1$.

Define an action of $SU(2)$ on $V_n(\mathbb{C}^2)$ by setting

$$(g \cdot P)(\eta) = P(g^{-1}\eta)$$

for $g \in SU(2)$, $P \in V_n(\mathbb{C}^2)$, and $\eta \in \mathbb{C}^2$. To verify that this is indeed a representation, calculate that

$$[g_1 \cdot (g_2 \cdot P)](\eta) = (g_2 \cdot P)(g_1^{-1}\eta) = P(g_2^{-1}g_1^{-1}\eta) = P((g_1 g_2)^{-1}\eta)$$
$$= [(g_1 g_2) \cdot P](\eta)$$

so that $g_1 \cdot (g_2 \cdot P) = (g_1 g_2) \cdot P$. Since smoothness and invertibility are clear, this action yields an $n + 1$-dimensional representation of $SU(2)$ on $V_n(\mathbb{C}^2)$.

Although these representations are fairly simple, they turn out to play an extremely important role as a building blocks in representation theory. With this in mind, we write them out in all their glory. If $g = \begin{pmatrix} a & -\overline{b} \\ b & \overline{a} \end{pmatrix} \in SU(2)$, then

$g^{-1} = \begin{pmatrix} \overline{a} & \overline{b} \\ -b & a \end{pmatrix}$, so that $g^{-1}\eta = (\overline{a}\eta_1 + \overline{b}\eta_2, -b\eta_1 + a\eta_2)$ where $\eta = (\eta_1, \eta_2)$. In particular, if $P = z_1^k z_2^{n-k}$, then $(g \cdot P)(\eta) = (\overline{a}\eta_1 + \overline{b}\eta_2)^k(-b\eta_1 + a\eta_2)^{n-k}$, so that

$$(2.3) \qquad \begin{pmatrix} a & -\overline{b} \\ b & \overline{a} \end{pmatrix} \cdot (z_1^k z_2^{n-k}) = (\overline{a}z_1 + \overline{b}z_2)^k(-bz_1 + az_2)^{n-k}.$$

Let us now consider another family of representations of $SU(2)$. Define

$$V_n'$$

to be the vector space of holomorphic functions in one variable of degree less than or equal to n. As such, V_n' has a basis consisting of $\{z^k \mid 0 \leq k \leq n\}$, so V_n' is also $n + 1$-dimensional. In this case, define an action of $SU(2)$ on V_n' by

$$(2.4) \qquad (g \cdot Q)(u) = (-bu + a)^n Q\left(\frac{\overline{a}u + \overline{b}}{-bu + a}\right)$$

for $g = \begin{pmatrix} a & -\overline{b} \\ b & \overline{a} \end{pmatrix} \in SU(2)$, $Q \in V_n'$, and $u \in \mathbb{C}$. It is easy to see that (Exercise 2.1) this yields a representation of $SU(2)$.

In fact, this apparently new representation is old news since it turns out that $V_n' \cong V_n(\mathbb{C}^2)$. To see this, we need to construct a bijective intertwining operator from $V_n(\mathbb{C}^2)$ to V_n'. Let $T : V_n(\mathbb{C}^2) \to V_n'$ be given by $(TP)(u) = P(u, 1)$ for $P \in V_n(\mathbb{C}^2)$ and $u \in \mathbb{C}$. This map is clearly bijective. To see that T is a G-map, use the definitions to calculate that

$$[T(g \cdot P)](u) = (g \cdot P)(u, 1) = P(\overline{a}u + \overline{b}, -bu + a)$$

$$= (-bu + a)^n P\left(\frac{\overline{a}u + \overline{b}}{-bu + a}, 1\right)$$

$$= (-bu + a)^n (TP)(u) = [g \cdot (TP)](u),$$

so $T(g \cdot P) = g \cdot (TP)$ as desired.

2.1.2.3 $O(n)$ and Harmonic Polynomials Let

$$V_m(\mathbb{R}^n)$$

be the vector space of complex-valued polynomials on $\mathbb{R}^n$ that are homogeneous of degree m. Since $V_m(\mathbb{R}^n)$ has a basis consisting of $\{x_1^{k_1} x_2^{k_2} \cdots x_n^{k_n} \mid k_i \in \mathbb{N}$ and $k_1 + k_2 + \cdots + k_n = m\}$, $\dim V_m(\mathbb{R}^n) = \binom{m+n-1}{m}$ (Exercise 2.4). Define an action of $O(n)$ on $V_m(\mathbb{R}^n)$ by

$$(g \cdot P)(x) = P(g^{-1}x)$$

for $g \in O(n)$, $P \in V_m(\mathbb{R}^n)$, and $x \in \mathbb{R}^n$. As in §2.1.2.2, this defines a representation. As fine and natural as this representation is, it actually contains a smaller, even nicer, representation.

Write $\Delta = \partial_{x_1}^2 + \cdots + \partial_{x_n}^2$ for the *Laplacian* on $\mathbb{R}^n$. It is a well-known corollary of the chain rule and the definition of $O(n)$ that Δ commutes with this action, i.e., $\Delta(g \cdot P) = g \cdot (\Delta P)$ (Exercise 2.5).

Definition 2.5. Let $\mathcal{H}_m(\mathbb{R}^n)$ be the subspace of all *harmonic polynomials* of degree m, i.e., $\mathcal{H}_m(\mathbb{R}^n) = \{P \in V_m(\mathbb{R}^n) \mid \Delta P = 0\}$.

If $P \in \mathcal{H}_m(\mathbb{R}^n)$ and $g \in O(n)$, then $\Delta(g \cdot P) = g \cdot (\Delta P) = 0$ so that $g \cdot P \in \mathcal{H}_m(\mathbb{R}^n)$. In particular, the action of $O(n)$ on $V_m(\mathbb{R}^n)$ descends to a representation of $O(n)$ (or $SO(n)$, of course) on $\mathcal{H}_m(\mathbb{R}^n)$. It will turn out that these representations do not break into any smaller pieces.

2.1.2.4 Spin and Half-Spin Representations Any representation (π, V) of $SO(n)$ automatically yields a representation of $\mathrm{Spin}_n(\mathbb{R})$ by looking at $(\pi \circ \mathcal{A}, V)$ where $\mathcal{A}$ is the covering map from $\mathrm{Spin}_n(\mathbb{R})$ to $SO(n)$. The set of representations of $\mathrm{Spin}_n(\mathbb{R})$ constructed this way is exactly the set of representations in which $-1 \in \mathrm{Spin}_n(\mathbb{R})$ acts as the identity operator. In this section we construct an important representation, called the spin representation, of $\mathrm{Spin}_n(\mathbb{R})$ that is *genuine*, i.e., one that does not originate from a representation of $SO(n)$ in this manner.

Let $(\cdot, \cdot)$ be the symmetric bilinear form on $\mathbb{C}^n$ given by the dot product. Write $n = 2m$ when n is even and write $n = 2m + 1$ when n is odd. Recall a subspace $W \subseteq \mathbb{C}^n$ is called *isotropic* if $(\cdot, \cdot)$ vanishes on W. It is well known that $\mathbb{C}^n$ can be written as a direct sum

$$(2.6) \qquad \mathbb{C}^n = \begin{cases} W \oplus W' & n \text{ even} \\ W \oplus W' \oplus \mathbb{C}e_0 & n \text{ odd} \end{cases}$$

for W, W' maximal isotropic subspaces (of dimension m) and e_0 a vector that is perpendicular to $W \oplus W'$ and satisfies $(e_0, e_0) = 1$. Thus, when n is even, take $W = \{(z_1, \ldots, z_m, iz_1, \ldots, iz_m) \mid z_k \in \mathbb{C}\}$ and $W' = \{(z_1, \ldots, z_m, -iz_1, \ldots, -iz_m) \mid z_k \in \mathbb{C}\}$. For n odd, take $W = \{(z_1, \ldots, z_m, iz_1, \ldots, iz_m, 0) \mid z_k \in \mathbb{C}\}$, $W' = \{(z_1, \ldots, z_m, -iz_1, \ldots, -iz_m, 0) \mid z_k \in \mathbb{C}\}$, and $e_0 = (0, \ldots, 0, 1)$.

Compared to our previous representations, the action of the spin representation is fairly complicated. We state the necessary definition below, although it will take some work to provide appropriate motivation and to show that everything is well defined. Recall (Lemma 1.39) that one realization of $\mathrm{Spin}_n(\mathbb{R})$ is $\{x_1 x_2 \cdots x_{2k} \mid x_i \in S^{n-1}$ for $2 \le 2k \le 2n\}$.

Definition 2.7. (1) The elements of $S = \bigwedge W$ are called *spinors* and $\mathrm{Spin}_n(\mathbb{R})$ has a representation on S called the *spin representation*.
(2) For *n even*, the action for the spin representation of $\mathrm{Spin}_n(\mathbb{R})$ on S is induced by the map

$$x \to \epsilon(w) - 2\iota(w'),$$

where $x \in S^{n-1}$ is uniquely written as $x = w + w'$ according to the decomposition $\mathbb{R}^n \subseteq \mathbb{C}^n = W \oplus W'$.

(3) Let $S^+ = \bigwedge^+ W = \bigoplus_k \bigwedge^{2k} W$ and $S^- = \bigwedge^- W = \bigoplus_k \bigwedge^{2k+1} W$. As vector spaces $S = S^+ \oplus S^-$.

(4) For n even, the spin representation action of $\mathrm{Spin}_n(\mathbb{R})$ on S preserves the subspaces S^+ and S^-. These two spaces are therefore representations of $\mathrm{Spin}_n(\mathbb{R})$ in their own right and called the *half-spin representations*.

(5) For n odd, the action for the spin representation of $\mathrm{Spin}_n(\mathbb{R})$ on S is induced by the map

$$x \to \epsilon(w) - 2\iota(w') + (-1)^{\deg} m_{i\zeta},$$

where $x \in S^{n-1}$ is uniquely written as $x = w + w' + \zeta e_0$ according to the decomposition $\mathbb{R}^n \subseteq \mathbb{C}^n = W \oplus W' \oplus \mathbb{C}e_0$, $(-1)^{\deg}$ is the linear operator acting by ± 1 on $\bigwedge^{\pm} W$, and $m_{i\zeta}$ is multiplication by $i\zeta$.

To start making proper sense of this definition, let $C_n(\mathbb{C}) = C_n(\mathbb{R}) \otimes_{\mathbb{R}} \mathbb{C}$. From the definition of $C_n(\mathbb{R})$, it is easy to see that $C_n(\mathbb{C})$ is simply $\mathcal{T}(\mathbb{C}^n)$ modulo the ideal generated by either $\{(z \otimes z + (z, z)) \mid z \in \mathbb{C}^n\}$ or equivalently by $\{(z_1 \otimes z_2 + z_2 \otimes z_1 + 2(z_1, z_2)) \mid z_i \in \mathbb{C}^n\}$ (c.f. Exercise 1.30).

Since $\mathrm{Spin}_n(\mathbb{R}) \subseteq C_n(\mathbb{C})$, $C_n(\mathbb{C})$ itself becomes a representation for $\mathrm{Spin}_n(\mathbb{R})$ under left multiplication. Under this action, $-1 \in \mathrm{Spin}_n(\mathbb{R})$ acts as m_{-1}, and so this representation is genuine. However, the spin representations turn out to be much smaller than $C_n(\mathbb{C})$. One way to find these smaller representations is to restrict left multiplication of $\mathrm{Spin}_n(\mathbb{R})$ to certain left ideals in $C_n(\mathbb{C})$. While this method works (e.g., Exercise 2.12), we take an equivalent path that realizes $C_n(\mathbb{C})$ as a certain endomorphism ring.

Theorem 2.8. *As algebras,*

$$C_n(\mathbb{C}) \cong \begin{cases} \mathrm{End} \bigwedge W & n \text{ even} \\ (\mathrm{End} \bigwedge W) \oplus (\mathrm{End} \bigwedge W) & n \text{ odd.} \end{cases}$$

Proof. n even: For $z = w + w' \in \mathbb{C}^n$, define $\widetilde{\Phi} : \mathbb{C}^n \to \mathrm{End} \bigwedge W$ by

$$\widetilde{\Phi}(z) = \epsilon(w) - 2\iota(w').$$

As an algebra map, extend $\widetilde{\Phi}$ to $\widetilde{\Phi} : \mathcal{T}_n(\mathbb{C}) \to \mathrm{End} \bigwedge W$. A simple calculation (Exercise 2.6) shows $\widetilde{\Phi}(z)^2 = m_{-2(w,w')} = m_{-(z,z)}$ so that $\widetilde{\Phi}$ descends to a map $\Phi : C_n(\mathbb{C}) \to \mathrm{End} \bigwedge W$.

To see that $\widetilde{\Phi}$ is an isomorphism, it suffices to check that $\widetilde{\Phi}$ is surjective since $C_n(\mathbb{C})$ and $\mathrm{End} \bigwedge W$ both have dimension 2^n. Pick a basis $\{w_1, \dots, w_m\}$ of W and let $\{w'_1, \dots, w'_m\}$ be the dual basis for W, i.e., (w_i, w'_j) is 0 when $i \neq j$ and 1 when $i = j$. With respect to this basis, $\widetilde{\Phi}$ acts in a particularly simple fashion. If $1 \leq i_1 < \cdots < i_k \leq m$, then $\widetilde{\Phi}(w_{i_1} \cdots w_{i_k} w'_{i_1} \cdots w'_{i_k})$ kills $\bigwedge^p W$ for $p < k$, maps

$\bigwedge^k W$ onto $\mathbb{C} w_{i_1} \wedge \cdots \wedge w_{i_k}$, and preserves $\bigwedge^p W$ for $p > k$. An inductive argument on $n - k$ therefore shows that the image of $\widetilde{\Phi}$ contains each projection of $\bigwedge W$ onto $\mathbb{C} w_{i_1} \wedge \cdots \wedge w_{i_k}$. Successive use of the operators $\widetilde{\Phi}(w_i)$ and $\widetilde{\Phi}(w_j')$ can then be used to map $w_{i_1} \wedge \cdots \wedge w_{i_k}$ to any other $w_{j_1} \wedge \cdots \wedge w_{j_l}$. This implies that $\widetilde{\Phi}$ is surjective (in familiar matrix notation, this shows that the image of $\widetilde{\Phi}$ contains all endomorphisms corresponding to each matrix basis element $E_{i,j}$).

n odd: For $z = w + w' + \zeta e_0 \in \mathbb{C}^n$, let $\widetilde{\Phi}^{\pm} : \mathbb{C}^n \to \text{End} \bigwedge W$ by

$$\widetilde{\Phi}^{\pm}(z) = \epsilon(w) - 2\iota(w') \pm (-1)^{\deg} m_{i\zeta}.$$

As an algebra map, extend $\widetilde{\Phi}^{\pm}$ to $\widetilde{\Phi}^{\pm} : T_n(\mathbb{C}) \to \text{End} \bigwedge W$. A simple calculation (Exercise 2.6) shows that $\widetilde{\Phi}^{\pm}(z)^2 = m_{-(z,z)}$ so that $\widetilde{\Phi}^{\pm}$ descends to a map $\widetilde{\Phi}^{\pm} : C_n(\mathbb{C}) \to \text{End} \bigwedge W$. Thus the map $\widetilde{\Phi} : C_n(\mathbb{C}) \to (\text{End} \bigwedge W) \oplus (\text{End} \bigwedge W)$ given by $\widetilde{\Phi}(v) = (\widetilde{\Phi}^+(v), \widetilde{\Phi}^-(v))$ is well defined.

To see that $\widetilde{\Phi}$ is an isomorphism, it suffices to verify that $\widetilde{\Phi}$ is surjective since $C_n(\mathbb{C})$ and $(\text{End} \bigwedge W) \oplus (\text{End} \bigwedge W)$ both have dimension 2^n. The argument is similar to the one given for the even case and left as an exercise (Exercise 2.7). $\quad \square$

Theorem 2.9. *As algebras,*

$$C_n^+(\mathbb{C}) \cong \begin{cases} (\text{End} \bigwedge^+ W) \oplus (\text{End} \bigwedge^- W) & n \text{ even} \\ (\text{End} \bigwedge W) & n \text{ odd}. \end{cases}$$

Proof. n even: From the definition of $\widetilde{\Phi}$ in the proof of Theorem 2.8, it is clear that the operators in $\widetilde{\Phi}(C_n^+(\mathbb{C}))$ preserve $\bigwedge^{\pm} W$. Thus restricted to $C_n^+(\mathbb{C})$, $\widetilde{\Phi}$ may be viewed as a map to $(\text{End} \bigwedge^+ W) \oplus (\text{End} \bigwedge^- W)$. Since this map is already known to be injective, it suffices to show that $\dim [(\text{End} \bigwedge^+ W) \oplus (\text{End} \bigwedge^- W)] = \dim C_n^+(\mathbb{C})$. In fact, it is a simple task (Exercise 2.9) to see that both dimensions are 2^{n-1}. For instance, Equation 1.34 and Theorem 1.35 show $\dim C_n^+(\mathbb{C}) = 2^{n-1}$. Alternatively, use Exercise 1.3.1 to show that $C_{n-1}(\mathbb{R}) \cong C_n^+(\mathbb{R})$.

n odd: In this case, operators in $\widetilde{\Phi}(C_n^+(\mathbb{C}))$ no longer have to preserve $\bigwedge^{\pm} W$. However, restriction of the map $\widetilde{\Phi}^+$ to $C_n^+(\mathbb{C})$ yields a map of $C_n^+(\mathbb{C})$ to $\text{End} \bigwedge W$ (alternatively, $\widetilde{\Phi}^-$ could have been used). By construction, this map is known to be surjective. Thus to see that the map is an isomorphism, it again suffices to show that $\dim (\text{End} \bigwedge W) = \dim C_n^+(\mathbb{C})$. As before, it is simple (Exercise 2.9) to see that both dimensions are 2^{n-1}. $\quad \square$

At long last, the origin of the spin representations can be untangled. Since $\text{Spin}_n(\mathbb{R}) \subseteq C_n^+(\mathbb{C})$, Definition 2.7 uses the homomorphism $\widetilde{\Phi}$ from Theorem 2.9 for n even and the homomorphism $\widetilde{\Phi}^+$ for n odd and restricts the action to $\text{Spin}_n(\mathbb{R})$. In the case of n even, $\widetilde{\Phi}$ can be further restricted to either $\text{End} \bigwedge^{\pm} W$ to construct the two half-spin representations. Finally, $-1 \in \text{Spin}_n(\mathbb{R})$ acts by m_{-1}, so the spin representations are genuine as claimed.

2.1.3 Exercises

Exercise 2.1 Show that Equation 2.4 defines a representation of $SU(2)$ on V_n'.

Exercise 2.2 **(a)** Find the joint eigenspaces for the action of $\{\text{diag}(e^{i\theta}, e^{-i\theta}) \mid \theta \in \mathbb{R}\} \subseteq SU(2)$ on $V_n(\mathbb{C}^2)$. That is, find all nonzero $P \in V_n(\mathbb{C}^2)$, so that $(\text{diag}(e^{i\theta}, e^{-i\theta})) \cdot P = \lambda_\theta P$ for all $\theta \in \mathbb{R}$ and some $\lambda_\theta \in \mathbb{C}$.
(b) Find the joint eigenspaces for the action of $SO(2)$ on $V_m(\mathbb{R}^2)$ and on $\mathcal{H}_m(\mathbb{R}^2)$.
(c) Find the joint eigenspaces for the action of

$$\{(\cos\theta_1 + e_1 e_2 \sin\theta_1)(\cos\theta_2 + e_3 e_4 \sin\theta_2) \mid \theta_i \in \mathbb{R}\} \subseteq \text{Spin}(4)$$

on the half-spin representations $S^\pm$.

Exercise 2.3 Define a Hermitian inner product on $V_n(\mathbb{C}^2)$ by

$$\left(\sum_k a_k z_1^k z_2^{n-k}, \sum_k b_k z_1^k z_2^{n-k} \right) = \sum k!(n-k)! \, a_k \overline{b_k}.$$

For $g \in SU(2)$, show that $(gP, gP') = (P, P')$ for $P, P' \in V_n(\mathbb{C}^2)$.

Exercise 2.4 Show that $| \{x_1^{k_1} x_2^{k_2} \cdots x_n^{k_n} \mid k_i \in \mathbb{N} \text{ and } k_1 + k_2 + \cdots + k_n = m\} | = \binom{m+n-1}{m}$.

Exercise 2.5 For $g \in O(n)$ and f a smooth function on $\mathbb{R}^n$, show that $\Delta(f \circ l_g) = (\Delta f) \circ l_g$ where $l_g(x) = gx$.

Exercise 2.6 **(a)** For n even and $\widetilde{\Phi}(z) = \epsilon(w) - 2\iota(w')$ for $z = w + w' \in \mathbb{C}^n$ with $w \in W$ and $w' \in W'$, show $\widetilde{\Phi}(z)^2 = m_{-2(w,w')} = m_{-(z,z)}$.
(b) For n odd and $\widetilde{\Phi}(z) = \epsilon(w) - 2\iota(w') \pm (-1)^{\deg} m_{i\zeta}$ for $z = w + w' + \zeta e_0 \in \mathbb{C}^n$ with $w \in W$, $w' \in W'$, and $\zeta \in \mathbb{C}$, show that $\widetilde{\Phi}(z)^2 = m_{-2(w,w') - \zeta^2} = m_{-(z,z)}$.

Exercise 2.7 In the proof of Theorem 2.8, show that the map $\widetilde{\Phi}$ is surjective when n is odd.

Exercise 2.8 Use Theorem 2.8 to compute the center of $C_n(\mathbb{C})$.

Exercise 2.9 From Theorem 2.9, show directly that $\dim C_n^+(\mathbb{C})$ and

$$\dim \left[\left(\text{End} \bigwedge^+ W \right) \oplus \left(\text{End} \bigwedge^- W \right) \right]$$

are both 2^{n-1} for n even, and $\dim \left(\text{End} \bigwedge W \right) = 2^{n-1}$ for n odd.

Exercise 2.10 Up to equivalence, show that the spin and half-spin representations are independent of the choice of the maximal isotropic decomposition for $\mathbb{C}^n$ (as in Equation 2.6).

Exercise 2.11 For n odd, $\widetilde{\Phi}^+$ was used to define the spin representation of $\text{Spin}_n(\mathbb{R})$ on $\bigwedge W$. Show that an equivalent representation is constructed by using $\widetilde{\Phi}^-$ in place of $\widetilde{\Phi}^+$.

Exercise 2.12 (a) Use the same notation as in the proof of Theorem 2.8. For n even, let $w_0' = w_1' \cdots w_m' \in C_n(\mathbb{C})$, let $\mathcal{J}$ be the left ideal of $C_n(\mathbb{C})$ generated by w_0', and let $T : \bigwedge W \to \mathcal{J}$ be the linear map satisfying $T(w_{i_1} \wedge \cdots \wedge w_{i_k}) = w_{i_1} \cdots w_{i_k} w_0'$. Show that T is a well-defined and $\mathrm{Spin}_n(\mathbb{R})$-intertwining isomorphism with respect to the spin action on $\bigwedge W$ and left Clifford multiplication on $\mathcal{J}$.
(b) For n odd, let $w_0' = (1 - ie_0)w_1' \cdots w_m'$. Show that there is an analogous $\mathrm{Spin}_n(\mathbb{R})$-intertwining isomorphism with respect to the spin action on $\bigwedge W$ and left Clifford multiplication on the appropriate left ideal of $C_n(\mathbb{C})$.

Exercise 2.13 (a) Define a nondegenerate bilinear form $(\cdot, \cdot)$ on $\bigwedge W$ by setting $\left(\bigwedge^k W, \bigwedge^l W \right) = 0$ when $k + l \neq m$ and requiring $\alpha(u^*) \wedge v = (u, v) \, w_1 \wedge \cdots \wedge w_m$ for $u \in \bigwedge^k W$ and $v \in \bigwedge^{m-k} W$ (see §1.3.2 for notation). Show that the form is symmetric when $m \equiv 0, 3 \bmod(4)$ and that it is skew-symmetric when $m \equiv 1, 2, \bmod(4)$.
(b) With respect to the spin representation action, show that $(g \cdot u, g \cdot v) = (u, v)$ for $u, v \in S = \bigwedge W$ and $g \in \mathrm{Spin}_n(\mathbb{R})$.
(c) For n even, show that $(\cdot, \cdot)$ restricts to a nondegenerate form on $S^{\pm} = \bigwedge^{\pm} W$ when m is even, but restricts to zero when m is odd.

2.2 Operations on Representations

2.2.1 Constructing New Representations

Given one or two representations, it is possible to form many new representations using standard constructions from linear algebra. For instance, if V and W are vector spaces, one can form new vector spaces via the direct sum, $V \oplus W$, the tensor product, $V \otimes W$, or the set of linear maps from V to W, $\mathrm{Hom}(V, W)$. The tensor product leads to the construction of the tensor algebra, $T(V) = \bigoplus_{k=0}^{\infty} \left(\bigotimes^k V \right)$, and its quotients, the *exterior algebra*, $\bigwedge(V) = \bigoplus_{k=0}^{\dim V} \bigwedge^k V$, and the *symmetric algebra*, $S(V) = \bigoplus_{k=0}^{\infty} S^k(V)$. Further constructions include the *dual* (or *contragradient*) space, $V^* = \mathrm{Hom}(V, \mathbb{C})$, and the *conjugate* space, $\overline{V}$, which has the same underlying additive structure as V, but is equipped with a new scalar multiplication structure, $\cdot'$, given by $z \cdot' v = \overline{z} v$ for $z \in \mathbb{C}$ and $v \in V$. Each of these new vector spaces also carries a representation as defined below.

Definition 2.10. Let V and W be finite-dimensional representations of a Lie group G.
(1) G acts on $V \oplus W$ by $g(v, w) = (gv, gw)$.
(2) G acts on $V \otimes W$ by $g \sum v_i \otimes w_j = \sum gv_i \otimes gw_j$.
(3) G acts on $\mathrm{Hom}(V, W)$ by $(gT)(v) = g\left[T\left(g^{-1}v \right) \right]$.
(4) G acts on $\bigotimes^k V$ by $g \sum v_{i_1} \otimes \cdots \otimes v_{i_k} = \sum (gv_{i_1}) \otimes \cdots \otimes (gv_{i_k})$.
(5) G acts on $\bigwedge^k V$ by $g \sum v_{i_1} \wedge \cdots \wedge v_{i_k} = \sum (gv_{i_1}) \wedge \cdots \wedge (gv_{i_k})$.
(6) G acts on $S^k(V)$ by $g \sum v_{i_1} \cdots v_{i_k} = \sum (gv_{i_1}) \cdots (gv_{i_k})$.

(7) G acts on V^* by $(gT)(v) = T\left(g^{-1}v\right)$.
(8) G acts on $\overline{V}$ by the same action as it does on V.

It needs to be verified that each of these actions define a representation. All are simple. We check numbers (3) and (5) and leave the rest for Exercise 2.14. For number (3), smoothness and invertibility are clear. It remains to verify the homomorphism property so we calculate

$$[g_1(g_2T)](v) = g_1\left[(g_2T)\left(g_1^{-1}v\right)\right] = g_1g_2\left[T(g_2^{-1}g_1^{-1}v)\right] = [(g_1g_2)T](v)$$

for $g_i \in G$, $T \in \text{Hom}(V, W)$, and $v \in V$. For number (5), recall that $\bigwedge^k V$ is simply $\bigotimes^k V$ modulo $\mathcal{I}_k$, where $\mathcal{I}_k$ is $\bigotimes^k V$ intersect the ideal generated by $\{v \otimes v \mid v \in V\}$. Since number (4) is a representation, it therefore suffices to show that the action of G on $\bigotimes^k V$ preserves $\mathcal{I}_k$—but this is clear.

Some special notes are in order. For number (1) dealing with $V \oplus W$, choose in the obvious way a basis for $V \oplus W$ that is constructed from a basis for V and a basis for W. With respect to this basis, the action of G can be realized on $V \oplus W$ by multiplication by a matrix of the form $\begin{pmatrix} * & 0 \\ 0 & * \end{pmatrix}$ where the upper left block is given by the action of G on V and the lower right block is given by the action of G on W.

For number (7) dealing with V^*, fix a basis $\mathcal{B} = \{v_i\}_{i=1}^n$ for V and let $\mathcal{B}^* = \{v_i^*\}_{i=1}^n$ be the *dual basis* for V^*, i.e., $v_i^*(v_j)$ is 1 when $i = j$ and is 0 when $i \neq j$. Using these bases, identify V and V^* with $\mathbb{C}^n$ by the coordinate maps $\left[\sum_i c_i v_i\right]_{\mathcal{B}} = (c_1, \dots, c_n)$ and $\left[\sum_i c_i v_i^*\right]_{\mathcal{B}^*} = (c_1, \dots, c_n)$. With respect to these bases, realize the action of g on V and V^* by a matrices M_g and M_g' so that $[g \cdot v]_{\mathcal{B}} = M_g [v]_{\mathcal{B}}$ and $[g \cdot T]_{\mathcal{B}^*} = M_g' [T]_{\mathcal{B}^*}$ for $v \in V$ and $T \in V^*$. In particular, $\left[M_g\right]_{i,j} = v_i^* (g \cdot v_j)$ and $\left[M_g'\right]_{i,j} = \left(g \cdot v_j^*\right)(v_i)$. Thus $\left[M_g'\right]_{i,j} = v_j^*(g^{-1} \cdot v_i) = \left[M_{g^{-1}}\right]_{j,i}$ so that $M_g' = M_g^{-1,t}$. In other words, once appropriate bases are chosen and the G action is realized by matrix multiplication, the action of G on V^* is obtained from the action of G on V simply by taking the *inverse transpose* of the matrix.

For number (8) dealing with $\overline{V}$, fix a basis for V and realize the action of g by a matrix M_g as above. To examine the action of g on $v \in \overline{V}$, recall that scalar multiplication is the conjugate of the original scalar multiplication in V. In particular, in $\overline{V}$, $g \cdot v$ is therefore realized by the matrix $\overline{M_g}$. In other words, once a basis is chosen and the G action is realized by matrix multiplication, the action of G on $\overline{V}$ is obtained from the action of G on V simply by taking the *conjugate* of the matrix.

It should also be noted that few of these constructions are independent of each other. For instance, the action in number (7) on V^* is just the special case of the action in number (3) on $\text{Hom}(V, W)$ in which $W = \mathbb{C}$ is the trivial representation. Also the actions in (4), (5), and (6) really only make repeated use of number (2). Moreover, as representations, it is the case that $V^* \otimes W \cong \text{Hom}(V, W)$ (Exercise 2.15) and, for compact G, $V^* \cong \overline{V}$ (Corollary 2.20).

2.2.2 Irreducibility and Schur's Lemma

Now that we have many ways to glue representations together, it makes sense to seek some sort of classification. For this to be successful, it is necessary to examine the smallest possible building blocks.

Definition 2.11. Let G be a Lie group and V a finite-dimensional representation of G.
(1) A subspace $U \subseteq V$ is *G-invariant* (also called a *submodule* or a *subrepresentation*) if $gU \subseteq U$ for $g \in G$. Thus U is a representation of G in its own right.
(2) A nonzero representation V is *irreducible* if the only G-invariant subspaces are $\{0\}$ and V. A nonzero representation is called *reducible* if there is a proper (i.e., neither zero nor all of V) G-invariant subspace of V.

It follows that a nonzero finite-dimensional representation V is irreducible if and only if

$$V = \text{span}_{\mathbb{C}}\{gv \mid g \in G\}$$

for each nonzero $v \in V$, since this property is equivalent to excluding proper G-invariant subspaces. For example, it is well known from linear algebra that this condition is satisfied for each of the standard representations in §2.1.2.1 and so each is irreducible.

For more general representations, this approach is often impossible to carry out. In those cases, other tools are needed. One important tool is based on the next result.

Theorem 2.12 (Schur's Lemma). *Let V and W be finite-dimensional representations of a Lie group G. If V and W are irreducible, then*

$$\dim \text{Hom}_G(V, W) = \begin{cases} 1 & \text{if } V \cong W \\ 0 & \text{if } V \not\cong W. \end{cases}$$

Proof. If nonzero $T \in \text{Hom}_G(V, W)$, then $\ker T$ is not all of V and G-invariant so irreducibility implies T is injective. Similarly, the image of T is nonzero and G-invariant, so irreducibility implies T is surjective and therefore a bijection. Thus there exists a nonzero $T \in \text{Hom}_G(V, W)$ if and only if $V \cong W$.

In the case $V \cong W$, fix a bijective $T_0 \in \text{Hom}_G(V, W)$. If also $T \in \text{Hom}_G(V, W)$, then $T \circ T_0^{-1} \in \text{Hom}_G(V, V)$. Since V is a finite-dimensional vector space over $\mathbb{C}$, there exists an eigenvalue λ for $T \circ T_0^{-1}$. As $\ker(T \circ T_0^{-1} - \lambda I)$ is nonzero and G-invariant, irreducibility implies $T \circ T_0^{-1} - \lambda I = 0$, and so $\text{Hom}_G(V, W) = \mathbb{C}T_0$. □

Note Schur's Lemma implies that

(2.13) $$\text{Hom}_G(V, V) = \mathbb{C}I$$

for irreducible V.

2.2.3 Unitarity

Definition 2.14. (1) Let V be a representation of a Lie group G. A form $(\cdot, \cdot)$: $V \times V \to \mathbb{C}$ is called *G-invariant* if $(gv, gv') = (v, v')$ for $g \in G$ and $v, v' \in V$.
(2) A representation V of a Lie group G is called *unitary* if there exists a G-invariant (Hermitian) inner product on V.

Noncompact groups abound with nonunitary representations (Exercise 2.18). However, compact groups are much more nicely behaved.

Theorem 2.15. *Every representation of a compact Lie group is unitary.*

Proof. Begin with any inner product $\langle \cdot, \cdot \rangle$ on V and define

$$(v, v') = \int_G \langle gv, gv' \rangle \, dg.$$

This is well defined since G is compact and $g \to \langle gv, gv' \rangle$ is continuous. The new form is clearly Hermitian and it is G-invariant since dg is right invariant. It remains only to see it is definite, but by definition, $(v, v) = \int_G \langle gv, gv \rangle \, dg$ which is positive for $v \neq 0$ since $\langle gv, gv \rangle > 0$. $\qquad\square$

Theorem 2.15 provides the underpinning for much of the representation theory of compact Lie groups. It also says a representation (π, V) of a compact Lie group is better than a homomorphism $\pi : G \to GL(V)$; it is a homomorphism to the *unitary group* on V with respect to the G-invariant inner product (Exercise 2.20).

Definition 2.16. A finite-dimensional representation of a Lie group is called *completely reducible* if it is a direct sum of irreducible submodules.

Reducible but not completely reducible representations show up frequently for noncompact groups (Exercise 2.18), but again, compact groups are much simpler. We note that an analogous result will hold even in the infinite-dimensional setting of unitary representations of compact groups (Corollary 3.15).

Corollary 2.17. *Finite-dimensional representations of compact Lie groups are completely reducible.*

Proof. Suppose V is a representation of a compact Lie group G that is reducible. Let $(\cdot, \cdot)$ be a G-invariant inner product. If $W \subseteq V$ is a proper G-invariant subspace, then $V = W \oplus W^\perp$. Moreover, $W^\perp$ is also a proper G-invariant subspace since $(gw', w) = (w', g^{-1}w) = 0$ for $w' \in W^\perp$ and $w \in W$. By the finite dimensionality of V and induction, the proof is finished. $\qquad\square$

As a result, any representation V of a compact Lie group G may be written as

$$(2.18) \qquad\qquad V \cong \bigoplus_{i=1}^{N} n_i V_i,$$

where $\{V_i \mid 1 \leq i \leq N\}$ is a collection of inequivalent irreducible representations of G and $n_i V_i = V_i \oplus \cdots \oplus V_i$ (n_i copies). To study any representation of G, it therefore suffices to understand each irreducible representation and to know how to compute the n_i. In §2.2.4 we will find a formula for n_i.

Understanding the set of irreducible representations will take much more work. The bulk of the remaining text is, in one way or another, devoted to answering this question. In §3.3 we will derive a large amount of information on the set of all irreducible representations by studying functions on G. However, we will not be able to classify and construct all irreducible representations individually until §7.3.5 and §?? where we study highest weights and associated structures.

Corollary 2.19. *If V is a finite-dimensional representation of a compact Lie group G, V is irreducible if and only if* $\dim \operatorname{Hom}_G(V, V) = 1$.

Proof.   If V is irreducible, then Schur's Lemma (Theorem 2.12) implies that $\dim \operatorname{Hom}_G(V, V) = 1$. On the other hand, if V is reducible, then $V = W \oplus W'$ for proper submodules W, W' of V. In particular, this shows that $\dim \operatorname{Hom}_G(V, V) \geq 2$ since it contains the projection onto either summand. Hence $\dim \operatorname{Hom}_G(V, V) = 1$ implies that V is irreducible. $\square$

The above result also has a corresponding version that holds even in the infinite-dimensional setting of unitary representations of compact groups (Theorem 3.12).

Corollary 2.20. *(1) If V is a finite-dimensional representation of a compact Lie group G, then $\overline{V} \cong V^*$.*
(2) If V is irreducible, then the G-invariant inner product is unique up to scalar multiplication by a positive real number.

Proof. For part (1), let $(\cdot, \cdot)$ be a G-invariant inner product on V. Define the bijective linear map $T : \overline{V} \to V^*$ by $Tv = (\cdot, v)$ for $v \in V$. To see that it is a G-map, calculate that $g(Tv) = (g^{-1}\cdot, v) = (\cdot, gv) = T(gv)$.

For part (2), assume V is irreducible. If $(\cdot, \cdot)'$ is another G-invariant inner product on V, define a second bijective linear map $T' : \overline{V} \to V^*$ by $T'v = (\cdot, v)'$. Schur's Lemma (Theorem 2.12) shows that $\dim \operatorname{Hom}_G(\overline{V}, V^*) = 1$. Since $T, T' \in \operatorname{Hom}_G(\overline{V}, V^*)$, there exists $c \in \mathbb{C}$, so $T' = cT$. Thus $(\cdot, v)' = c(\cdot, v)$ for all $v \in V$. It is clear that c must be in $\mathbb{R}$ and positive. $\square$

Corollary 2.21. *Let V be a finite-dimensional representation of a compact Lie group G with a G-invariant inner product $(\cdot, \overset{\bullet}{\cdot})$. If V_1, V_2 are inequivalent irreducible submodules of V, then $V_1 \perp V_2$, i.e., $(V_1, V_2) = 0$.*

Proof. Consider $W = \{v_1 \in V_1 \mid (v_1, V_2) = 0\}$. Since $(\cdot, \cdot)$ is G-invariant, W is a submodule of V_1. If $(V_1, V_2) \neq 0$, i.e., $W \neq V_1$, then irreducibility implies that $W = \{0\}$ and $(\cdot, \cdot)$ yields a nondegenerate pairing of V_1 and V_2. Thus the map $v_1 \to (\cdot, v_1)$ exhibits an equivalence $\overline{V_1} \cong V_2^*$. This implies that $V_1 \cong \overline{V_2}^* \cong V_2$. $\square$

2.2.4 Canonical Decomposition

Definition 2.22. (1) Let G be a compact Lie group. Denote the set of equivalence classes of irreducible (unitary) representations of G by $\widehat{G}$. When needed, choose a representative representation (π, E_π) for each $[\pi] \in \widehat{G}$.
(2) Let V be a finite-dimensional representation of G. For $[\pi] \in \widehat{G}$, let $V_{[\pi]}$ be the largest subspace of V that is a direct sum of irreducible submodules equivalent to E_π. The submodule $V_{[\pi]}$ is called the π-*isotypic component* of V.
(3) The *multiplicity* of π in V, m_π, is $\frac{\dim V_{[\pi]}}{\dim E_\pi}$, i.e., $V_{[\pi]} \cong m_\pi E_\pi$.

First, we verify that $V_{[\pi]}$ is well defined. The following lemma does that as well as showing that $V_{[\pi]}$ is the sum of *all* submodules of V equivalent to E_π.

Lemma 2.23. *If V_1, V_2 are direct sums of irreducible submodules isomorphic to E_π, then so is $V_1 + V_2$.*

Proof. By finite dimensionality, it suffices to check the following: if $\{W_i\}$ are G-submodules of a representation and W_1 is irreducible satisfying $W_1 \not\subseteq W_2 \oplus \cdots \oplus W_n$, then $W_1 \cap (W_2 \oplus \cdots \oplus W_n) = \{0\}$. However, $W_1 \cap (W_2 \oplus \cdots \oplus W_n)$ is a G-invariant submodule of W_1, so the initial hypothesis and irreducibility finish the argument. □

If V, W are representations of a Lie group G and $V \cong W \oplus W$, note this decomposition is not canonical. For example, if $c \in \mathbb{C}\backslash\{0\}$, then $W' = \{(w, cw) \mid w \in W\}$ and $W'' = \{(w, -cw) \mid w \in W\}$ are two other submodules both equivalent to W and satisfying $V \cong W' \oplus W''$. The following result gives a uniform method of handling this ambiguity as well as giving a formula for the n_i in Equation 2.18.

Theorem 2.24 (Canonical Decomposition). *Let V be a finite-dimensional representation of a compact Lie group G.*
(1) There is a G-intertwining isomorphism ι_π

$$\mathrm{Hom}_G(E_\pi, V) \otimes E_\pi \xrightarrow{\cong} V_{[\pi]}$$

induced by mapping $T \otimes v \to T(v)$ for $T \in \mathrm{Hom}_G(E_\pi, V)$ and $v \in V$. In particular, the multiplicity of π is

$$m_\pi = \dim \mathrm{Hom}_G(E_\pi, V).$$

(2) There is a G-intertwining isomorphism

$$\bigoplus_{[\pi] \in \widehat{G}} \mathrm{Hom}_G(E_\pi, V) \otimes E_\pi \xrightarrow{\cong} V = \bigoplus_{[\pi] \in \widehat{G}} V_{[\pi]}.$$

Proof. For part (1), let $T \in \mathrm{Hom}_G(E_\pi, V)$ be nonzero. Then $\ker T = \{0\}$ by the irreducibility of E_π. Thus T is an equivalence of E_π with $T(E_\pi)$, and so $T(E_\pi) \subseteq V_{[\pi]}$. Thus ι_π is well defined. Next, by the definition of the G-action on $\mathrm{Hom}(E_\pi, V)$

and the definition of $\text{Hom}_G(E_\pi, V)$, it follows that G acts trivially on $\text{Hom}_G(E_\pi, V)$. Thus $g(T \otimes v) = T \otimes (gv)$, so $\iota_\pi(g(T \otimes v)) = T(gv) = gT(v) = g\iota_\pi(T \otimes v)$, and so ι_π is a G-map. To see that ι_π is surjective, let $V_1 \cong E_\pi$ be a direct summand in $V_{[\pi]}$ with equivalence given by $T : E_\pi \to V_1$. Then $T \in \text{Hom}_G(E_\pi, V)$ and V_1 clearly lies in the image of ι_π. Finally, make use of a dimension count to show that ι_π is injective. Write $V_{[\pi]} = V_1 \oplus \cdots \oplus V_{m_\pi}$ with $V_i \cong E_\pi$. Then

$$\dim \text{Hom}_G(E_\pi, V) = \dim \text{Hom}_G(E_\pi, V_{[\pi]}) = \dim \text{Hom}_G(E_\pi, V_1 \oplus \cdots \oplus V_{m_\pi})$$

$$= \sum_{i=1}^{m_\pi} \dim \text{Hom}_G(E_\pi, V_i) = m_\pi$$

by Schur's Lemma (Theorem 2.12). Thus $\dim \text{Hom}_G(E_\pi, V) \otimes E_\pi = m_\pi \dim E_\pi = \dim V_{[\pi]}$.

For part (2), it only remains to show that $V = \bigoplus_{[\pi] \in \widehat{G}} V_{[\pi]}$. By Equation 2.18, $V = \sum_{[\pi] \in \widehat{G}} V_{[\pi]}$ and by Corollary 2.21 the sum is direct. $\square$

See Theorem 3.19 for the generalization to the infinite-dimensional setting of unitary representations of compact groups.

2.2.5 Exercises

Exercise 2.14 Verify that the actions given in Definition 2.10 are representations.

Exercise 2.15 (a) Let V and W be finite-dimensional representations of a Lie group G. Show that $V^* \otimes W$ is equivalent to $\text{Hom}(V, W)$ by mapping $T \otimes w$ to the linear map $wT(\cdot)$.
(b) Show, as representations, that $V \otimes V \cong S^2(V) \oplus \bigwedge^2(V)$.

Exercise 2.16 If V is an irreducible finite-dimensional representation of a Lie group G, show that V^* is also irreducible.

Exercise 2.17 This exercise considers a natural generalization of $V_n(\mathbb{C}^2)$. Let W be a representation of a Lie group G. Define $V_n(W)$ to be the space of holomorphic polynomials on W that are homogeneous of degree n and let $(gP)(\eta) = P(g^{-1}\eta)$. Show that there is an equivalence of representations $S^n(W^*) \cong V_n(W)$ induced by viewing $T_1 \cdots T_n, T_i \in W^*$, as a function on W.

Exercise 2.18 (a) Show that the map $\pi : t \to \begin{pmatrix} 1 & t \\ 0 & 1 \end{pmatrix}$ produces a representation of $\mathbb{R}$ on $\mathbb{C}^2$.
(b) Show that this representation is not unitary.
(c) Find all invariant submodules.
(b) Show that the representation is reducible and yet not completely reducible.

Exercise 2.19 Use Schur's Lemma (Theorem 2.12) to quickly calculate the centers of the groups having standard representations listed in §2.1.2.1.

Exercise 2.20 Let $(\cdot, \cdot)$ be an inner product on $\mathbb{C}^n$. Show that $U(n) \cong \{g \in GL(n, \mathbb{C}) \mid (gv, gv') = (v, v') \text{ for } v, v' \in \mathbb{C}^n\}$.

Exercise 2.21 **(a)** Use Equation 2.13 to show that all irreducible finite-dimensional representations of an Abelian Lie group are 1-dimensional (c.f. Exercise 3.18).
(b) Classify all irreducible representations of S^1 and show that $\widehat{S^1} \cong \mathbb{Z}$.
(c) Find the irreducible summands of the representation of S^1 on $\mathbb{C}^2$ generated by the isomorphism $S^1 \cong SO(2)$.
(d) Show that a smooth homomorphism $\varphi : \mathbb{R} \to \mathbb{C}$ satisfies the differential equation $\varphi' = [\varphi'(0)]\varphi$. Use this to show that the set of irreducible representations of $\mathbb{R}$ is indexed by $\mathbb{C}$ and that the unitary ones are indexed by $i\mathbb{R}$.
(e) Use part (d) to show that the set of irreducible representations of $\mathbb{R}^+$ under its multiplicative structure is indexed by $\mathbb{C}$ and that the unitary ones are indexed by $i\mathbb{R}$.
(f) Classify all irreducible representations of $\mathbb{C} \cong \mathbb{R}^2$ under its additive structure and of $\mathbb{C}\backslash\{0\}$ under its multiplicative structure.

Exercise 2.22 Let V be a finite-dimensional representation of a compact Lie group G. Show the set of G-invariant inner products on G is isomorphic to $\text{Hom}_G(V^*, V^*)$.

Exercise 2.23 **(a)** Let $\pi_i : V_i \to U(n)$ be two (unitary) equivalent irreducible representations of a compact Lie group G. Use Corollary 2.20 to show that there exists a *unitary* transformation intertwining π_1 and π_2.
(b) Repeat part (a) without the hypothesis of irreducibility.

Exercise 2.24 Let V be a finite-dimensional representation of a compact Lie group G and let $W \subseteq V$ be a subrepresentation. Show that $W_{[\pi]} \subseteq V_{[\pi]}$ for $[\pi] \in \widehat{G}$.

Exercise 2.25 Suppose V is a finite-dimensional representation of a compact Lie group G. Show that the set of G-intertwining automorphisms of V is isomorphic to $\prod_{[\pi]\in\widehat{G}} GL(m_\pi, \mathbb{C})$ where m_π is the multiplicity of the isotypic component $V_{[\pi]}$.

2.3 Examples of Irreducibility

2.3.1 $SU(2)$ and $V_n(\mathbb{C}^2)$

In this section we show that the representation $V_n(\mathbb{C}^2)$ from §2.1.2.2 of $SU(2)$ is irreducible. In fact, we will later see (Theorem 3.32) these are, up to equivalence, the only irreducible representations of $SU(2)$. The trick employed here points towards the powerful techniques that will be developed in §4 where derivatives, i.e., the tangent space of G, are studied systematically (c.f. Lemma 6.6).

Let $H \subseteq V_n(\mathbb{C}^2)$ be a nonzero invariant subspace. From Equation 2.3,

$$(2.25) \qquad \text{diag}(e^{i\theta}, e^{-i\theta}) \cdot (z_1^k z_2^{n-k}) = e^{i(n-2k)\theta} z_1^k z_2^{n-k}.$$

As the joint eigenvalues $e^{i(n-2k)\theta}$ are distinct and since H is preserved by $\{\text{diag}(e^{i\theta}, e^{-i\theta})\}$, H is spanned by some of the joint eigenvectors $z_1^k z_2^{n-k}$. In particular, there is a k_0, so $z_1^{k_0} z_2^{n-k_0} \in H$.

Let $K_t = \begin{pmatrix} \cos t & -\sin t \\ \sin t & \cos t \end{pmatrix} \in SU(2)$ and let $\eta_t = \begin{pmatrix} \cos t & i \sin t \\ i \sin t & \cos t \end{pmatrix} \in SU(2)$.

Since H is $SU(2)$ invariant, $\frac{1}{2}(K_t \pm i\eta_t) z_1^{k_0} z_2^{n-k_0} \in H$. Thus, when the limits exist, $\frac{d}{dt}\left[\frac{1}{2}(K_t \pm i\eta_t) z_1^{k_0} z_2^{n-k_0}\right]|_{t=0} \in H$. Using Equation 2.3, a simple calculation (Exercise 2.26) shows that

$$(2.26) \qquad \frac{1}{2}\frac{d}{dt}\left[(K_t \pm i\eta_t) z_1^{k_0} z_2^{n-k_0}\right]|_{t=0} = \begin{cases} k_0 \, z_1^{k_0-1} z_2^{n-k_0+1} & \text{for} + \\ (k_0 - n) \, z_1^{k_0+1} z_2^{n-k_0-1} & \text{for} - . \end{cases}$$

Induction therefore implies that $V_n(\mathbb{C}) \subseteq H$, and so $V_n(\mathbb{C})$ is irreducible.

2.3.2 $SO(n)$ and Harmonic Polynomials

In this section we show that the representation of $SO(n)$ on the harmonic polynomials $\mathcal{H}_m(\mathbb{R}^n) \subseteq V_m(\mathbb{R}^n)$ is irreducible (see §2.1.2.3 for notation). Let $D_m(\mathbb{R}^n)$ be the space of complex constant coefficient differential operators on $\mathbb{R}^n$ of degree m. Recall that the algebra isomorphism from $\bigoplus_m V_m(\mathbb{R}^n)$ to $\bigoplus_m D_m(\mathbb{R}^n)$ is generated by mapping $x_i \to \partial_{x_i}$. In general, if $q \in \bigoplus_m V_m(\mathbb{R}^n)$, write ∂_q for the corresponding element of $\bigoplus_m D_m(\mathbb{R}^n)$.

Define $\langle \cdot, \cdot \rangle$ a Hermitian form on $V_m(\mathbb{R}^n)$ by $\langle p, q \rangle = \partial_{\bar{q}}(p) \in \mathbb{C}$ for $p, q \in V_m(\mathbb{R}^n)$. Since $\{x_1^{k_1} x_2^{k_2} \ldots x_n^{k_n} \mid k_i \in \mathbb{N}$ and $k_1 + k_2 + \cdots + k_n = m\}$ turns out to be an orthogonal basis for $V_m(\mathbb{R}^n)$, it is easy to see that $\langle \cdot, \cdot \rangle$ is an inner product. In fact, $\langle \cdot, \cdot \rangle$ is actually $O(n)$-invariant (Exercise 2.27), although we will not need this fact.

Lemma 2.27. *With respect to the inner product* $\langle \cdot, \cdot \rangle$ *on* $V_m(\mathbb{R}^n)$, $\mathcal{H}_m(\mathbb{R}^n)^{\perp} = |x|^2 V_{m-2}(\mathbb{R}^n)$ *where* $|x|^2 = \sum_{i=1}^n x_i^2 \in V_2(\mathbb{R}^n)$. *As* $O(n)$-*modules,*

$$V_m(\mathbb{R}^n) \cong \mathcal{H}_m(\mathbb{R}^n) \oplus \mathcal{H}_{m-2}(\mathbb{R}^n) \oplus \mathcal{H}_{m-4}(\mathbb{R}^n) \oplus \cdots.$$

Proof. Let $p \in V_m(\mathbb{R}^n)$ and $q \in V_{m-2}(\mathbb{R}^n)$. Then $\langle p, |x|^2 q \rangle = \partial_{|x|^2 \bar{q}} p = \partial_{\bar{q}} \Delta p = \langle \Delta p, q \rangle$. Thus $\left[|x|^2 V_{m-2}(\mathbb{R}^n)\right]^{\perp} = \mathcal{H}_m(\mathbb{R}^n)$ so that

$$(2.28) \qquad V_m(\mathbb{R}^n) = \mathcal{H}_m(\mathbb{R}^n) \oplus |x|^2 V_{m-2}(\mathbb{R}^n).$$

Induction therefore shows that

$$V_m(\mathbb{R}^n) = \mathcal{H}_m(\mathbb{R}^n) \oplus |x|^2 \mathcal{H}_{m-2}(\mathbb{R}^n) \oplus |x|^4 \mathcal{H}_{m-4}(\mathbb{R}^n) \oplus \cdots.$$

The last statement of the lemma follows by observing that $O(n)$ fixes $|x|^{2k}$. $\qquad \square$

By direct calculation, $\dim \mathcal{H}_m(\mathbb{R}^1) = 0$ for $m \geq 2$. For $n \geq 2$, however, it is clear that $\dim V_m(\mathbb{R}^n) > \dim V_{m-1}(\mathbb{R}^n)$ so that $\dim \mathcal{H}_m(\mathbb{R}^n) \geq 1$.

Lemma 2.29. *If G is a compact Lie group with finite-dimensional representations U, V, W satisfying $U \oplus V \cong U \oplus W$, then $V \cong W$.*

Proof. Using Equation 2.18, decompose $U \cong \bigoplus_{[\pi] \in \widehat{G}} m_\pi E_\pi$, $V \cong \bigoplus_{[\pi] \in \widehat{G}} m'_\pi E_\pi$, and $W \cong \bigoplus_{[\pi] \in \widehat{G}} m''_\pi E_\pi$. The condition $U \oplus V \cong U \oplus W$ therefore implies that $m_\pi + m'_\pi = m_\pi + m''_\pi$ so that $m'_\pi = m''_\pi$ and $V \cong W$.   □

Definition 2.30. If H is a Lie subgroup of a Lie group G and V is a representation of G, write $V|_H$ for the representation of H on V given by restricting the action of G to H.

For the remainder of this section, view $O(n-1)$ as a Lie subgroup of $O(n)$ via the embedding $g \to \begin{pmatrix} 1 & 0 \\ 0 & g \end{pmatrix}$.

Lemma 2.31.

$$\mathcal{H}_m(\mathbb{R}^n)|_{O(n-1)} \cong \mathcal{H}_m(\mathbb{R}^{n-1}) \oplus \mathcal{H}_{m-1}(\mathbb{R}^{n-1}) \oplus \cdots \oplus \mathcal{H}_0(\mathbb{R}^{n-1}).$$

Proof. Any $p \in V_m(\mathbb{R}^n)$ may be uniquely written as $p = \sum_{k=0}^m x_1^k p_k$ with $p_k \in V_{m-k}(\mathbb{R}^{n-1})$ where $\mathbb{R}^n$ is viewed as $\mathbb{R} \times \mathbb{R}^{n-1}$. Since $O(n-1)$ acts trivially on x_1^k,

$$(2.32) \qquad V_m(\mathbb{R}^n)|_{O(n-1)} \cong \bigoplus_{k=0}^m V_{m-k}(\mathbb{R}^{n-1}).$$

Applying Equation 2.28 first (restricted to $O(n-1)$) and then Equation 2.32, we get

$$V_m(\mathbb{R}^n)|_{O(n-1)} \cong \mathcal{H}_m(\mathbb{R}^n)|_{O(n-1)} \oplus \bigoplus_{k=0}^{m-2} V_{m-2-k}(\mathbb{R}^{n-1}).$$

Applying Equation 2.32 first and then Equation 2.28 yields

$$V_m(\mathbb{R}^n)|_{O(n-1)} \cong \bigoplus_{k=0}^m \left[\mathcal{H}_{m-k}(\mathbb{R}^{n-1}) \oplus V_{m-2-k}(\mathbb{R}^{n-1}) \right]$$
$$= \left[\bigoplus_{k=0}^m \mathcal{H}_{m-k}(\mathbb{R}^{n-1}) \right] \oplus \left[\bigoplus_{k=0}^{m-2} V_{m-2-k}(\mathbb{R}^{n-1}) \right].$$

The proof is now finished by Lemma 2.29.   □

Theorem 2.33. $\mathcal{H}_m(\mathbb{R}^n)$ *is an irreducible $O(n)$-module and, in fact, is irreducible under $SO(n)$ for $n \geq 3$.*

Proof. See Exercise 2.31 for the case of $n = 2$. In this proof assume $n \geq 3$.

$\mathcal{H}_m(\mathbb{R}^n)|_{SO(n-1)}$ *contains, up to scalar multiplication, a unique $SO(n-1)$-invariant function:* If $f \in \mathcal{H}_m(\mathbb{R}^n)$ is nonzero and $SO(n)$-invariant, then it is constant on each sphere in $\mathbb{R}^n$ and thus a function of the radius. Homogeneity implies that $f(x) = C |x|^m$ for some nonzero constant. It is trivial to check the condition that $\Delta f = 0$ now forces $m = 0$. Thus only $\mathcal{H}_0(\mathbb{R}^n)$ contains a nonzero $SO(n)$-invariant function. The desired result now follows from the previous observation and Lemma 2.31.

If V is a finite-dimensional $SO(n)$-invariant subspace of continuous functions on S^{n-1}, then V contains a nonzero $SO(n-1)$-invariant function: Here the action of $SO(n)$ on V is, as usual, given by $(gf)(s) = f(g^{-1}s)$. Since $SO(n)$ acts transitively on S^{n-1} and V is nonzero invariant, there exists $f \in V$, so $f(1, 0, \ldots, 0) \neq 0$.

Define $\widetilde{f}(s) = \int_{SO(n-1)} f(gs)\,dg$. If $\{f_i\}$ is a basis of V, then $f(gs) = (g^{-1}f)(s)$ and so may be written as $f(gs) = \sum_i c_i(g) f_i(s)$ for some smooth functions c_i. By integrating, it follows that $\widetilde{f} \in V$. From the definition, it is clear that $\widetilde{f}$ is $SO(n-1)$-invariant. It is nonzero since $\widetilde{f}(1, 0, \dots, 0) = f(1, 0, \dots, 0)$.

$\mathcal{H}_m(\mathbb{R}^n)$ *is an irreducible $SO(n)$-module:* Suppose $\mathcal{H}_m(\mathbb{R}^n) = V_1 \oplus V_2$ for proper $SO(n)$-invariant subspaces. By homogeneity, restricting functions in V_i from $\mathbb{R}^n$ to S^{n-1} is injective. Hence, both V_1 and V_2 contain independent $SO(n-1)$-invariant functions. But this contradicts the fact that $\mathcal{H}_m(\mathbb{R}^n)$ has only one independent $SO(n-1)$-invariant function. $\qquad\square$

A relatively small dose of functional analysis (Exercise 3.14) can be used to further show that $L^2(S^{n-1}) = \bigoplus_{m=0}^{\infty} \mathcal{H}_m(\mathbb{R}^n)|_{S^{n-1}}$ (Hilbert space direct sum) and that $\mathcal{H}_m(\mathbb{R}^n)|_{S^{n-1}}$ is the eigenspace of the Laplacian on S^{n-1} with eigenvalue $-m(n+m-2)$.

2.3.3 Spin and Half-Spin Representations

The spin representation $S = \bigwedge W$ of $\mathrm{Spin}_n(\mathbb{R})$ for n odd and the half-spin representations $S^{\pm} = \bigwedge^{\pm} W$ for n even were constructed in §2.1.2.4, where W is a maximal isotropic subspace of $\mathbb{C}^n$. This section shows that these representations are irreducible.

For n even with $n = 2m$, let $W = \{(z_1, \dots, z_m, iz_1, \dots, iz_m) \mid z_k \in \mathbb{C}\}$ and $W' = \{(z_1, \dots, z_m, -iz_1, \dots, -iz_m) \mid z_k \in \mathbb{C}\}$. Identify W with $\mathbb{C}^m$ by projecting onto the first m coordinates. For $x = (x_1, \dots, x_m)$ and $y = (y_1, \dots, y_m)$ in $\mathbb{R}^m$, let $(x, y) = (x_1, \dots, x_m, y_1, \dots, y_m) \in \mathbb{R}^n$. In particular, $(x, y) = \frac{1}{2}(x - iy, i(x - iy)) + \frac{1}{2}(x + iy, -i(x + iy))$. Using Definition 2.7, the identification of $\mathbb{C}^m$ with W, and noting $((a, -ia), (b, ib)) = 2(a, b)$, the spin action of $\mathrm{Spin}_{2m}(\mathbb{R})$ on $\bigwedge^{\pm} \mathbb{C}^m \cong S^{\pm}$ is induced by having (x, y) act as

$$(2.34) \qquad \frac{1}{2}\epsilon(x - iy) - 2\iota(x + iy).$$

For n odd with $n = 2m + 1$, take $W = \{(z_1, \dots, z_m, iz_1, \dots, iz_m, 0) \mid z_k \in \mathbb{C}\}$, $W' = \{(z_1, \dots, z_m, -iz_1, \dots, -iz_m, 0) \mid z_k \in \mathbb{C}\}$, and $e_0 = (0, \dots, 0, 1)$. As above, identify W with $\mathbb{C}^m$ by projecting onto the first m coordinates. For $x = (x_1, \dots, x_m)$ and $y = (y_1, \dots, y_m)$ in $\mathbb{R}^m$ and $u \in \mathbb{R}$, let $(x, y, u) = (x_1, \dots, x_m, y_1, \dots, y_m, u) \in \mathbb{R}^n$. In particular, $(x, y, u) = \frac{1}{2}(x - iy, i(x - iy), 0) + \frac{1}{2}(x + iy, -i(x + iy), 0) + (0, 0, u)$. Using Definition 2.7 and the identification of $\mathbb{C}^m$ with W, the spin action of $\mathrm{Spin}_{2m+1}(\mathbb{R})$ on $\bigwedge \mathbb{C}^m \cong S$ is induced by having (x, y, u) act as

$$(2.35) \qquad \frac{1}{2}\epsilon(x - iy) - 2\iota(x + iy) + (-1)^{\deg} m_{iu}.$$

Theorem 2.36. *For n even, the half-spin representations $S^{\pm}$ of $\mathrm{Spin}_n(\mathbb{R})$ are irreducible. For n odd, the spin representation S of $\mathrm{Spin}_n(\mathbb{R})$ is irreducible.*

Proof. Using the standard basis $\{e_j\}_{j=1}^n$, calculate

$$(e_j \pm ie_{j+m})(e_k \pm ie_{k+m}) = e_je_k \pm i(e_je_{k+m} + e_{j+m}e_k) - e_{j+m}e_{k+m}$$

for $1 \le j, k \le m$. Since e_je_k, e_je_{k+m}, $e_{j+m}e_k$, and $e_{j+m}e_{k+m}$ lie in $\mathrm{Spin}_n(\mathbb{R})$, Equations 2.34 and 2.35 imply that the operators $\epsilon(e_j)\epsilon(e_k)$ and $\iota(e_j)\iota(e_k)$ on $\bigwedge \mathbb{C}^m$ are achieved by linear combinations of the action of elements of $\mathrm{Spin}_n(\mathbb{R})$ on $\bigwedge \mathbb{C}^m$.

For n *even*, let W be a nonzero $\mathrm{Spin}_n(\mathbb{R})$-invariant subspace contained in either $S^+ \cong \bigwedge^+ \mathbb{C}^m$ or $S^- \cong \bigwedge^- \mathbb{C}^m$. The operators $\epsilon(e_j)\epsilon(e_k)$ can be used to show that W contains a nonzero element in either $\bigwedge^m \mathbb{C}^m$ or $\bigwedge^{m-1} \mathbb{C}^m$, depending on the parity of m. In the first case, since $\dim \bigwedge^m \mathbb{C}^m = 1$, the operators $\iota(e_j)\iota(e_k)$ can be used to generate all of $\bigwedge^\pm \mathbb{C}^m$. In the second case, the operators $\iota(e_j)\iota(e_k)$ and $\epsilon(e_{j'})\epsilon(e_k)$ can be used to generate all of $\bigwedge^{m-1} \mathbb{C}^m$ after which the operators $\iota(e_j)\iota(e_k)$ can be used to generate all of $S^\pm$. Thus both half-spin representation are irreducible.

Similarly, for n *odd*, examination of the element $(e_j \pm ie_{j+m})e_n$ shows that the operators $\epsilon(e_j)(-1)^{\deg}$ and $\iota(e_j)(-1)^{\deg}$ are obtainable as linear combinations of the action of elements of $\mathrm{Spin}_n(\mathbb{R})$ on $\bigwedge \mathbb{C}^m$. Hence any nonzero $\mathrm{Spin}_n(\mathbb{R})$-invariant subspace W of $\bigwedge \mathbb{C}^m$ contains $\bigwedge^m \mathbb{C}^m$ by use of the operators $\epsilon(e_j)(-1)^{\deg}$. Finally, the operators $\iota(e_j)(-1)^{\deg}$ can then be used to show that $W = \bigwedge \mathbb{C}^m$ so that S is irreducible.   □

2.3.4 Exercises

Exercise 2.26 Verify Equation 2.26.

Exercise 2.27 (a) For $g \in O(n)$, use the chain rule to show that $\partial_{g \cdot x_i} f = g\left(\partial_{x_i} f\right)$ for smooth f on $\mathbb{R}^n$.
(b) For $g \in O(n)$, show that $\partial_{g \cdot p} f = g\left(\partial_p f\right)$ for $p \in V_m(\mathbb{R}^n)$.
(c) Show that $\langle \cdot, \cdot \rangle$ is $O(n)$-invariant on $V_m(\mathbb{R}^n)$.

Exercise 2.28 For $p \in V_m(\mathbb{R}^n)$ show that there exists a unique $h \in \bigoplus_k \mathcal{H}_{m-2k}(\mathbb{R}^n)$, so $p|_{S^{n-1}} = h|_{S^{n-1}}$.

Exercise 2.29 Show that Δ is an $O(n)$-map from $V_m(\mathbb{R}^n)$ onto $V_{m-2}(\mathbb{R}^n)$.

Exercise 2.30 Show that $\dim \mathcal{H}_0(\mathbb{R}^n) = 1$, $\dim \mathcal{H}_1(\mathbb{R}^n) = n$, and $\dim \mathcal{H}_m(\mathbb{R}^n) = \frac{(2m+n-2)(m+n-3)!}{m!(n-2)!}$ for $m \ge 2$.

Exercise 2.31 Show that $\mathcal{H}_m(\mathbb{R}^2)$ is $O(2)$-irreducible but not $SO(2)$-irreducible when $m \ge 2$.

Exercise 2.32 Exercises 2.32 through 2.34 outline an alternate method of proving irreducibility of $\mathcal{H}_m(\mathbb{R}^n)$ using reproducing kernels ([6]). Let $\mathcal{H}$ be a Hilbert space of functions on a space X that is closed under conjugation and such that evaluation at any $x \in X$ is a continuous operator on $\mathcal{H}$. Write $(\cdot, \cdot)$ for the inner product on $\mathcal{H}$. Then for $x \in X$, there exists a unique $\phi_x \in \mathcal{H}$, so $f(x) = (f, \phi_x)$ for $f \in \mathcal{H}$. The function $\Phi : X \times X \to \mathbb{C}$, given by $(x, y) \to (\phi_y, \phi_x)$, is called the *reproducing kernel*.

(a) Show that $\Phi(x, y) = \phi_y(x)$ for $x, y \in X$ and $f(x) = (f, \Phi(\cdot, x))$ for $f \in \mathcal{H}$.

(b) Show that span$\{\phi_x \mid x \in X\}$ is dense in $\mathcal{H}$.

(c) If $\{e_\alpha\}_{\alpha \in A}$ is an orthonormal basis of $\mathcal{H}$, then $\Phi(x, y) = \sum_\alpha e_\alpha(x) \overline{e_\alpha(y)}$.

(d) If there exists a measure μ on X such that $\mathcal{H}$ is a closed subspace in $L^2(X, d\mu)$, then $f(x) = \int_X \overline{\Phi(y, x)} f(y) \, d\mu(y)$.

Exercise 2.33 Suppose there is a Lie group G acting transitively on X. Fix $x_0 \in X$ so that $X \cong G/H$ where $H = G^{x_0}$. Let G act on functions by $(gf)(x) = f(g^{-1}x)$ for $g \in G$ and $x \in X$. Assume this action preserves $\mathcal{H}$, is unitary, and that $\mathcal{H}^H = \{f \in \mathcal{H} \mid hf = f \text{ for } h \in H\}$ is one dimensional.

(a) Show that $g\phi_x = \phi_{gx}$ and $\Phi(gx, gy) = \Phi(x, y)$ for $g \in G$ and $x, y \in X$.

(b) Let W be a nonzero closed G-invariant subspace of $\mathcal{H}$ and write Φ_W for its reproducing kernel. Show that the function $x \to \Phi_W(x, x_0)$ lies in $\mathcal{H}^H$.

(c) Show that G acts irreducibly on $\mathcal{H}$.

Exercise 2.34 Let $\mathcal{H} = \mathcal{H}_m(\mathbb{R}^n) \subseteq V_m(\mathbb{R}^n)$, where $V_m(\mathbb{R}^n)$ is viewed as sitting in $L^2(S^{n-1})$ by restriction to S^{n-1}. Let $p_0 = (1, 0, \dots, 0)$.

(a) Show that $V_m(\mathbb{R}^n)^{O(n-1)}$ consists of all functions of the form

$$x \to \sum_{j=0}^{\lfloor \frac{m}{2} \rfloor} (-1)^j c_j (x, x)^j (x, p_0)^{k-2j}$$

for constants $c_j \in \mathbb{C}$.

(b) Find a linear recurrence formula on the c_j to show that $\dim \mathcal{H}_m(\mathbb{R}^n)^{O(n-1)} = 1$.

(c) Show that $\mathcal{H}_m(\mathbb{R}^n)$ is irreducible under $O(n)$.

(d) Show that $\mathcal{H}_m(\mathbb{R}^n)$ is still irreducible under restriction to $SO(n)$ for $n \geq 3$.

Exercise 2.35 Let $G = U(n)$, $V_{p,q}(\mathbb{C}^n)$ be the set of complex polynomials homogeneous of degree p in $z_1, \dots, z_n$ and homogeneous of degree q in $\overline{z_1}, \dots, \overline{z_n}$ equipped with the typical action of G, $\Delta_{p,q} = \sum_j \partial_{z_j} \partial_{\overline{z_j}}$, and $\mathcal{H}_{p,q}(\mathbb{C}^n) = V_{p,q}(\mathbb{C}^n) \cap \ker \Delta_{p,q}$. Use restriction to S^{2n-1} and techniques similar to those found in Exercises 2.32 through 2.34 to demonstrate the following.

(a) Show that $\Delta_{p,q}$ is a G-map from $V_{p,q}(\mathbb{C}^n)$ onto $V_{p-1,q-1}(\mathbb{C}^n)$.

(b) Show $\mathcal{H}_m(\mathbb{R}^{2n}) \cong \bigoplus_{p+q=m} \mathcal{H}_{p,q}(\mathbb{C}^n)$.

(c) Show that $\mathcal{H}_{p,q}(\mathbb{C}^n)$ is $U(n)$-irreducible.

(d) Show that $\mathcal{H}_{p,q}(\mathbb{C}^n)$ is still irreducible under restriction to $SU(n)$.

Exercise 2.36 Show that S^+ and S^- are inequivalent representations of $\text{Spin}_n(\mathbb{R})$ for n even.

3

Harmonic Analysis

Throughout this chapter let G be a *compact* Lie group. This chapter studies a number of function spaces on G such as the set of continuous functions on G, $C(G)$, or the set of square integrable functions on G, $L^2(G)$, with respect to the Haar measure dg. These function spaces are examined in the light of their behavior under left and right translation by G.

3.1 Matrix Coefficients

3.1.1 Schur Orthogonality

Let (π, V) be a finite-dimensional unitary representation of a compact Lie group G with G-invariant inner product $(\cdot, \cdot)$. If $\{v_i\}$ is a basis for V, let $\{v_i^*\}$ be the *dual basis* for V, i.e., $(v_i, v_j^*) = \delta_{i,j}$ where $\delta_{i,j}$ is 1 when $i = j$ and 0 when $i \neq j$. With respect to this basis, the linear transformation $\pi(g) : V \to V$, $g \in G$, can be realized as matrix multiplication by the matrix whose entry in the $(i, j)^{\text{th}}$ position is

$$(gv_j, v_i^*).$$

The function $g \to (gv_j, v_i^*)$ is a smooth complex-valued function on G. The study of linear combinations of such functions turns out to be quite profitable.

Definition 3.1. Any function on a compact Lie group G of the form $f_{u,v}^V(g) = (gu, v)$ for a finite-dimensional unitary representation V of G with $u, v \in V$ and G-invariant inner product $(\cdot, \cdot)$ is called a *matrix coefficient* of G. The collection of all matrix coefficients is denoted $MC(G)$.

Lemma 3.2. $MC(G)$ *is a subalgebra of the set of smooth functions on G and contains the constant functions. If $\{v_i^\pi\}_{i=1}^{n_\pi}$ is a basis for E_π, $[\pi] \in \widehat{G}$, then $\{f_{v_i^\pi, v_j^\pi}^{E_\pi} \mid [\pi] \in \widehat{G}$ and $1 \leq i, j \leq n_\pi\}$ span $MC(G)$.*

Proof. By definition, a matrix coefficient is clearly a smooth function on G. If V, V' are unitary representations of G with G-invariant inner products $(\cdot, \cdot)_V$ and $(\cdot, \cdot)_{V'}$, then $U \oplus V$ is unitary with respect to the inner product $\left((u, v), (u', v')\right)_{V \oplus V'} = (u, u')_V + (v, v')_{V'}$ and $V \otimes V'$ is unitary with respect to the inner product

$$\left(\sum_i u_i \otimes v_i, \sum_j u'_j \otimes v'_j \right)_{V \otimes V'} = \sum_{i,j} (u_i, u'_j)_V \, (v_i, v'_j)_{V'}$$

(Exercise 3.1). Thus $cf^V_{u,u'} + f^{V'}_{v,v'} = f^{V \oplus V'}_{(cu,v),(u',v')}$, so $MC(G)$ is a subspace and $f^V_{u,u'} f^{V'}_{v,v'} = f^{V \otimes V'}_{u \otimes v, u' \otimes v'}$, so $MC(G)$ is an algebra. The constant functions are easily achieved as matrix coefficients of the trivial representation.

To verify the final statement of the lemma, first decompose V into irreducible mutually perpendicular summands (Exercise 3.2) as $V = \bigoplus_i V_i$ where each $V_i \cong E_{\pi_i}$. Any v, $v' \in V$ can be written $v = \sum_i v_i$ and $v' = \sum_i v'_i$ with $v_i, v'_i \in V_i$ so that $f^V_{v,v'} = \sum_i f^{V_i}_{v_i, v'_i}$. If $T_i : V_i \to E_{\pi_i}$ is an intertwining isomorphism, then $(T_i v_i, T_i v'_i)_{E_{\pi_i}} = (v_i, v'_i)_V$ defines a unitary structure on E_{π_i} so that $f^V_{v,v'} = \sum_i f^{E_{\pi_i}}_{T_i v_i, T_i v'_i}$. Expanding $T_i v_i$ and $T_i v'_i$ in terms of the basis for E_{π_i} finishes the proof. $\square$

The next theorem calculates the L^2 inner product of the matrix coefficients corresponding to irreducible representations.

Theorem 3.3 (Schur Orthogonality Relations). *Let U, V be irreducible finite-dimensional unitary representations of a compact Lie group G with G-invariant inner products $(\cdot, \cdot)_U$ and $(\cdot, \cdot)_V$. If $u_i \in U$ and $v_i \in V$,*

$$\int_G (gu_1, u_2)_U \, \overline{(gv_1, v_2)_V} \, dg = \begin{cases} 0 & \text{if } U \not\cong V \\ \frac{1}{\dim V} (u_1, v_1)_V \, \overline{(u_2, v_2)_V} & \text{if } U = V. \end{cases}$$

Proof. For $u \in U$ and $v \in V$, define $T_{u,v} : U \to V$ by $T_{u,v}(\cdot) = v \, (\cdot, u)_U$. For the sake of clarity, initially write the action of each representation as (π_U, U) and (π_V, V). Then the function $g \to \pi_U(g) \circ T_{u,v} \circ \pi_V^{-1}(g)$, $g \in G$, can be viewed, after choosing bases, as a matrix valued function. Integrating on each coordinate of the matrix (*c.f.* vector-valued integration in §3.2.2), define $\widetilde{T}_{u,v} : U \to V$ by

$$\widetilde{T}_{u,v} = \int_G \pi_U(g) \circ T_{u,v} \circ \pi_V^{-1}(g) \, dg.$$

For $h \in G$, the invariance of the measure implies that

$$\pi_U(h) \circ \widetilde{T}_{u,v} = \int_G \pi_U(hg) \circ T_{u,v} \circ \pi_V^{-1}(g) \, dg = \int_G \pi_U(g) \circ T_{u,v} \circ \pi_V^{-1}(h^{-1}g) \, dg$$

$$= \widetilde{T}_{u,v} \circ \pi_V(h),$$

so that $\widetilde{T}_{u,v} \in \mathrm{Hom}_G(U, V)$. Irreducibility and Schur's Lemma (Theorem 2.12) show that $\widetilde{T}_{u,v} = cI$ where $c = c(u, v) \in \mathbb{C}$ with $c = 0$ when $U \not\cong V$. Unwinding the definitions and using the change of variables $g \to g^{-1}$, calculate

$$c(u_1, v_1)_V = (\widetilde{T}_{u_2,v_2} u_1, v_1)_V = \int_G (g T_{u_2,v_2} g^{-1} u_1, v_1)_V \, dg$$

$$= \int_G ((g^{-1} u_1, u_2)_U \, g v_2, v_1)_V \, dg = \int_G (g u_1, u_2)_U \, (g^{-1} v_2, v_1)_V \, dg$$

$$= \int_G (g u_1, u_2)_U \, (v_2, g v_1)_V \, dg = \int_G (g u_1, u_2)_U \, \overline{(g v_1, v_2)}_V \, dg.$$

Thus the theorem is finished when $U \ncong V$. When $U = V$, it remains to calculate c. For this, take the trace of the identity $cI = \widetilde{T}_{u_2,v_2}$ to get

$$c \dim V = \operatorname{tr} \widetilde{T}_{u_2,v_2} = \int_G \operatorname{tr}\left[g \circ T_{u_2,v_2} \circ g^{-1} \right] dg$$

$$= \int_G \operatorname{tr} T_{u_2,v_2} \, dg = \operatorname{tr} T_{u_2,v_2}.$$

To quickly calculate $\operatorname{tr} T_{u_2,v_2}$ for nonzero u_2, choose a basis for $U = V$ with v_2 as the first element. Since $T_{u_2,v_2}(\cdot) = v_2(\cdot, u_2)_V$, $\operatorname{tr} T_{u_2,v_2} = (v_2, u_2)_V$, so that $c = \frac{1}{\dim V} \overline{(u_2, v_2)}_V$ which finishes the proof. $\qquad\square$

If $U \cong V$ and $T : U \to V$ is a G-intertwining isomorphism, Theorem 2.20 implies there is a positive constant $c \in \mathbb{R}$, so that $(u_1, u_2)_U = c(Tu_1, Tu_2)_V$. In this case, the Schur orthogonality relation becomes

$$\int_G (g u_1, u_2)_U \, \overline{(g v_1, v_2)}_V \, dg = \frac{c}{\dim V} (Tu_1, v_1)_V \, \overline{(Tu_2, v_2)}_V.$$

Of course, T can be scaled so that $c = 1$ by replacing T with $\sqrt{c} T$.

3.1.2 Characters

Definition 3.4. The *character* of a finite-dimensional representation (π, V) of a compact Lie group G is the function on G defined by $\chi_V(g) = \operatorname{tr} \pi(g)$.

It turns out that character theory provides a powerful tool for studying representations. In fact, we will see in Theorem 3.7 below that, up to equivalence, a character completely determines the representation. Note for $\dim V > 1$, a character in the above sense is usually not a homomorphism.

Theorem 3.5. *Let* V, V_i *be finite-dimensional representations of a compact Lie group* G.
(1) $\chi_V \in MC(G)$.
(2) $\chi_V(e) = \dim V$.
(3) *If* $V_1 \cong V_2$, *then* $\chi_{V_1} = \chi_{V_2}$.
(4) $\chi_V(hgh^{-1}) = \chi_V(g)$ *for* $g, h \in G$.
(5) $\chi_{V_1 \oplus V_2} = \chi_{V_1} + \chi_{V_2}$.
(6) $\chi_{V_1 \otimes V_2} = \chi_{V_1} \chi_{V_2}$.
(7) $\chi_{V^*}(g) = \chi_{\overline{V}}(g) = \overline{\chi_V(g)} = \chi_V(g^{-1})$.
(8) $\chi_{\mathbb{C}}(g) = 1$ *for the trivial representation* $\mathbb{C}$.

Proof. Each statement of the theorem is straightforward to prove. We prove parts (1), (4), (5), and (7) and leave the rest as an exercise (Exercise 3.3). For part (1), let $\{v_i\}$ be an orthonormal basis for V with respect to a G-invariant inner product $(\cdot, \cdot)$. Then $\chi_V(g) = \sum_i (gv_i, v_i)$ so that $\chi_V \in MC(G)$. For part (4), calculate

$$\chi_V(hgh^{-1}) = \operatorname{tr}\left[\pi(h)\pi(g)\pi(h)^{-1}\right] = \operatorname{tr}\pi(g) = \chi_V(g).$$

For part (5), §2.2.1 shows that the action of G on $V_1 \oplus V_2$ can be realized by a matrix of the form $\begin{pmatrix} * & 0 \\ 0 & * \end{pmatrix}$ where the upper left block is given by the action of G on V_1 and the lower right block is given by the action of G on V_2. Taking traces finishes the assertion. For part (7), the equivalence $V^* \cong \overline{V}$ shows $\chi_{V^*}(g) = \chi_{\overline{V}}(g)$. From the discussion in §2.2.1 on $\overline{V}$, the matrix realizing the action of g on $\overline{V}$ is the conjugate of the matrix realizing the action of g on V. Taking traces shows $\chi_{\overline{V}}(g) = \overline{\chi_V(g)}$. Similarly, from the discussion in §2.2.1 on V^*, the matrix realizing the action of g on V^* is the inverse transpose of the matrix realizing the action of g on V. Taking traces shows $\chi_{V^*}(g) = \chi_V(g^{-1})$. $\square$

Definition 3.6. If V is a finite-dimensional representation of a Lie group G, let $V^G = \{v \in V \mid gv = v \text{ for } g \in G\}$, i.e., V^G is the isotypic component of V corresponding to the trivial representation.

The next theorem calculates the L^2 inner product of characters corresponding to irreducible representations.

Theorem 3.7. *(1) Let V, W be finite-dimensional representations of a compact Lie group G. Then*

$$\int_G \chi_V(g)\,\overline{\chi_W(g)}\,dg = \dim \operatorname{Hom}_G(V, W).$$

In particular, $\int_G \chi_V(g)\,dg = \dim V^G$ and if V, W are irreducible, then

$$\int_G \chi_V(g)\,\overline{\chi_W(g)}\,dg = \begin{cases} 0 & \text{if } V \not\cong W \\ 1 & \text{if } U \cong V. \end{cases}$$

(2) Up to equivalence, V is completely determined by its character, i.e., $\chi_V = \chi_W$ if and only if $V \cong W$. In particular, if V_i are representations of G, then $V \cong \bigoplus_i n_i V_i$ if and only if $\chi_V = \sum_i n_i \chi_{V_i}$.
(3) V is irreducible if and only if $\int_G |\chi_V(g)|^2\,dg = 1$.

Proof. Begin with the assumption that V, W are irreducible. Let $\{v_i\}$ and $\{w_j\}$ be an orthonormal bases for V and W with respect to the G-invariant inner products $(\cdot, \cdot)_V$ and $(\cdot, \cdot)_W$. Then

$$\chi_V(g)\,\overline{\chi_W(g)} = \sum_{i,j} (gv_i, v_i)_V\,\overline{(gw_j, w_j)_W},$$

so Schur orthogonality (Theorem 3.3) implies that $\int_G \chi_V(g)\,\overline{\chi_W(g)}\,dg$ is 0 when $V \not\cong W$. When $U \cong V$, $\chi_W = \chi_V$, so Schur orthogonality implies that

$$\int_G \chi_V(g)\,\overline{\chi_V(g)}\,dg = \frac{1}{\dim V} \sum_{i,j} \left| (v_i, v_j)_V \right|^2 = 1.$$

For arbitrary V, W, decompose V and W into irreducible summands as $V \cong \bigoplus_{[\pi]\in\widehat{G}} m_\pi E_\pi$ and $W = \bigoplus_{[\pi]\in\widehat{G}} n_\pi E_\pi$. Hence

$$\int_G \chi_V(g)\,\overline{\chi_W(g)}\,dg = \sum_{[\pi],[\pi']\in\widehat{G}} m_\pi n_{\pi'} \int_G \chi_{E_\pi}(g)\,\overline{\chi_{E_{\pi'}}(g)}\,dg$$

$$= \sum_{[\pi]\in\widehat{G}} m_\pi n_\pi = \sum_{[\pi],[\pi']\in\widehat{G}} m_\pi n_{\pi'} \dim \mathrm{Hom}_G(E_\pi, E_{\pi'})$$

$$= \dim \mathrm{Hom}_G \Big(\bigoplus_{[\pi]\in\widehat{G}} m_\pi E_\pi, \bigoplus_{[\pi]\in\widehat{G}} n_\pi E_\pi \Big) = \dim \mathrm{Hom}_G(V, W).$$

The remaining statements follow easily from this result and the calculation of multiplicity in Theorem 2.24. In particular since V^G is the isotypic component of V corresponding to the trivial representation, $\dim \mathrm{Hom}_G(\mathbb{C}, V) = \dim V^G$ and thus $\dim V^G = \int_G \chi_{\mathbb{C}}(g)\,\overline{\chi_V(g)}\,dg = \int_G \overline{\chi_V(g)}\,dg$. Since $\dim V^G$ is a real number, the integrand may be conjugated with impunity and part (1) follows.

For part (2), V is completely determined by the multiplicities $m_\pi = \dim \mathrm{Hom}_G(E_\pi, V)$, $[\pi] \in \widehat{G}$. As this number is calculated by $\int_G \chi_{E_\pi}(g)\,\overline{\chi_V(g)}\,dg$, the representation is completely determined by χ_V. For part (3), V is irreducible if and only if $\dim \mathrm{Hom}_G(V, V) = 1$ by Corollary 2.19. In turn, this this is equivalent to $\int_G \chi_V(g)\,\overline{\chi_V(g)}\,dg = 1$.   □

As an application of the power of character theory, we prove a theorem classifying irreducible representations of the direct product of two compact Lie groups, $G_1 \times G_2$, in terms of the irreducible representations of G_1 and G_2. This allows us to eventually focus our study on compact Lie groups that are as small as possible.

Definition 3.8. If V_i is a finite-dimensional representation of a Lie group G_i, $V_1 \otimes V_2$ is a representation of $G_1 \times G_2$ with action given by $(g_1, g_2) \sum_i v_{i_1} \otimes v_{i_2} = \sum_i (g_1 v_{i_1}) \otimes (g_2 v_{i_2})$.

Theorem 3.9. *For compact Lie groups G_i, a finite-dimensional representation W of $G_1 \times G_2$ is irreducible if and only if $W \cong V_1 \otimes V_2$ for finite-dimensional irreducible representations V_i of G_i.*

Proof. If V_i are irreducible representations of G_i, then $\int_{G_i} \left| \chi_{V_i}(g) \right|^2 dg = 1$. Since $\chi_{V_1 \otimes V_2}(g_1, g_2) = \chi_{V_1}(g_1)\,\chi_{V_2}(g_2)$ (Exercise 3.3) and since Haar measure on $G_1 \times G_2$ is given by $dg_1\,dg_2$ by uniqueness,

$$\int_{G_1 \times G_2} \left| \chi_{\chi_{V_1 \otimes V_2}}(g_1, g_2) \right|^2 dg_1 \, dg_2$$

$$(3.10) \qquad\qquad = \left(\int_{G_1} \left| \chi_{V_1}(g_1) \right|^2 dg_1 \right) \left(\int_{G_2} \left| \chi_{V_2}(g_2) \right|^2 dg_2 \right)$$

$$= 1,$$

so that $V_1 \otimes V_2$ is $G_1 \times G_2$-irreducible.

Conversely, suppose W is $G_1 \times G_2$-irreducible. Identifying G_1 with $G_1 \times \{e\}$ and G_2 with $\{e\} \times G_2$, decompose W with respect to G_2 as

$$\bigoplus_{[\pi] \in \widehat{G_2}} \mathrm{Hom}_{G_2}(E_\pi, W) \otimes E_\pi$$

under the G_2-map Φ induced by $\Phi(T \otimes v) = T(v)$. Recall that G_2 acts trivially on $\mathrm{Hom}_{G_2}(E_\pi, W)$ and view $\mathrm{Hom}_{G_2}(E_\pi, W)$ as a representation of G_1 by setting $(g_1 T)(v) = (g_1, e)T(v)$. Thus $\bigoplus_{[\pi] \in \widehat{G_2}} \mathrm{Hom}_{G_2}(E_\pi, W) \otimes E_\pi$ is a representation of $G_1 \times G_2$ and, in fact, Φ is now a $G_1 \times G_2$-intertwining isomorphism to W since

$$(g_1, g_2)\Phi(T \otimes v) = (g_1, e)(e, g_2)\Phi(T \otimes v) = (g_1, e)\Phi(T \otimes g_2 v)$$

$$= (g_1, e)T(g_2 v) = (g_1 T)(g_2 v) = \Phi((g_1 T) \otimes (g_2 v)).$$

As W is irreducible, there exists exactly one $[\pi] \in \widehat{G_2}$ so that

$$W \cong \mathrm{Hom}_{G_2}(E_\pi, W) \otimes E_\pi.$$

Since E_π is G_2-irreducible, a calculation as in Equation 3.10 shows $\mathrm{Hom}_{G_2}(E_\pi, W)$ is G_1-irreducible as well. $\qquad\square$

As a corollary of Theorem 3.9 (Exercise 3.10), it easily follows that $\widehat{G_1 \times G_2} \cong \widehat{G_1} \times \widehat{G_2}$.

3.1.3 Exercises

Exercise 3.1 If V, V' are finite-dimensional unitary representations of a Lie group G with G-invariant inner products $(\cdot, \cdot)$ and $(\cdot, \cdot)'$, show the form $\big((u, u'), (v, v')\big) = (u, v) + (u', v')'$ on $V \oplus V'$ is a G-invariant inner product and the form

$$\left(\sum_i u_i \otimes u_i', \sum_j v_j \otimes v_j' \right) = \sum_{i,j} (u_i, v_j)(u_i', v_j')'$$

on $V \otimes V'$ is a G-invariant inner product.

Exercise 3.2 Show that any finite-dimensional unitary representation V of a compact Lie group G can be written as a direct sum of irreducible summands that are mutually perpendicular.

Exercise 3.3 Prove the remaining parts of Theorem 3.5. Also, if V_i are finite-dimensional representations of a compact Lie group G_i, show that $\chi_{V_1 \otimes V_2}(g_1, g_2) = \chi_{V_1}(g_1)\, \chi_{V_2}(g_2)$.

Exercise 3.4 Let G be a finite group acting on a finite set M. Define a representation of G on $C(M) = \{f : M \to \mathbb{C}\}$ by $(gf)(m) = f(g^{-1}m)$. Show that $\chi_{C(M)}(g) = |M^g|$ for $g \in G$ where $M^g = \{m \in M \mid gm = m\}$.

Exercise 3.5 (a) For the representation $V_n(\mathbb{C}^2)$ of $SU(2)$ from §2.1.2.2, calculate $\chi_{V_n(\mathbb{C}^2)}(g)$ for $g \in SU(2)$ in terms of the eigenvalues of g.
(b) Use a character computation to establish the *Clebsch–Gordan* formula:

$$V_n(\mathbb{C}^2) \otimes V_m(\mathbb{C}^2) \cong \bigoplus_{j=0}^{\min\{n,m\}} V_{n+m-2j}(\mathbb{C}^2).$$

Exercise 3.6 (a) For the representations $V_m(\mathbb{R}^3)$ and $\mathcal{H}_m(\mathbb{R}^3)$ of $SO(3)$ from §2.1.2.3, calculate $\chi_{V_m(\mathbb{R}^3)}(g)$ and $\chi_{\mathcal{H}_m(\mathbb{R}^3)}(g)$ for $g \in SO(3)$ of the form

$$\begin{pmatrix} 1 & 0 & 0 \\ 0 & \cos\theta & -\sin\theta \\ 0 & \sin\theta & \cos\theta \end{pmatrix}.$$

(b) For the half-spin representations $S^\pm$ of Spin(4) from §2.1.2.4, calculate $\chi_{S^\pm}(g)$ for $g \in$ Spin(4) of the form $(\cos\theta_1 + e_1 e_2 \sin\theta_1)(\cos\theta_2 + e_3 e_4 \sin\theta_2)$.

Exercise 3.7 Let V be a finite-dimensional representation of G. Show $\chi_{\bigwedge^2 V}(g) = \frac{1}{2}\left(\chi_V(g)^2 - \chi_V(g^2)\right)$ and $\chi_{S^2 V} = \frac{1}{2}(\chi_V(g)^2 + \chi_V(g^2))$. Use this to show that $V \otimes V \cong S^2 V \oplus \bigwedge^2 V$ (c.f., Exercise 2.15).

Exercise 3.8 A finite-dimensional representation (π, V) of a compact Lie group G is said to be of *real type* if there is a real vector space V_0 on which G acts that gives rise to the action on V by extension of scalars, i.e., by $V = V_0 \otimes_\mathbb{R} \mathbb{C}$. It is said to be of *quaternionic type* if there is a quaternionic vector space on which G acts that gives rise to the action on V by restriction of scalars. It is said to be of *complex type* if it is neither real nor quaternionic type.
(a) Show that V is of real type if and only if V possesses an invariant nondegenerate symmetric bilinear form. Show that V is of quaternionic type if and only if V possesses an invariant nondegenerate skew-symmetric bilinear form.
(b) Show that the set of G-invariant bilinear forms on V are given by $\mathrm{Hom}_G(V \otimes V, \mathbb{C}) \cong \mathrm{Hom}_G(V, V^*)$ (c.f., Exercise 2.15).
(c) For the remainder of the problem, let V be irreducible. Show that V is of complex type if and only if $V \not\cong V^*$. When $V \cong V^*$, use Exercise 3.7 to conclude that V is of real or quaternionic type, but not both.
(d) Using Theorem 3.7 and the character formulas in Exercise 3.7, show that

$$\int_G \chi_V(g^2)\, dg = \begin{cases} 1 & \text{if } V \text{ is of real type} \\ 0 & \text{if } V \text{ is of complex type} \\ -1 & \text{if } V \text{ is of quaternionic type.} \end{cases}$$

(e) If χ_V is real valued, show that V is of real or quaternionic type.

Exercise 3.9 Let (π, V) be a finite-dimensional representation of a compact Lie group G. Use unitarity and an eigenspace decomposition to show $|\chi_V(g)| \leq \dim V$ with equality if and only if $\pi(g)$ is multiplication by a scalar.

Exercise 3.10 Let $[\pi_i] \in \widehat{G_i}$ for compact Lie groups G_i. Now show that the map $(E_{\pi_1}, E_{\pi_2}) \to E_{\pi_1} \otimes E_{\pi_2}$ induces an isomorphism $\widehat{G_1} \times \widehat{G_2} \cong \widehat{G_1 \times G_2}$.

3.2 Infinite-Dimensional Representations

In many applications it is important to remove the finite-dimensional restriction from the definition of a representation. As infinite-dimensional spaces are a bit more tricky than finite-dimensional ones, this requires a slight reworking of a few definitions. None of these modifications affect the finite-dimensional setting. Once these adjustments are made, it is perhaps a bit disappointing that the infinite-representation theory for compact Lie groups reduces to the finite-dimensional theory.

3.2.1 Basic Definitions and Schur's Lemma

Recall that a *topological vector space* is a vector space equipped with a topology so that vector addition and scalar multiplication are continuous. If V and V' are topological vector spaces, write $\mathrm{Hom}(V, V')$ for the set of *continuous* linear transformations from V to V' and write $GL(V)$ for the set of invertible elements of $\mathrm{Hom}(V, V)$.

The following definition (c.f. Definitions 2.1, 2.2, and 2.11) provides the necessary modifications to allow the study of infinite-dimensional representations. As usual in infinite dimensional settings, the main additions consist of explicitly requiring the action of the Lie group to be continuous in both variables and liberal use of the adjectives continuous and closed.

Definition 3.11. **(1)** A *representation* of a Lie group G on a topological vector space V is a pair (π, V), where $\pi : G \to GL(V)$ is a homomorphism and the map $G \times V \to V$ given by $(g, v) \to \pi(g)v$ is *continuous*.
(2) If (π, V) and (π', V') are representations on topological vector spaces, $T \in \mathrm{Hom}(V, V')$ is called an *intertwining operator* or *G-map* if $T \circ \pi = \pi' \circ T$.
(3) The set of all G-maps is denoted by $\mathrm{Hom}_G(V, V')$.
(4) The representations V and V' are *equivalent*, $V \cong V'$, if there exists a bijective G-map from V to V'.
(5) A subspace $U \subseteq V$ is *G-invariant* if $gU \subseteq U$ for $g \in G$. Thus when U is closed, U is a representation of G in its own right and is also called a *submodule* or a *subrepresentation*.
(6) A nonzero representation V is *irreducible* if the only *closed* G-invariant subspaces are $\{0\}$ and V. A nonzero representation is called *reducible* if there is a proper *closed* G-invariant subspace of V.

For the most part, the interesting topological vector space representations we will examine will be *unitary representations* on *Hilbert spaces*, i.e., representations on complete inner product spaces where the inner product is invariant under the Lie group (Definition 2.14). More generally, many of the results are applicable to Hausdorff *locally convex topological spaces* and especially to *Fréchet spaces* (see [37]). Recall that locally convex topological spaces are topological vector spaces whose topology is defined by a family of seminorms. A Fréchet space is a complete locally convex Hausdorff topological spaces whose topology is defined by a countable family of seminorms.

As a first example of an infinite-dimensional unitary representation on a Hilbert space, consider the action of S^1 on $L^2(S^1)$ given by $\left(\pi(e^{i\theta})f\right)(e^{i\alpha}) = f(e^{i(\alpha-\theta)})$ for $e^{i\theta} \in S^1$ and $f \in L^2(S^1)$. We will soon see (Lemma 3.20) that this example generalizes to any compact Lie group.

Next we upgrade Schur's Lemma (Theorem 2.12) to handle unitary representations on Hilbert spaces.

Theorem 3.12 (Schur's Lemma). *Let V and W be unitary representations of a Lie group G on Hilbert spaces. If V and W are irreducible, then*

$$\dim \operatorname{Hom}_G(V, W) = \begin{cases} 1 & \text{if } V \cong W \\ 0 & \text{if } V \not\cong W. \end{cases}$$

In general, the representation V is irreducible if and only if $\operatorname{Hom}_G(V, V) = \mathbb{C}I$.

Proof. Start with V and W irreducible. If $T \in \operatorname{Hom}_G(V, W)$ is nonzero, then $\ker T$ is closed, not all of V, and G-invariant, so irreducibility implies $\ker T = \{0\}$. Similarly, the image of T is nonzero and G-invariant, so continuity and irreducibility imply that $\overline{\operatorname{range} T} = W$.

Using the definition of the adjoint map of T, $T^* : W \to V$, it immediately follows that $T^* \in \operatorname{Hom}_G(W, V)$ and that T^* is nonzero, injective, and has dense range (Exercise 3.11). Let $S = T^* \circ T \in \operatorname{Hom}_G(V, V)$ so that $S^* = S$. In the finite-dimensional case, we used the existence of an eigenvalue to finish the proof. In the infinite-dimensional setting however, eigenvalues (point spectrum) need not generally exist. To clear this hurdle, we invoke a standard theorem from a functional analysis course.

The Spectral Theorem for normal bounded operators (see [74] or [30] for details) says that there exists a projection valued measure E so that $S = \int_{\sigma(S)} \lambda \, dE$, where $\sigma(S)$ is the spectrum of S. It has the nice property that the only bounded endomorphisms of V commuting with S are the ones commuting with each self-adjoint projection $E(\Delta)$, Δ a Borel subset of $\sigma(S)$. In terms of understanding the notation $\int_{\sigma(S)} \lambda \, dE$, The Spectral Theorem also says that S is the limit, in the operator norm, of operators of the form $\sum_i \lambda_i E(\Delta_i)$ where $\{\Delta_i\}$ is a partition of $\sigma(S)$ and $\lambda_i \in \Delta_i$.

Since $S \in \operatorname{Hom}_G(V, V)$, $\pi(g)$ commutes with $E(\Delta)$ for each $g \in G$, so that $E(\Delta) \in \operatorname{Hom}_G(V, V)$. It has already been shown that nonzero elements of $\operatorname{Hom}_G(V, V)$ are injective. As $E(\Delta)$ is a projection, it must therefore be 0 or I. Thus $\sum_i \lambda_i E(\Delta_i) = kI$ for some (possibly zero) constant k. In particular, S is a multiple of the identity. Since S is injective, S is a nonzero multiple of the identity. It follows that T is invertible and that $V \cong W$.

Now suppose $T_i \in \mathrm{Hom}_G(V, W)$ are nonzero. Let $S = T_2^{-1} \circ T_1 \in \mathrm{Hom}_G(V, V)$ and write $S = \frac{1}{2}[(S + S^*) - i(iS - iS^*)]$. Using the same argument as above applied to the self-adjoint intertwining operators $S + S^*$ and $iS - iS^*$, it follows that S is a multiple of the identity. This proves the first statement of the theorem.

To prove the second statement, it only remains to show $\dim \mathrm{Hom}_G(V, V) \geq 2$ when V is not irreducible. If $U \subseteq V$ is a proper closed G-invariant subspace, then so is $U^\perp$ by unitarity. The two orthogonal projections onto U and $U^\perp$ do the trick.   □

3.2.2 G-Finite Vectors

Throughout the rest of the book there will be numerous occasions where *vector-valued integration* on compact sets is required. In a finite-dimensional vector space, a basis can be chosen and then integration can be done coordinate-by-coordinate. For instance, vector-valued integration in this setting was already used in the proof of Theorem 3.7 for the definition of $\widetilde{T}_{u,v}$. Obvious generalizations can be made to Hilbert spaces by tossing in limits. In any case, functional analysis provides a general framework for this type of operation which we recall now (see [74] for details). Remember that G is still a compact Lie group throughout this chapter.

Let V be a Hausdorff locally convex topological space and $F : G \to V$ a continuous function. Then there exists a unique element in V, called

$$\int_G f(g)\,dg,$$

so that $T\left(\int_G f(g)\,dg\right) = \int_G T(f(g))\,dg$ for each $T \in \mathrm{Hom}(V, \mathbb{C})$. If V is a Fréchet space, $\int_G f(g)\,dg$ is the limit of elements of the form

$$\sum_{i=1}^n f(g_i)\,dg(\Delta_i),$$

where $\{\Delta_i\}_{i=1}^n$ is a finite Borel partition of G, $g_i \in \Delta_i$, and $dg(\Delta_i)$ is the measure of Δ_i with respect to the invariant measure.

Recall that a linear map T on V is *positive* if $(Tv, v) \geq 0$ for all $v \in V$ and strictly greater than zero for some v. The linear map T is *compact* if the closure of the image of the unit ball under T is compact. It is a standard fact from functional analysis that the set of compact operators is a closed left and right ideal under composition within the set of bounded operators (e.g., [74] or [30]).

We now turn our attention to finding a *canonical decomposition* (Theorem 2.24) suitable for unitary representations on Hilbert spaces. The hardest part is getting started. In fact, the heart of the matter is really contained in Lemma 3.13 below.

Lemma 3.13. *Let (π, V) be a unitary representation of a compact Lie group G on a Hilbert space. There exists a nonzero finite-dimensional G-invariant (closed) subspace of V.*

Proof. Begin with any self-adjoint positive compact operator $T_0 \in \text{Hom}(V, V)$, e.g., any nonzero finite rank projection will work. Using vector-valued integration in $\text{Hom}(V, V)$, define

$$T = \int_G \pi(g) \circ T_0 \circ \pi(g)^{-1} \, dg.$$

Since T is the limit in norm of operators of the form $\sum_i dg(\Delta_i) \pi(g_i) \circ T_0 \circ \pi(g_i)^{-1}$ with $g_i \in \Delta_i \subseteq G$, T is still a compact operator. T is G-invariant since dg is left invariant (e.g., see the proof of Theorem 3.7 and the operator $\widetilde{T}_{u,v}$). Using the positivity of T_0, T is seen to be nonzero by calculating

$$(Tv, v) = \int_G (\pi(g)T_0\pi(g)^{-1}v, \, v) \, dg = \int_G (T_0\pi(g)^{-1}v, \, \pi(g)^{-1}v) \, dg,$$

where $(\cdot, \cdot)$ is the invariant inner product on V. Since V is unitary, the adjoint of $\pi(g)$ is $\pi(g)^{-1}$. Using the fact that T_0 is self-adjoint, it therefore follows that T is also self-adjoint.

An additional bit of functional analysis is needed to finish the proof. Use the Spectral Theorem for compact self-adjoint operators (see [74] or [30] for details) to see that T possesses a nonzero eigenvalue λ whose corresponding (nonzero) eigenspace is finite dimensional. This eigenspace, i.e., $\ker(T - \lambda I)$, is the desired nonzero finite-dimensional G-invariant subspace of V. $\qquad\square$

If $\{V_\alpha\}_{\alpha \in A}$ are Hilbert spaces with inner products $(\cdot, \cdot)_\alpha$, recall that the *Hilbert space direct sum* is

$$\widehat{\bigoplus_{\alpha \in A}} V_\alpha = \{(v_\alpha) \mid v_\alpha \in V_\alpha \text{ and } \sum_{\alpha \in A} \|v_\alpha\|_\alpha^2 < \infty\}.$$

$\widehat{\bigoplus}_\alpha V_\alpha$ is a Hilbert space with inner product $((v_\alpha), (v_\alpha')) = \sum_\alpha (v_\alpha, v_\alpha')_\alpha$ and contains $\bigoplus_\alpha V_\alpha$ as a dense subspace with $V_\alpha \perp V_\beta$ for distinct $\alpha, \beta \in A$.

Definition 3.14. If V is a representation of a Lie group G on a topological vector space, the set of *G-finite vectors* is the set of all $v \in V$ so that Gv generates a finite-dimensional subspace, i.e.,

$$V_{G\text{-fin}} = \{v \in V \mid \dim(\text{span}\{gv \mid g \in G\}) < \infty\}.$$

The next corollary shows that we do not really get anything new by allowing infinite-dimensional unitary Hilbert space representations.

Corollary 3.15. *Let (π, V) be a unitary representation of a compact Lie group G on a Hilbert space. There exists finite-dimensional irreducible G-submodules $V_\alpha \subseteq V$ so that*

$$V = \widehat{\bigoplus_\alpha} V_\alpha.$$

In particular, the irreducible unitary representations of G are all finite dimensional. Moreover, the set of G-finite vectors is dense in V.

Proof. Zorn's Lemma says that any partially ordered set has a maximal element if every linearly ordered subset has an upper bound. With this in mind, consider the collection of all sets $\{V_\alpha \mid \alpha \in \mathcal{A}\}$ satisfying the properties: (1) each V_α is finite-dimensional, G-invariant, and irreducible; and (2) $V_\alpha \perp V_\beta$ for distinct $\alpha, \beta \in \mathcal{A}$. Partially order this collection by inclusion. By taking a union, every linearly ordered subset clearly has an upper bound. Let $\{V_\alpha \mid \alpha \in \mathcal{A}\}$ be a maximal element. If $\bigoplus_\alpha V_\alpha \neq V$, then $\left(\bigoplus_\alpha V_\alpha\right)^\perp$ is closed, nonempty, and G-invariant, and so a unitary Hilbert space representation in its own right. In particular, Lemma 3.13 and Corollary 2.17 imply that there exists a finite dimensional, G-invariant, irreducible submodule $V_\gamma \subseteq \left(\bigoplus_\alpha V_\alpha\right)^\perp$. This, however, violates maximality and the corollary is finished. □

As was the case in §2.2.4, the above decomposition is not canonical. This situation will be remedied next in §3.2.3 below.

3.2.3 Canonical Decomposition

First, we update the notion of isotypic component from Definition 2.22 in order to handle infinite-dimensional unitary representations. The only real change replaces direct sums with Hilbert space direct sums.

Definition 3.16. Let V be a unitary representation of a compact Lie group G on a Hilbert space. For $[\pi] \in \widehat{G}$, let $V_{[\pi]}$ be the largest subspace of V that is a Hilbert space direct sum of irreducible submodules equivalent to E_π. The submodule $V_{[\pi]}$ is called the π-*isotypic component* of V.

As in the finite-dimensional case, the above definition of the isotypic component $V_{[\pi]}$ is well defined and $V_{[\pi]}$ is the closure of the sum of all submodules of V equivalent to E_π These statements are verified using Zorn's Lemma in a fashion similar to the proof of Corollary 3.15 (Exercise 3.12).

Lemma 3.17. *Let V be a unitary representation of a compact Lie group G on a Hilbert space with invariant inner product $(\cdot, \cdot)_V$ and let E_π, $[\pi] \in \widehat{G}$, be an irreducible representation of G with invariant inner product $(\cdot, \cdot)_{E_\pi}$. Then $\mathrm{Hom}_G(E_\pi, V)$ is a Hilbert space with a G-invariant inner product $(\cdot, \cdot)_{\mathrm{Hom}}$ defined by $(T_1, T_2)_{\mathrm{Hom}} I = T_2^* \circ T_1$. It satisfies*

$$(3.18) \qquad (T_1, T_2)_{\mathrm{Hom}} (x_1, x_2)_{E_\pi} = (T_1 x_1, T_2 x_2)_V$$

for $T_i \in \mathrm{Hom}_G(E_\pi, V)$ and $x_i \in E_\pi$. Moreover, $\|T\|_{\mathrm{Hom}}$ is the same as the operator norm of T.

Proof. The adjoint of T_2, $T_2^* \in \mathrm{Hom}(V, E_\pi)$, is still a G-map since

$$(T_2^*(gv), x)_{E_\pi} = (gv, T_2 x)_V = (v, T_2(g^{-1}x))_V = (T_2^* v, g^{-1}x)_{E_\pi}$$
$$= (g T_2^* v, x)_{E_\pi}$$

for $x \in E_\pi$ and $v \in V$. Thus $T_2^* \circ T_1 \in \text{Hom}(E_\pi, E_\pi)$. Schur's Lemma implies that there is a scalar $(T_1, T_2)_{\text{Hom}} \in \mathbb{C}$, so that $(T_1, T_2)_{\text{Hom}} I = T_2^* \circ T_1$.

By definition, $(\cdot, \cdot)_{\text{Hom}}$ is clearly a Hermitian form on $\text{Hom}_G(E_\pi, V)$ and

$$(T_1 x_1, T_2 x_2)_V = (T_2^*(T_1 x_1), x_2)_{E_\pi} = ((T_1, T_2)_{\text{Hom}} x_1, x_2)_{E_\pi}$$
$$= (T_1, T_2)_{\text{Hom}} (x_1, x_2)_{E_\pi}.$$

In particular, for $T \in \text{Hom}_G(E_\pi, V)$, $\|T\|_{\text{Hom}}$ is the quotient of $\|Tx\|_V$ and $\|x\|_{E_\pi}$ for any nonzero $x \in E_\pi$. Thus $\|T\|_{\text{Hom}}$ is the same as the operator norm of T viewed as an element of $\text{Hom}(E_\pi, V)$. Hence $(\cdot, \cdot)_{\text{Hom}}$ is an inner product making $\text{Hom}_G(E_\pi, V)$ into a Hilbert space. □

Note Equation 3.18 is independent of the choice of invariant inner product on E_π. To see this directly, observe that scaling $(\cdot, \cdot)_{E_\pi}$ scales T_2^*, and therefore $(\cdot, \cdot)_{\text{Hom}_G(E_\pi, V)}$, by the inverse scalar so that the product of $(\cdot, \cdot)_{E_\pi}$ and $(\cdot, \cdot)_{\text{Hom}_G(E_\pi, V)}$ remains unchanged.

If V_i are Hilbert spaces with inner products $(\cdot, \cdot)_i$, recall that the *Hilbert space tensor product*, $V_1 \widehat{\otimes} V_2$, is the completion of $V_1 \otimes V_2$ with respect to the inner product generated by $(v_1 \otimes v_2, v_1' \otimes v_2') = (v_1, v_1')(v_2, v_2')$ (c.f. Exercise 3.1).

Theorem 3.19 (Canonical Decomposition). *Let V be a unitary representation of a compact Lie group G on a Hilbert space.*
(1) There is a G-intertwining unitary isomorphism ι_π

$$\text{Hom}_G(E_\pi, V) \widehat{\otimes} E_\pi \xrightarrow{\cong} V_{[\pi]}$$

induced by $\iota_\pi (T \otimes v) = T(v)$ for $T \in \text{Hom}_G(E_\pi, V)$ and $v \in V$.
(2) There is a G-intertwining unitary isomorphism

$$\widehat{\bigoplus_{[\pi] \in \widehat{G}}} \text{Hom}_G(E_\pi, V) \widehat{\otimes} E_\pi \xrightarrow{\cong} V = \widehat{\bigoplus_{[\pi] \in \widehat{G}}} V_{[\pi]}.$$

Proof. As in the proof of Theorem 2.24, ι_π is a well-defined G-map from $\text{Hom}_G(E_\pi, V) \otimes E_\pi$ to $V_{[\pi]}$ with dense range (since $V_{[\pi]}$ is a Hilbert space direct sum of irreducible submodules instead of finite direct sum as in Theorem 2.24). As Lemma 3.17 implies ι_π is unitary on $\text{Hom}_G(E_\pi, V) \otimes E_\pi$, it follows that ι_π is injective and uniquely extends by continuity to a G-intertwining unitary isomorphism from $\text{Hom}_G(E_\pi, V) \widehat{\otimes} E_\pi$ to $V_{[\pi]}$. Finally, V is the closure of $\sum_{[\pi] \in \widehat{G}} V_{[\pi]}$ by Corollary 3.15 and the sum is orthogonal by Corollary 2.21. □

3.2.4 Exercises

Exercise 3.11 Let V and W be unitary representations of a compact Lie group G on Hilbert spaces and let $T \in \text{Hom}_G(V, W)$ be injective with dense range. Show that $T^* \in \text{Hom}_G(W, V)$, T^* is injective, and that T^* has dense range.

Exercise 3.12 Let V be a unitary representation of a compact Lie group G on a Hilbert space and let $[\pi] \in \widehat{G}$.

(a) Consider the collection of all sets $\{V_\alpha \mid \alpha \in \mathcal{A}\}$ satisfying the properties: (1) each V_α is a submodule of V isomorphic to E_π and (2) $V_\alpha \perp V_\beta$ for distinct $\alpha, \beta \in \mathcal{A}$. Partially order this collection by inclusion and use Zorn's Lemma to show that there is a maximal element.

(b) Write $\{V_\alpha \mid \alpha \in \mathcal{A}\}$ for the maximal element. Show that the orthogonal projection $P : V \to \left(\widehat{\bigoplus}_{\alpha \in \mathcal{A}} V_\alpha\right)^\perp$ is a G-map. If $V_\gamma \subseteq V$ is any submodule equivalent to E_π, use irreducibility and maximality to show that $P V_\gamma = \{0\}$.

(c) Show that the definition of the isotypic component $V_{[\pi]}$ in Definition 3.16 is well defined and that $V_{[\pi]}$ is the closure of the sum of all submodules of V equivalent to E_π.

Exercise 3.13 Recall that $\widehat{S^1} \cong \mathbb{Z}$ via the one-dimensional representations $\pi_n(e^{i\theta}) = e^{in\theta}$ for $n \in \mathbb{Z}$ (Exercise 2.21). View $L^2(S^1)$ as a unitary representation of S^1 under the action $(e^{i\theta} \cdot f)(e^{i\alpha}) = f(e^{i(\alpha-\theta)})$ for $f \in L^2(S^1)$. Calculate $\mathrm{Hom}_{S^1}(\pi_n, L^2(S^1))$ and conclude that $L^2(S^1) = \widehat{\bigoplus}_{n \in \mathbb{Z}} \mathbb{C} e^{in\theta}$.

Exercise 3.14 Use Exercise 2.28 and Theorem 2.33 to show that

$$L^2(S^{n-1}) = \widehat{\bigoplus}_{m \in \mathbb{N}} \mathcal{H}_m(\mathbb{R}^n)|_{S^{n-1}}, \quad n \geq 2,$$

is the canonical decomposition of $L^2(S^{n-1})$ under $O(n)$ (or $SO(n)$ for $n \geq 3$) with respect to usual action $(gf)(v) = f(g^{-1}v)$.

Exercise 3.15 Recall that the irreducible unitary representations of $\mathbb{R}$ are given by the one-dimensional representations $\pi_r(x) = e^{irx}$ for $r \in \mathbb{R}$ (Exercise 2.21) and consider the unitary representation of $\mathbb{R}$ on $L^2(\mathbb{R})$ under the action $(x \cdot f)(y) = f(x - y)$ for $f \in L^2(\mathbb{R})$. Show $L^2(\mathbb{R}) \neq \widehat{\bigoplus}_{r \in \mathbb{R}} L^2(\mathbb{R})_{\pi_r}$ by showing $L^2(\mathbb{R})_{\pi_r} = \{0\}$.

3.3 The Peter–Weyl Theorem

Let G be a compact Lie group. In this section we decompose $L^2(G)$ under left and right translation of functions. The canonical decomposition reduces the work to calculating $\mathrm{Hom}_G(E_\pi, L^2(G))$. Instead of attacking this problem directly, it turns out to be easy (Lemma 3.23) to calculate that $\mathrm{Hom}_G(E_\pi, C(G)_{G\text{-fin}})$. Using the Stone–Weierstrass Theorem (Theorem 3.25), it is shown that $C(G)_{G\text{-fin}}$ is dense in $L^2(G)$. In turn, this density result allows the calculation of $\mathrm{Hom}_G(E_\pi, L^2(G))$.

3.3.1 The Left and Right Regular Representation

The set of continuous functions on a compact Lie group G, $C(G)$, is a Banach space with respect to the norm $\|f\|_{C(G)} = \sup_{g \in G} |f(g)|$ and the set of square integrable functions, $L^2(G)$, is a Hilbert space with respect to the norm $\|f\|_{L^2(G)} = \int_G |f(g)|^2 \, dg$. Both spaces carry a left and right action l_g and r_g of G given by

$$(l_g f)(h) = f(g^{-1}h)$$
$$(r_g f)(h) = f(hg)$$

which, as the next theorem shows, are representations. They are called the *left* and *right regular representations*.

Lemma 3.20. *The left and right actions of a compact Lie group G on $C(G)$ and $L^2(G)$ are representations and norm preserving.*

Proof. The only statement from Definition 3.11 that still requires checking is continuity of the map $(g, f) \to l_g f$ (since r_g is handled similarly). Working in $C(G)$ first, calculate

$$\left| f_1(g_1^{-1}h) - f_2(g_2^{-1}h) \right| \le \left| f_1(g_1^{-1}h) - f_1(g_2^{-1}h) \right| + \left| f_1(g_2^{-1}h) - f_2(g_2^{-1}h) \right|$$
$$\le \left| f_1(g_1^{-1}h) - f_1(g_2^{-1}h) \right| + \| f_1 - f_2 \|_{C(G)}.$$

Since f_1 is continuous on compact G and since the map $g \to g^{-1}h$ is continuous, it follows that $\left\| l_{g_1} f_1 - l_{g_2} f_2 \right\|_{C(G)}$ can be made arbitrarily small by choosing (g_1, f_1) sufficiently close to (g_2, f_2).

Next, working with $f_i \in L^2(G)$, choose $f \in C(G)$ and calculate the following:

$$\left\| l_{g_1} f_1 - l_{g_2} f_2 \right\|_{L^2(G)} = \left\| f_1 - l_{g_1^{-1}g_2} f_2 \right\|_{L^2(G)}$$
$$\le \| f_1 - f_2 \|_{L^2(G)} + \left\| f_2 - l_{g_1^{-1}g_2} f_2 \right\|_{L^2(G)}$$
$$= \| f_1 - f_2 \|_{L^2(G)} + \left\| l_{g_1} f_2 - l_{g_2} f_2 \right\|_{L^2(G)}$$
$$\le \| f_1 - f_2 \|_{L^2(G)} + \left\| l_{g_1} f_2 - l_{g_1} f \right\|_{L^2(G)}$$
$$+ \left\| l_{g_1} f - l_{g_2} f \right\|_{L^2(G)} + \left\| l_{g_2} f - l_{g_2} f_2 \right\|_{L^2(G)}$$
$$= \| f_1 - f_2 \|_{L^2(G)} + 2 \| f_2 - f \|_{L^2(G)} + \left\| l_{g_1} f - l_{g_2} f \right\|_{L^2(G)}$$
$$\le \| f_1 - f_2 \|_{L^2(G)} + 2 \| f_2 - f \|_{L^2(G)} + \left\| l_{g_1} f - l_{g_2} f \right\|_{C(G)}.$$

Since f may be chosen arbitrarily close to f_2 in the L^2 norm and since G already acts continuously on $C(G)$, the result follows. □

The first important theorem identifies the G-finite vectors of $C(G)$ with the set of matrix coefficients, $MC(G)$. Even though there are two actions of G on $C(G)$, i.e., l_g and r_g, it turns out that both actions produce the same set of G-finite vectors (Theorem 3.21). As a result, write $C(G)_{G\text{-fin}}$ unambiguously for the set of G-finite vectors with respect to either action.

Theorem 3.21. *(1) For a compact Lie group G, the set of G-finite vectors of $C(G)$ with respect to left action, l_g, coincides with set of G-finite vectors of $C(G)$ with respect to right action, r_g.*
(2) $C(G)_{G\text{-fin}} = MC(G)$.

Proof. We first show that $C(G)_{G\text{-fin}}$, with respect to left action, is the set of matrix coefficients. Let $f^V_{u,v}(g) = (gu, v)$ be a matrix coefficient for a finite-dimensional unitary representation V of G with $u, v \in V$ and G-invariant inner product $(\cdot, \cdot)$. Then $(l_g f^V_{u,v})(h) = (g^{-1}hu, v) = (hu, gv)$ so that $l_g f^V_{u,v} = f^V_{u,gv}$. Hence $\{l_g f^V_{u,v} \mid g \in G\} \subseteq \{f^V_{u,v'} \mid v' \in V\}$. Since V is finite dimensional, $f^V_{u,v} \in C(G)_{G\text{-fin}}$, and thus $MC(G) \subseteq C(G)_{G\text{-fin}}$.

Conversely, let $f \in C(G)_{G\text{-fin}}$. By definition, there is a finite-dimensional submodule, $V \subseteq C(G)$, with respect to the left action so that $f \in V$. Since $\overline{gf} = g\overline{f}$, $\overline{V} = \{\overline{v} \mid v \in V\}$ is also a finite-dimensional submodule of $C(G)$. Write $(\cdot, \cdot)$ for the L^2 norm restricted to $\overline{V}$. The linear functional on $\overline{V}$ that evaluates functions at e is continuous, so there exists $\overline{v}_0 \in \overline{V}$ so that $\overline{v}(e) = (\overline{v}, \overline{v}_0)$ for $\overline{v} \in \overline{V}$. In particular, $\overline{f}(g) = l_{g^{-1}}\overline{f}(e) = (l_{g^{-1}}\overline{f}, \overline{v}_0) = (\overline{f}, l_g\overline{v}_0)$. In particular, $f = f^{\overline{V}}_{\overline{v}_0, \overline{f}} \in MC(G)$. Thus $C(G)_{G\text{-fin}} \subseteq MC(G)$ and part (2) is done (with respect to the left action).

For part (1), let f be a left G-finite vector in $C(G)$. By the above paragraph, there is a matrix coefficient, so $f = f^V_{u,v}$. Thus $(r_g f)(h) = (hgu, v)$ so that $r_g f = f^V_{gu,v}$. Since $\{gu \mid g \in G\}$ is contained in the finite-dimensional space V, it follows that the set of left G-finite vectors are contained in the set of right G-finite vectors.

Conversely, let f be a right G-finite vector. As before, pick a finite-dimensional submodule, $V \subseteq C(G)$, with respect to the left action so that $f \in V$. Write $(\cdot, \cdot)$ for the L^2 norm restricted to V. The linear functional on V that evaluates functions at e is continuous so there exists $v_0 \in V$, so that $v(e) = (v, v_0)$ for $v \in V$. In particular, $f(g) = r_g f(e) = (r_g f, v_0)$. In particular, $f = f^V_{f,v_0} \in MC(G)$, so that the set of right G-finite vectors is contained in the set of left G-finite vectors. $\square$

Based on our experience with the canonical decomposition, we hope $C(G)_{G\text{-fin}}$ decomposes under the *left* action into terms isomorphic to

$$\mathrm{Hom}_G(E_\pi, C(G)_{G\text{-fin}}) \otimes E_\pi$$

for $[\pi] \in \widehat{G}$. In this case, l_g acts trivially on $\mathrm{Hom}_G(E_\pi, C(G)_{G\text{-fin}})$ so that E_π carries the entire left action. However, Theorem 3.21 says that $C(G)_{G\text{-fin}}$ is actually a $G \times G$-module under the action $((g_1, g_2)f)(g) = (r_{g_1}l_{g_2}f)(g) = f(g_2^{-1}gg_1)$. In light of Theorem 3.9, it is therefore reasonable to hope $\mathrm{Hom}_G(E_\pi, C(G)_{G\text{-fin}})$ will carry the *right* action. This, of course, requires a different action on $\mathrm{Hom}_G(E_\pi, C(G)_{G\text{-fin}})$ than the trivial action defined in §2.2.1. Towards this end and with respect to the left action on $C(G)_{G\text{-fin}}$, define a second action of G on $\mathrm{Hom}_G(E_\pi, C(G)_{G\text{-fin}})$ and $\mathrm{Hom}_G(E_\pi, C(G))$ by

$$(3.22) \qquad\qquad (gT)(x) = r_g(Tx)$$

for $g \in G$, $x \in E_\pi$, and $T \in \mathrm{Hom}_G(E_\pi, C(G))$. To verify this is well defined, calculate

$$l_{g_1}((g_2T)(x)) = l_{g_1}r_{g_2}(Tx) = r_{g_2}l_{g_1}(Tx) = r_{g_2}(T(g_1x)) = ((g_2T)(g_1x)),$$

so that $g_2T \in \mathrm{Hom}_G(E_\pi, C(G))$. If $T \in \mathrm{Hom}_G(E_\pi, C(G)_{G\text{-fin}})$, then $g_2T \in \mathrm{Hom}_G(E_\pi, C(G)_{G\text{-fin}})$ as well by Theorem 3.21.

The next lemma is a special case of *Frobenius Reciprocity* in §7.4.1. It does not depend on the fact that E_π is irreducible.

Lemma 3.23. *With respect to the left action of a compact Lie group G on $C(G)_{G\text{-fin}}$ and the action on $\mathrm{Hom}_G(E_\pi, C(G)_{G\text{-fin}})$ given by Equation 3.22,*

$$\mathrm{Hom}_G(E_\pi, C(G)) = \mathrm{Hom}_G(E_\pi, C(G)_{G\text{-fin}}) \cong E_\pi^*$$

as G-modules. The intertwining map is induced by mapping $T \in \mathrm{Hom}_G(E_\pi, C(G)_{G\text{-fin}})$ to $\lambda_T \in E_\pi^*$ where

$$\lambda_T(x) = (T(x))(e)$$

for $x \in E_\pi$.

Proof. Let $T \in \mathrm{Hom}_G(E_\pi, C(G)_{G\text{-fin}})$ and define λ_T as in the statement of the lemma. This is a G-map since

$$(g\lambda_T)(x) = \lambda_T(g^{-1}x) = (T(g^{-1}x))(e) = (l_{g^{-1}}(Tx))(e) = (Tx)(g)$$
$$= (r_g(Tx))(e) = ((gT)(x))(e) = \lambda_{gT}(x),$$

so $g\lambda_T = \lambda_{gT}$ for $g \in G$.

We claim that the inverse map is obtained by mapping $\lambda \in E_\pi^*$ to

$$T_\lambda \in \mathrm{Hom}_G(E_\pi, C(G)_{G\text{-fin}})$$

by

$$(T_\lambda(x))(h) = \lambda(h^{-1}x)$$

for $h \in G$. To see that this is well defined, calculate

$$(l_g(T_\lambda(x)))(h) = (T_\lambda(x))(g^{-1}h) = \lambda(h^{-1}gx) = (T_\lambda(gx))(h)$$

so that $l_g(T_\lambda(x)) = T_\lambda(gx)$. This shows that T_λ is a G-map and, since E_π is finite dimensional, $T_\lambda(x) \in C(G)_{G\text{-fin}}$. To see that this operation is the desired inverse, calculate

$$\lambda_{T_\lambda}(x) = (T_\lambda(x))(e) = \lambda(x)$$

and

$$(T_{\lambda_T}(x))(h) = \lambda_T(h^{-1}x) = (T(h^{-1}x))(e) = (l_{h^{-1}}(Tx))(e) = (Tx)(h).$$

Hence $\mathrm{Hom}_G(E_\pi, C(G)_{G\text{-fin}}) \cong E_\pi^*$.

To see $\mathrm{Hom}_G(E_\pi, C(G)_{G\text{-fin}}) = \mathrm{Hom}_G(E_\pi, C(G))$, observe that the map $T \to \lambda_T$ is actually a well-defined map from $\mathrm{Hom}_G(E_\pi, C(G))$ to E_π^*. Since the inverse is still given by $\lambda \to T_\lambda$ and $T_\lambda \in \mathrm{Hom}_G(E_\pi, C(G)_{G\text{-fin}})$, the proof is finished. $\square$

Note that if E_π^* inherits an invariant inner product from E_π in the usual fashion, the above isomorphism need not be unitary with respect to the inner product on $\mathrm{Hom}_G(E_\pi, C(G))$ given in Lemma 3.17. In fact, they can be off by a scalar multiple determined by $\dim E_\pi$. The exact relationship will be made clear in §3.4.

3.3.2 Main Result

For $n \in \mathbb{Z}$, consider the representation (π_n, E_{π_n}) of S^1 where $E_{\pi_n} = \mathbb{C}$ and $\pi_n :$ $S^1 \to GL(1, \mathbb{C})$ is given by $(\pi_n(g))(x) = g^n x$ for $g \in S^1$ and $x \in E_{\pi_n}$. In so doing, we realize the isomorphism $\mathbb{Z} \cong \widehat{S^1}$ (c.f. Exercise 3.13). Define the function $f_n : S^1 \to \mathbb{C}$ by $f_n(g) = g^n$. Standard results from Fourier analysis show that $\{f_n \mid n \in \mathbb{Z}\}$ is an orthonormal basis for $L^2(S^1)$. By mapping $1 \in E_{\pi_n} \to f_n$, we could say that there is an is an induced isomorphism $\bigoplus_{n \in \mathbb{Z}} E_{\pi_n} \cong L^2(S^1)$. This map even intertwines with the right regular action of $L^2(S^1)$.

In order to generalize to groups that are not Abelian and to accommodate both the left and right regular actions, we will phrase the result a bit differently. Consider the map from $E_{\pi_n}^* \otimes E_{\pi_n}$ to $L^2(S^1)$ induced by mapping $\lambda \otimes x \in E_{\pi_n}^* \otimes E_{\pi_n}$ to the function $f_{\lambda \otimes x}$ where $f_{\lambda \otimes x}(g) = \lambda(\pi_n(g^{-1})x)$ for $g \in S^1$. If $1^* \in E_{\pi_n}^*$ maps 1 to 1, notice $f_{1^* \otimes 1} = f_{-n}$, so there is still an induced isomorphism

$$\widehat{\bigoplus}_{\pi_n \in \widehat{S^1}} E_{\pi_n}^* \otimes E_{\pi_n} \cong L^2(S^1).$$

Moreover, it is easy to check that this isomorphism is an $S^1 \times S^1$-intertwining map with $(g_1, g_2) \in S^1 \times S^1$ acting on on $L^2(S^1)$ by $r_{g_1} \circ l_{g_2}$. Thus the results of Fourier analysis on S^1 can be thought of as arising directly from the representation theory of S^1. This result will generalize to all compact Lie groups.

Theorem 3.24. *Let G be a compact Lie group. As a $G \times G$-module with $(g_1, g_2) \in$ $G \times G$ acting as $r_{g_1} \circ l_{g_2} = l_{g_2} \circ r_{g_1}$ on $C(G)_{G\text{-fin}}$,*

$$C(G)_{G\text{-fin}} \cong \bigoplus_{[\pi] \in \widehat{G}} E_\pi^* \otimes E_\pi.$$

The intertwining isomorphism is induced by mapping $\lambda \otimes x \in E_\pi^* \otimes E_\pi$ to $f_{\lambda \otimes x} \in$ $C(G)_{G\text{-fin}}$ where $f_{\lambda \otimes x}(g) = \lambda(g^{-1}x)$ for $g \in G$.

Proof. The proof of this theorem is really not much more than the proof of Theorem 2.24 coupled with Lemma 3.23 and Theorem 3.21. To see that the given map is a G-map, calculate

$$((g_1, g_2)f_{\lambda \otimes x})(g) = \lambda(g_1^{-1}g^{-1}g_2 x) = (g_1 \lambda)(g^{-1}g_2 x) = f_{g_1 \lambda \otimes g_2 x}.$$

To see that the map is surjective, Lemma 3.2 shows that it suffices to verify that each matrix coefficient of the form $f_{u,v}^{E_\pi}(g) = (gu, v)$ is achieved where $[\pi] \in \widehat{G}$, $(\cdot, \cdot)$ is a G-invariant inner product on E_π, and $u, v \in E_\pi$. Since $C(G)_{G\text{-fin}}$ is closed under complex conjugation, it suffices to show $\overline{f_{u,v}^{E_\pi}(g)} = (v, gu) = (g^{-1}v, u)$ is achieved. For this, take $\lambda = (\cdot, u)$ so that $f_{\lambda \otimes v} = f_{u,v}^{E_\pi}$.

It remains to see that the map is injective. Any element of the kernel lies in a finite sum of $W = \bigoplus_{i=1}^N E_{\pi_i}^* \otimes E_{\pi_i}$. Restricted to W, the kernel is $G \times G$-invariant. Since the kernel's isotypic components are contained in the isotypic components of W, it follows that the kernel is either $\{0\}$ or a direct sum of certain of the $E_{\pi_i}^* \otimes E_{\pi_i}$. As $f_{\lambda \otimes x}(g)$ is clearly nonzero for nonzero $\lambda \otimes x \in E_{\pi_i}^* \otimes E_{\pi_i}$, the kernel must be $\{0\}$. $\square$

Theorem 3.25 (Peter–Weyl). *Let G be a compact Lie group. $C(G)_{G\text{-fin}}$ is dense in $C(G)$ and in $L^2(G)$.*

Proof. Since $C(G)$ is dense in $L^2(G)$, it suffices to prove the first statement. For this, recall that $C(G)_{G\text{-fin}}$ is an algebra that is closed under complex conjugation and contains 1. By the Stone–Weierstrass Theorem, it only remains to show that $C(G)_{G\text{-fin}}$ separates points. For this, using left translation, it is enough to show that for any $g_0 \in G$, $g_0 \neq e$, there exists $f \in C(G)_{G\text{-fin}}$ so that $f(g_0) \neq f(e)$.

By the Hausdorff condition and continuity of left translation, choose an open neighborhood U of e so that $U \cap (g_0 U) = \emptyset$. The characteristic function for U, χ_U, is a nonzero function in $L^2(G)$. Since $l_{g_0}\chi_U = \chi_{g_0 U}$, $(l_{g_0}\chi_U, \chi_U) = 0$. Because $(\chi_U, \chi_U) > 0$, l_{g_0} cannot be the identity operator on $L^2(G)$. By Corollary 3.15 and with respect to the left action of G on $L^2(G)$, there exist finite-dimensional irreducible G-submodules $V_\alpha \subseteq L^2(G)$ so that $L^2(G) = \widehat{\bigoplus}_\alpha V_\alpha$. In particular, there is an α_0 so that l_{g_0} does not act by the identity on V_{α_0}. Thus there exists $x \in V_{\alpha_0}$ so that $l_{g_0}x \neq x$, and so there is a $y \in V_{\alpha_0}$, so that $(l_{g_0}x, y) \neq (x, y)$. The matrix coefficient $f = f_{x,y}^{V_\alpha}$ is therefore the desired function. $\square$

Coupling this density result with the canonical decomposition and the version of Frobenius reciprocity contained in Lemma 3.23, it is now possible to decompose $L^2(G)$. Since the two results are so linked, the following corollary is also often referred to as the Peter–Weyl Theorem.

Corollary 3.26. *Let G be a compact Lie group. As a $G \times G$-module with $(g_1, g_2) \in G \times G$ acting as $r_{g_1} \circ l_{g_2}$ on $L^2(G)$,*

$$L^2(G) \cong \widehat{\bigoplus_{[\pi] \in \widehat{G}}} E_\pi^* \otimes E_\pi.$$

The intertwining isomorphism is induced by mapping $\lambda \otimes v \in E_\pi^* \otimes E_\pi$ to $f_{\lambda \otimes v}$ where $f_{\lambda \otimes v}(g) = \lambda(g^{-1}v)$ for $g \in G$. With respect to the same conventions as in Lemma 3.23, $\mathrm{Hom}_G(E_\pi, L^2(G)) = \mathrm{Hom}_G(E_\pi, C(G)) \cong E_\pi^*$ as G-modules.

Proof. With respect to the left action, the canonical decomposition says that there is an intertwining isomorphism

$$\iota : \widehat{\bigoplus_{[\pi] \in \widehat{G}}} \mathrm{Hom}_G(E_\pi, L^2(G)) \,\widehat{\otimes}\, E_\pi \to L^2(G)$$

induced by $\iota(T \otimes v) = T(v)$ for $T \in \mathrm{Hom}_G(E_\pi, L^2(G))$ and $v \in L^2(G)$. Using the natural inclusion $\mathrm{Hom}_G(E_\pi, C(G)) \hookrightarrow \mathrm{Hom}_G(E_\pi, L^2(G))$ and Lemma 3.23, there is an injective map $\kappa : E_\pi^* \hookrightarrow \mathrm{Hom}_G(E_\pi, L^2(G))$ induced by mapping $\lambda \in E_\pi^*$ to $T_\lambda \in \mathrm{Hom}_G(E_\pi, L^2(G))$ via $(T_\lambda(v))(g) = \lambda(g^{-1}v)$. We first show that κ is an isomorphism.

Argue by contradiction. Suppose $\kappa(E_\pi^*)$ is a proper subset of $\mathrm{Hom}_G(E_\pi, L^2(G))$. Then, since ι is an isomorphism and E_π^* is finite dimensional, $\iota(\kappa(E_\pi^*) \otimes E_\pi)$ is a proper closed subset of $\iota(\mathrm{Hom}_G(E_\pi, L^2(G)) \otimes E_\pi)$. Choose a nonzero $f \in \iota(\mathrm{Hom}_G(E_\pi, L^2(G)) \otimes E_\pi)$ that is perpendicular to $\iota(\kappa(E_\pi^*) \otimes E_\pi)$. By virtue of the fact that $\iota(\mathrm{Hom}_G(E_\pi, L^2(G)) \otimes E_\pi)$ is the π-isotypic component of $L^2(G)$ for the left action and by Corollary 2.21, it follows that f is perpendicular to $\iota(\bigoplus_{[\pi] \in \widehat{G}} \kappa(E_\pi^*) \otimes E_\pi)$. Since $\iota(T_\lambda \otimes v) = T_\lambda(v) = f_{\lambda \otimes v}$, Theorem 3.24 shows f is perpendicular to $C(G)_{G\text{-fin}}$. By the Peter–Weyl Theorem, this is a contradiction, and so $E_\pi^* \cong \mathrm{Hom}_G(E_\pi, L^2(G))$.

Hence there is an isomorphism $\widehat{\bigoplus}_{[\pi] \in \widehat{G}} E_\pi^* \otimes E_\pi \to L^2(G)$ induced by mapping $\lambda \otimes v$ to $f_{\lambda \otimes v}$. The calculation given in the proof of Theorem 3.24 shows that this map is a $G \times G$-map when restricted to the subspace $\bigoplus_{[\pi] \in \widehat{G}} E_\pi^* \otimes E_\pi$. Since this subspace is dense, continuity finishes the proof. $\qquad\square$

By Lemma 3.17, $E_\pi^* \cong \mathrm{Hom}_G(E_\pi, L^2(G))$ is equipped with a natural inner product. In §3.4 we will see how to rescale the above isomorphism on each component $E_\pi^* \otimes E_\pi$, so that the resulting map is unitary.

3.3.3 Applications

3.3.3.1 Orthonormal Basis for $L^2(G)$ and Faithful Representations

Corollary 3.27. *Let G be a compact Lie group. If $\{v_i^\pi\}_{i=1}^{n_\pi}$ is an orthonormal basis for E_π, $[\pi] \in \widehat{G}$, then $\{(\dim E_\pi)^{\frac{1}{2}} f_{v_i^\pi, v_j^\pi}^{E_\pi} \mid [\pi] \in \widehat{G} \text{ and } 1 \le i, j \le n_\pi\}$ is an orthonormal basis for $L^2(G)$.*

Proof. This follows immediately from Lemma 3.2, Theorem 3.21, the Schur orthogonality relations, and the Peter–Weyl Theorem. $\qquad\square$

Theorem 3.28. *A compact Lie group G possesses a faithful representation, i.e., there exists a (finite-dimensional representation) (π, V) of G for which π is injective.*

Proof. By the proof of the Peter–Weyl Theorem, for $g_1 \in G^0$, $g_1 \ne e$, there exists a finite-dimensional representation (π_1, V_1) of G, so that $\pi_1(g_1)$ is not the identity operator. Thus $\ker \pi_1$ is a closed proper Lie subgroup of G, and so a compact Lie group in its own right. Since $\ker \pi_1$ is a regular submanifold that does not contain a neighborhood of e, it follows that $\dim \ker \pi_1 < \dim G$. If $\dim \ker \pi_1 > 0$, choose $g_2 \in (\ker \pi_1)^0$, $g_2 \ne e$, and let (π_2, V_2) be a representation of G, so that $\pi_2(g_2)$ is not the identity. Then $\ker(\pi_1 \oplus \pi_2)$ is a compact Lie group with $\ker(\pi_1 \oplus \pi_2) < \dim \ker \pi_1$.

Continuing in this manner, there are representations (π_i, V_i), $1 \le i \le N$, of G, so that $\dim \ker(\pi_1 \oplus \cdots \oplus \pi_N) = 0$. Since G is compact, $\ker(\pi_1 \oplus \cdots \oplus \pi_N) = \{h_1, h_2, \ldots, h_M\}$ for $h_i \in G$. Choose representations (π_{N+i}, V_{N+i}), $1 \le i \le M$, of G, so that $\pi_{N+i}(h_i)$ is not the identity. The representation $\pi_1 \oplus \cdots \oplus \pi_{N+M}$ does the trick. $\qquad\square$

Thus compact groups fall in the category of *linear groups* since each is now seen to be isomorphic to a closed subgroup of $GL(n, \mathbb{C})$. Even better, since compact, each is isomorphic to a closed subgroup of $U(n)$ by Theorem 2.15.

3.3.3.2 Class Functions

Definition 3.29. Let G be a Lie group. A function $f \in C(G)$ is called a *continuous class function* if $f(ghg^{-1}) = f(h)$ for all $g, h \in G$. Similarly, a function $f \in L^2(G)$ is called an L^2 *class function* if for each $g \in G$, $f(ghg^{-1}) = f(h)$ for almost all $h \in G$.

Theorem 3.30. *Let G be a compact Lie group and let χ be the set of irreducible characters, i.e., $\chi = \{\chi_{E_\pi} \mid [\pi] \in \widehat{G}\}$.*
(1) The span of χ equals the set of continuous class functions in $C(G)_{G\text{-fin}}$.
(2) The span of χ is dense in the set of continuous class functions.
(3) The set χ is an orthonormal basis for the set of L^2 class functions. In particular, if f is an L^2 class function, then

$$f = \sum_{[\pi] \in \widehat{G}} (f, \chi_{E_\pi})_{L^2(G)} \, \chi_{E_\pi}$$

as an L^2 function with respect to L^2 convergence and

$$\|f\|^2_{L^2(G)} = \sum_{[\pi] \in \widehat{G}} |(f, \chi_{E_\pi})_{L^2(G)}|^2.$$

Proof. For part (1), recall from Theorem 3.24 that $C(G)_{G\text{-fin}} \cong \bigoplus_{[\pi] \in \widehat{G}} E_\pi^* \otimes E_\pi$ as a $G \times G$-module. View $C(G)_{G\text{-fin}}$ and $E_\pi^* \otimes E_\pi$ as G-modules via the diagonal embedding $G \hookrightarrow G \times G$ given by $g \to (g, g)$. In particular, $(gf)(h) = f(g^{-1}hg)$ for $f \in C(G)_{G\text{-fin}}$, so that f is a class function if and only if $gf = f$ for all $g \in G$.

Also recall that the isomorphism of G-modules $E_\pi^* \otimes E_\pi \cong \text{Hom}(E_\pi, E_\pi)$ from Exercise 2.15 is induced by mapping $\lambda \otimes v$ to the linear map $v\lambda(\cdot)$ for $\lambda \in E_\pi^*$ and $v \in E_\pi$. Using this isomorphism,

$$(3.31) \qquad\qquad C(G)_{G\text{-fin}} \cong \bigoplus_{[\pi] \in \widehat{G}} \text{Hom}(E_\pi, E_\pi)$$

as a G-module under the diagonal action. For $T \in \text{Hom}(E_\pi, E_\pi)$, T satisfies $gT = T$ for all $g \in G$ if and only if $T \in \text{Hom}_G(E_\pi, E_\pi)$. By Schur's Lemma, this is if and only if $T = \mathbb{C}I_{E_\pi}$ where I_{E_π} is the identity operator. Thus the set of class functions in $C(G)_{G\text{-fin}}$ is isomorphic to $\bigoplus_{[\pi] \in \widehat{G}} \mathbb{C}I_{E_\pi}$.

If $\{x_i\}$ is an orthonormal basis for E_π and $(\cdot, \cdot)$ is a G-invariant inner product, then $I_{E_\pi} = \sum_i (\cdot, x_i)x_i$. Tracing the definitions back, the corresponding element in $E_\pi^* \otimes E_\pi$ is $\sum_i (\cdot, x_i) \otimes x_i$ and the corresponding function in $C(G)_{G\text{-fin}}$ is $g \to \sum_i (g^{-1}x_i, x_i)$. Since $(g^{-1}x_i, x_i) = \overline{(gx_i, x_i)}$, this means the class function corresponding to I_{E_π} under Equation 3.31 is exactly $\overline{\chi_{E_\pi}}$. In light of Lemma 3.2 and Theorem 3.21, part (1) is finished .

For part (2), let f be a continuous class function. By the Peter–Weyl Theorem, for $\epsilon > 0$ choose $\varphi \in C(G)_{G\text{-fin}}$, so that $\|f - \varphi\|_{C(G)} < \epsilon$. Define $\widetilde{\varphi}(h) = \int_G \varphi(g^{-1}hg) \, dg$, so that $\widetilde{\varphi}$ is a continuous class function. Using the fact that f is a class function,

$$\|f - \widetilde{\varphi}\|_{C(G)} = \sup_{h \in G} |f(h) - \widetilde{\varphi}(h)| = \sup_{h \in G} \left| \int_G \left(f(g^{-1}hg) - \varphi(g^{-1}hg) \right) dg \right|$$

$$\leq \sup_{h \in G} \int_G \left| f(g^{-1}hg) - \varphi(g^{-1}hg) \right| dg \leq \|f - \varphi\|_{C(G)} < \epsilon.$$

It therefore suffices to show that $\widetilde{\varphi} \in \text{span} \, \chi$.

For this, use Theorem 3.24 to write $\varphi(g) = \sum_i (gx_i, y_i)$ for $x_i, y_i \in E_{\pi_i}$. Thus $\widetilde{\varphi}(h) = \sum_i (\int_G g^{-1}hgx_i \, dg, y_i)$. However, on E_{π_i}, the operator $\int_G g^{-1}hg \, dg$ is a G-map and therefore acts as a scalar c_i by Schur's Lemma. Taking traces on E_{π_i},

$$\chi_{E_{\pi_i}}(h) = \text{tr}\left(\int_G g^{-1}hg \, dg \right) = \text{tr}\left(c_i I_{E_{\pi_i}} \right) = c_i \dim E_{\pi_i},$$

so that $\widetilde{\varphi}(h) = \sum_i \frac{(x_i, y_i)}{\dim E_{\pi_i}} \chi_{E_{\pi_i}}(h)$ which finishes (2).

For part (3), let f be an L^2 class function. By the Peter–Weyl Theorem, choose $\varphi \in C(G)_{G\text{-fin}}$ so that $\|f - \varphi\|_{L^2(G)} < \epsilon$. Then $\widetilde{\varphi} \in \text{span} \, \chi$. Using the integral form of the Minkowski integral inequality and invariant integration,

$$\|f - \widetilde{\varphi}\|_{L^2(G)} = \left(\int_G |f(h) - \widetilde{\varphi}(h)|^2 \, dh \right)^{\frac{1}{2}}$$

$$= \left(\int_G \left| \int_G \left(f(g^{-1}hg) - \varphi(g^{-1}hg) \right) dg \right|^2 dh \right)^{\frac{1}{2}}$$

$$\leq \int_G \left(\int_G \left| f(g^{-1}hg) - \varphi(g^{-1}hg) \right|^2 dh \right)^{\frac{1}{2}} dg$$

$$= \int_G \left(\int_G |f(h) - \varphi(h)|^2 \, dh \right)^{\frac{1}{2}} dg = \|f - \varphi\|_{L^2(G)} < \epsilon.$$

The proof is finished by the Schur orthogonality relations and elementary Hilbert space theory.   □

3.3.3.3 Classification of Irreducible Representation of $SU(2)$.
From §2.1.2.2, recall that the representations $V_n(\mathbb{C}^2)$ of $SU(2)$ were shown to be irreducible in §2.3.1. By dimension, each is obviously inequivalent to the others. In fact, they are the only irreducible representations up to isomorphism (c.f. Exercise 6.8 for a purely algebraic proof).

Theorem 3.32. *The map $n \to V_n(\mathbb{C}^2)$ establishes an isomorphism $\mathbb{N} \cong \widehat{SU(2)}$.*

Proof. Viewing S^1 as a subgroup of $SU(2)$ via the inclusion $e^{i\theta} \to \text{diag}(e^{i\theta}, e^{-i\theta})$, Equation 2.25 calculates the character of $V_n(\mathbb{C}^2)$ restricted to S^1 to be

$$(3.33) \qquad \chi_{V_n(\mathbb{C}^2)}(e^{i\theta}) = \sum_{k=0}^{n} e^{i(n-2k)\theta}.$$

A simple inductive argument (Exercise 3.21) using Equation 3.33 shows that span$\{\chi_{V_n(\mathbb{C}^2)}(e^{i\theta}) \mid n \in \mathbb{N}\}$ equals span$\{\cos n\theta \mid n \in \mathbb{N}\}$.

Since every element of $SU(2)$ is uniquely diagonalizable to elements of the form $e^{\pm i\theta} \in S^1$, it is easy to see (Exercise 3.21) that restriction to S^1 establishes a norm preserving bijection from the set of continuous class functions on $SU(2)$ to the set of even continuous functions on S^1.

From elementary Fourier analysis, span$\{\cos n\theta \mid n \in \mathbb{N}\}$ is dense in the set of even continuous functions on S^1. Thus span$\{\chi_{V_n(\mathbb{C}^2)}(e^{i\theta}) \mid n \in \mathbb{N}\}$ is dense within the set of continuous class functions on $SU(2)$ and therefore dense within the set of L^2 class functions. Part (3) of Theorem 3.30 therefore shows that there are no other irreducible characters. Since a representation is determined by its character, Theorem 3.7, the proof is finished. $\square$

Notice dim $V_n(\mathbb{C}^2) = n + 1$, so that the dimension is a complete invariant for irreducible representations of $SU(2)$.

3.3.4 Exercises

Exercise 3.16 Recall that $\widehat{S^1} \cong \mathbb{Z}$ (c.f. Exercise 3.13). Use the theorems of this chapter to recover the standard results of Fourier analysis on S^1. Namely, show that the trigonometric polynomials, span$\{e^{in\theta} \mid n \in \mathbb{Z}\}$, are dense in $C(S^1)$ and that $\{e^{in\theta} \mid n \in \mathbb{Z}\}$ is an orthonormal basis for $L^2(S^1)$.

Exercise 3.17 (a) Let G be a compact Lie group. Use the fact that $\widehat{G \times G} \cong \widehat{G} \times \widehat{G}$ (Exercise 3.10) and the nature of G-finite vectors to show that any $G \times G$-submodule of $C(G)_{G\text{-fin}}$ corresponds to $\bigoplus_{[\pi] \in \mathcal{A}} E_\pi^* \otimes E_\pi$ for some $\mathcal{A} \subseteq \widehat{G}$ under the correspondence $C(G)_{G\text{-fin}} \cong \bigoplus_{[\pi] \in \widehat{G}} E_\pi^* \otimes E_\pi$.
(b) Let $\pi : G \to GL(n, \mathbb{C})$ be a faithful representation of G with $\pi_{i,j}(g)$, denoting the $(i, j)^{\text{th}}$ entry of the matrix $\pi(g)$ for $g \in G$. Show that the set of functions $\{\pi_{i,j}, \overline{\pi_{i,j}} \mid 1 \le i, j, \le n\}$ generate $MC(G) = C(G)_{G\text{-fin}}$ as an algebra over $\mathbb{C}$. In particular, $C(G)_{G\text{-fin}}$ is finitely generated.
(c) Let V be a faithful representation of G. Show that each irreducible representation of G is a submodule of $(\bigotimes^n V) \oplus (\bigotimes^m \overline{V})$ for some $n, m \in \mathbb{N}$.

Exercise 3.18 Let G be a compact Lie group. The *commutator subgroup* of G, G', is the subgroup generated by $\{g_1 g_2 g_1^{-1} g_2^{-1} \mid g_i \in G\}$ and G is Abelian if and only if $G' = \{e\}$. Use the fact that G' acts trivially on 1-dimensional representations to show that all irreducible representations of a compact Lie group are one-dimensional if and only if G is Abelian (c.f. Exercise 2.21).

Exercise 3.19 Let G be a finite group.
(a) Show that $\int_G f(g)\, dg = \frac{1}{|G|} \sum_{g \in G} f(g)$.
(b) Use character theory to show that the number of inequivalent irreducible representations is the number of conjugacy classes in G.
(c) Show that $|G|$ equals the sum of the squares of the dimensions of its irreducible representations.

Exercise 3.20 If a compact Lie group G is not finite, show that $\widehat{G}$ is countably infinite.

Exercise 3.21 (a) Viewing $S^1 \hookrightarrow SU(2)$ via the inclusion $e^{i\theta} \to \text{diag}(e^{i\theta}, e^{-i\theta})$, show that the span$\{\chi_{V_n(\mathbb{C}^2)}(e^{i\theta}) \mid n \in \mathbb{N}\}$ equals the span$\{\cos n\theta \mid n \in \mathbb{N}\}$.
(b) Show restriction to S^1 establishes a norm preserving bijection from the set of continuous class functions on $SU(2)$ to the set of even continuous functions on S^1 (c.f. §7.3.1 for a general statement).

Exercise 3.22 (a) Continue to view $S^1 \hookrightarrow SU(2)$. For the representations $V_n(\mathbb{C}^2)$ of $SU(2)$, show $\chi_{V_n(\mathbb{C}^2)}(e^{i\theta}) = \frac{\sin(n+1)\theta}{\sin\theta}$ when $\theta \notin \pi\mathbb{Z}$.
(b) Let f be a continuous class function on $SU(2)$. Show that

$$\int_{SU(2)} f(g)\, dg = \frac{2}{\pi} \int_0^\pi f(\text{diag}(e^{i\theta}, e^{-i\theta})) \sin^2\theta\, d\theta$$

by first showing the above integral equation holds when $f = \chi_{V_n(\mathbb{C}^2)}$, c.f. Exercise 7.9.

Exercise 3.23 (a) Let V be an irreducible representation of a compact Lie group G. Show that

$$\dim V \int_G \chi_V(g^{-1}hgk)\, dg = \chi_V(h)\chi_V(k)$$

for $h, k \in G$.
(b) Conversely, if $f \in C(G)$ satisfies $\int_G f(g^{-1}hgk)\, dg = f(h)f(k)$ for all $h, k \in G$, show that there is an irreducible representation V of G, so that $f = (\dim V)^{-1}\chi_V$.

Exercise 3.24 (a) Use the isomorphism $SO(3) \cong SU(2)/\{\pm I\}$, Lemma 1.23, to show that the set of inequivalent irreducible representations of $SO(3)$ can be indexed by $\{V_{2n}(\mathbb{C}^2) \mid n \in \mathbb{N}\}$.
(b) Using a dimension count, Theorem 2.33, and Exercise 2.30, show that $V_{2n}(\mathbb{C}^2) \cong \mathcal{H}_n(\mathbb{R}^3)$ as $SO(3)$-modules. Conclude that $\{\mathcal{H}_n(\mathbb{R}^3) \mid n \in \mathbb{N}\}$ comprises a complete set of inequivalent irreducible representations for $SO(3)$.
(c) Use Exercise 3.5 to show that

$$\mathcal{H}_n(\mathbb{R}^3) \otimes \mathcal{H}_m(\mathbb{R}^3) \cong \bigoplus_{j=0}^{\min\{n,m\}} \mathcal{H}_{n+m-j}(\mathbb{R}^3).$$

3.4 Fourier Theory

Recall that the *Fourier transform* on S^1 can be thought of as an isomorphism $\wedge : L^2(S^1) \to l^2(\mathbb{Z})$, where

$$\widehat{f}(n) = \int_{S^1} f(e^{i\theta}) e^{-in\theta} \frac{d\theta}{2\pi}$$

with $\|f\| = \|\widehat{f}\|$. The inverse is given by the *Fourier series*

$$f(\theta) = \sum_{n \in \mathbb{Z}} \widehat{f}(n)e^{in\theta},$$

where convergence is as $L^2(S^1)$ functions. It is well known that even when $f \in C(S^1)$, the Fourier series may not converge pointwise to f. However continuity and any positive Lipschitz condition will guarantee uniform convergence.

Since we recognize $\frac{d\theta}{2\pi}$ as the invariant measure on S^1 and $\mathbb{Z}$ as parametrizing $\widehat{S^1}$ with n corresponding to the (one-dimensional) representation $e^{i\theta} \to e^{in\theta}$, it seems likely this result can be generalized to any compact Lie group G. In fact, the scalar valued Fourier transform in Theorem 3.43 will establish a unitary isomorphism

$$\{L^2(G) \text{ class functions}\} \cong l^2(\widehat{G}).$$

Note in the case of $G = S^1$, the class function assumption is vacuous since S^1 is Abelian.

In order to handle all L^2 functions when G is not Abelian, the operator valued Fourier transform in the Plancherel Theorem (Theorem 3.38) will establish a unitary isomorphism

$$L^2(G) \cong \widehat{\bigoplus}_{[\pi] \in \widehat{G}} \text{End}(E_\pi).$$

Remarkably, this isomorphism will also preserve the natural algebra structure of both sides. Note that for $G = S^1$, the right-hand side in the above equation reduces to $l^2(\widehat{G})$ since $\text{End}(E_\pi) \cong \mathbb{C}$.

In terms of proofs, most of the work needed for the general case is already done in Corollary 3.26. In essence, only some bookkeeping and definition chasing is required to appropriately rescale existing maps.

3.4.1 Convolution

Let G be a compact Lie group. Write $\text{End}(V) = \text{Hom}(V, V)$ for the set of endomorphisms on a vector space V. Since G has finite volume, $L^2(G) \subseteq L^1(G)$, so that the following definition makes sense.

Definition 3.34. (1) For $[\pi] \in \widehat{G}$, define $\pi : L^2(G) \to \text{End}(E_\pi)$ by

$$(\pi(f))(v) = \int_G f(g)gv\, dg$$

for $f \in L^2(G)$ and $v \in E_\pi$.
(2) Define $\widetilde{f} \in L^2(G)$ by $\widetilde{f}(g) = \overline{f(g^{-1})}$.

From a standard analysis course (Exercise 3.25 or see [37] or [73]), recall that the *convolution* operator $* : L^2(G) \times L^2(G) \to C(G)$ is given by

$$(f_1 * f_2)(g) = \int_G f_1(gh^{-1})f_2(h)\, dh$$

for $f_i \in L^2(G)$ and $g \in G$.

Lemma 3.35. *Let G be a compact Lie group, $[\pi] \in \widehat{G}$ with G-invariant inner product $(\cdot, \cdot)$ on E_π, f_i, $f \in L^2(G)$, and $v_i \in E_\pi$.*
(1) $\pi(f_1 * f_2) = \pi(f_1) \circ \pi(f_2)$.
(2) $(\pi(f)v_1, v_2) = (v_1, \pi(\tilde{f})v_2)$, i.e., $\pi(f)^* = \pi(\tilde{f})$.

Proof. For part (1) with $v \in E_\pi$, use Fubini's Theorem and a change of variables $g \to gh$ to calculate

$$\pi(f_1 * f_2)(v) = \int_G \int_G f_1(gh^{-1})f_2(h)gv\,dh\,dg$$
$$= \int_G \int_G f_1(g)f_2(h)ghv\,dg\,dh$$
$$= \int_G f_1(g)g\left(\int_G f_2(h)hv\,dh\right)dg = \pi(f_1)\,(\pi(f_2)(v))\,.$$

For part (2), calculate the following:

$$(\pi(f)v_1, v_2) = \int_G f(g)(gv_1, v_2)\,dg = \int_G (v_1, \overline{f(g)}g^{-1}v_2)\,dg$$
$$= \int_G (v_1, \tilde{f}(g)gv_2)\,dg = (v_1, \pi(\tilde{f})v_2)\,. \qquad \square$$

3.4.2 Plancherel Theorem

The motivation for the next definition comes from Corollary 3.26 and the decomposition $L^2(G) \cong \widehat{\bigoplus}_{[\pi]\in\widehat{G}} E_\pi^* \otimes E_\pi$ coupled with the isomorphism $E_\pi^* \otimes E_\pi \cong \text{End}(E_\pi)$.

Definition 3.36. (1) Let G be a compact Lie group and $[\pi] \in \widehat{G}$ with a G-invariant inner product $(\cdot, \cdot)$ on E_π. Then $\text{End}(E_\pi)$ is a Hilbert space with respect to the *Hilbert–Schmidt* inner product

$$(T, S)_{HS} = \text{tr}(S^* \circ T) = \sum_i (Tv_i, Sv_i)$$

with $T, S \in \text{End}(E_\pi)$, S^* the adjoint of S with respect to $(\cdot, \cdot)$, and $\{v_i\}$ an orthonormal basis for E_π. The corresponding *Hilbert–Schmidt* norm is

$$\|T\|_{HS} = \text{tr}(T^*T)^{\frac{1}{2}} = \left(\sum_i \|Tv_i\|^2\right)^{\frac{1}{2}}.$$

Write $\text{End}(E_\pi)_{HS}$ when viewing $\text{End}(E_\pi)$ as a Hilbert space equipped with the Hilbert–Schmidt inner product.
(2) Let $\text{Op}(\widehat{G})$ be the Hilbert space

$$\text{Op}(\widehat{G}) = \widehat{\bigoplus_{[\pi]\in\widehat{G}}} \text{End}(E_\pi)_{HS}.$$

Equip $\mathrm{Op}(\widehat{G})$ with the algebra structure

$$(T_\pi)_{[\pi]\in\widehat{G}}(S_\pi)_{[\pi]\in\widehat{G}} = ((\dim E_\pi)^{-\frac{1}{2}} T_\pi \circ S_\pi)_{[\pi]\in\widehat{G}}$$

and the $G \times G$-module structure

$$(g_1, g_2)(T_\pi)_{[\pi]\in\widehat{G}} = \left(\pi(g_2) \circ T \circ \pi(g_1^{-1})\right)_{[\pi]\in\widehat{G}}$$

for $g_i \in G$ and $T \in \mathrm{End}(E_\pi)$.

Some comments are in order. First, note that the inner product on $\mathrm{End}(E_\pi)_{HS}$ is independent of the choice of invariant inner product on E_π since scaling the inner product on E_π does not change S^*. Secondly, it must be verified that the algebra structure and $G \times G$-module structure on $\mathrm{Op}(\widehat{G})$ are well defined. Since these are straightforward exercises, they are left to the reader (Exercise 3.26).

Definition 3.37. (1) Let G be a compact Lie group. The *operator valued Fourier transform*, $\mathcal{F} : L^2(G) \to \mathrm{Op}(\widehat{G})$, is defined by

$$\mathcal{F}f = \left((\dim E_\pi)^{\frac{1}{2}} \pi(f)\right)_{[\pi]\in\widehat{G}}.$$

(2) For $T_\pi \in \mathrm{End}(E_\pi)$, write $\mathrm{tr}(T_\pi \circ g^{-1})$ for the smooth function on G defined by $g \to \mathrm{tr}(T_\pi \circ \pi(g^{-1}))$. The *inverse operator valued Fourier transform*, $\mathcal{I} : \mathrm{Op}(\widehat{G}) \to L^2(G)$, is given by

$$\mathcal{I}(T_\pi)_{[\pi]\in\widehat{G}} = \sum_{[\pi]\in\widehat{G}} (\dim E_\pi)^{\frac{1}{2}} \mathrm{tr}(T_\pi \circ g^{-1})$$

with respect to L^2 convergence.

It is necessary to check that $\mathcal{F}$ and $\mathcal{I}$ are well defined and inverses of each other. These details will be checked in the proof below. In the following theorem, view $L^2(G)$ as an algebra with respect to convolution and remember that $L^2(G)$ is a $G \times G$-module with $(g_1, g_2) \in G \times G$ acting as $r_{g_1} \circ l_{g_2}$ so $((g_1, g_2)f)(g) = f(g_2^{-1}gg_1)$ for $f \in L^2(G)$ and $g_i, g \in G$.

Theorem 3.38 (Plancherel Theorem). *Let G be a compact Lie group. The maps $\mathcal{F}$ and $\mathcal{I}$ are well defined unitary, algebra, $G \times G$-intertwining isomorphisms and inverse to each other so that*

$$\mathcal{F} : L^2(G) \xrightarrow{\cong} \mathrm{Op}(\widehat{G})$$

with $\|f\|_{L^2(G)} = \|\mathcal{F}f\|_{\mathrm{Op}(\widehat{G})}$, $\mathcal{F}(f_1 * f_2) = (\mathcal{F}f_1)(\mathcal{F}f_2)$, $\mathcal{F}((g_1, g_2)f) = (g_1, g_2)(\mathcal{F}f)$, *and* $\mathcal{F}^{-1} = \mathcal{I}$ *for* $f \in L^2(G)$ *and* $g_i \in G$.

Proof. Recall the decomposition $L^2(G) \cong \widehat{\bigoplus}_{[\pi] \in \widehat{G}} E_\pi^* \otimes E_\pi$ from Corollary 3.26 that maps $\lambda \otimes v \in E_\pi^* \otimes E_\pi$ to $f_{\lambda \otimes v}$ where $f_{\lambda \otimes v}(g) = \lambda(g^{-1}v)$ for $g \in G$. Since $\mathrm{Op}(\widehat{G}) = \widehat{\bigoplus}_{[\pi] \in \widehat{G}} \mathrm{End}(E_\pi)_{HS}$ and since isometries on dense sets uniquely extend by continuity, it suffices to check that $\mathcal{F}$ restricts to a unitary, algebra, $G \times G$-intertwining isomorphism from $\mathrm{span}\{f_{\lambda \otimes v} \mid \lambda \otimes v \in E_\pi^* \otimes E_\pi\}$ to $\mathrm{End}(E_\pi)$ with inverse $\mathcal{I}$. Here $\mathrm{End}(E_\pi)$ is viewed as a subspace of $\mathrm{Op}(\widehat{G})$ under the natural inclusion $\mathrm{End}(E_\pi) \hookrightarrow \mathrm{Op}(\widehat{G})$.

Write $(\cdot, \cdot)$ for a G-invariant inner product on E_π. Any $\lambda \in E_\pi^*$ may be uniquely written as $\lambda = (\cdot, v)$ for some $v \in E_\pi$. Thus the main problem revolves around evaluating $\pi'(f_{(\cdot, v_1) \otimes v_2})$ for $[\pi'] \in \widehat{G}$ and $v_i \in E_\pi$. Therefore choose $w_i \in E_{\pi'}$ and a G-invariant inner product $(\cdot, \cdot)'$ on $E_{\pi'}$ and calculate

$$(\pi'(f_{(\cdot, v_1) \otimes v_2})(w_1), w_2)' = \int_G (\pi(g^{-1})v_2, v_1)(\pi'(g)w_1, w_2)' \, dg$$

$$= \int_G (\pi'(g)w_1, w_2)' \overline{(\pi(g)v_1, v_2)} \, dg.$$

If $\pi' \not\cong \pi$, the Schur orthogonality relations imply that $(\pi'(f_{(\cdot, v_1) \otimes v_2})(w_1), w_2)' = 0$, so that $\pi'(f_{(\cdot, v_1) \otimes v_2}) = 0$. Thus $\mathcal{F}$ maps $\mathrm{span}\{f_{\lambda \otimes v} \mid \lambda \otimes v \in E_\pi^* \otimes E_\pi\}$ to $\mathrm{End}(E_\pi)$. On the other hand, if $\pi' = \pi$, the Schur orthogonality relations imply that

$$(\pi(f_{(\cdot, v_1) \otimes v_2})(w_1), w_2) = (\dim E_\pi)^{-1} (w_1, v_1)\overline{(w_2, v_2)}.$$

In particular, $\pi(f_{(\cdot, v_1) \otimes v_2}) = (\dim E_\pi)^{-1} (\cdot, v_1)v_2$, so

$$\mathcal{F} f_{(\cdot, v_1) \otimes v_2} = (\dim E_\pi)^{-\frac{1}{2}} (\cdot, v_1)v_2.$$

Viewed as a map from $\mathrm{span}\{f_{\lambda \otimes v} \mid \lambda \otimes v \in E_\pi^* \otimes E_\pi\}$ to $\mathrm{End}(E_\pi)$, this shows that $\mathcal{F}$ is surjective and, by dimension count, an isomorphism.

To see that $\mathcal{I}$ is the inverse of $\mathcal{F}$, calculate the trace using an orthonormal basis that starts with $\|v_2\|^{-1} v_2$:

$$\mathrm{tr}([(\cdot, v_1)v_2] \circ \pi(g^{-1})) = \left([[(\cdot, v_1)v_2] \circ \pi(g^{-1})] (\frac{v_2}{\|v_2\|}), \frac{v_2}{\|v_2\|} \right)$$

$$= (\pi(g^{-1})v_2, v_1) = f_{(\cdot, v_1) \otimes v_2}(g).$$

Thus

(3.39) $$\mathcal{I}\left((\dim E_\pi)^{-\frac{1}{2}} (\cdot, v_1)v_2\right) = f_{(\cdot, v_1) \otimes v_2}$$

and $\mathcal{I} = \mathcal{F}^{-1}$.

To check unitarity, use the Schur orthogonality relations to calculate

$$\left(f_{(\cdot, v_1) \otimes v_2}, f_{(\cdot, v_3) \otimes v_4}\right)_{L^2(G)} = \int_G (g^{-1}v_2, v_1) \overline{(g^{-1}v_4, v_3)} \, dg$$

$$= (\dim E_\pi)^{-1} (v_2, v_4)\overline{(v_1, v_3)}.$$

To calculate a Hilbert–Schmidt norm, first observe that the adjoint of $(\cdot, v_3)v_4 \in \text{End}(E_\pi)_{HS}$ is $(\cdot, v_4)v_3$ since

$$((v_5, v_3)v_4, v_6) = (v_5, v_3)(v_4, v_6) = (v_5, (v_6, v_4)v_3).$$

Hence

$$
\begin{aligned}
\left(\mathcal{F}f_{(\cdot, v_1) \otimes v_2}, \mathcal{F}f_{(\cdot, v_3) \otimes v_4}\right)_{HS} &= (\dim E_\pi)^{-1} \left((\cdot, v_1)v_2, (\cdot, v_3)v_4\right)_{HS} \\
&= (\dim E_\pi)^{-1} \operatorname{tr}\left[((\cdot, v_1)v_2, v_4)v_3\right] \\
&= (\dim E_\pi)^{-1} (v_2, v_4) \operatorname{tr}\left[(\cdot, v_1)v_3\right] \\
&= (\dim E_\pi)^{-1} (v_2, v_4) \left((\frac{v_3}{\|v_3\|}, v_1)v_3, \frac{v_3}{\|v_3\|}\right) \\
&= (\dim E_\pi)^{-1} (v_2, v_4)(v_3, v_1) \\
&= \left(f_{(\cdot, v_1) \otimes v_2}, f_{(\cdot, v_3) \otimes v_4}\right)_{L^2(G)},
\end{aligned}
$$

and so $\mathcal{F}$ is unitary.

To check that the algebra structures are preserved, simply use Lemma 3.35 to observe that $\pi(f_1 * f_2) = \pi(f_1) \circ \pi(f_2)$. Thus

$$
\begin{aligned}
\mathcal{F}(f_1 * f_2) &= (\dim E_\pi)^{\frac{1}{2}} \pi(f_1) \circ \pi(f_2) \\
&= (\dim E_\pi)^{-\frac{1}{2}} \mathcal{F}f_1 \circ \mathcal{F}f_2 = (\mathcal{F}f_1)(\mathcal{F}f_2),
\end{aligned}
$$

as desired.

Finally, to see $\mathcal{F}$ is a $G \times G$-map, first observe that

$$
\begin{aligned}
\left((g_1, g_2)f_{(\cdot, v_1) \otimes v_2}\right)(g) &= f_{(\cdot, v_1) \otimes v_2}(g_2^{-1}gg_1) = (g_1^{-1}g^{-1}g_2v_2, v_1) \\
&= (g^{-1}g_2v_2, g_1v_1) = f_{(\cdot, g_1 v_1) \otimes g_2 v_2}(g).
\end{aligned}
$$

Thus

$$
\begin{aligned}
\mathcal{F}((g_1, g_2)f_{(\cdot, v_1) \otimes v_2}) &= \mathcal{F}f_{(\cdot, g_1 v_1) \otimes g_2 v_2} = (\dim E_\pi)^{-\frac{1}{2}} (\cdot, g_1v_1)g_2v_2 \\
&= \pi(g_2) \circ (\cdot, v_1)v_2 \circ \pi(g_1^{-1}) = (g_1, g_2)(\mathcal{F}f),
\end{aligned}
$$

which finishes the proof. $\qquad\square$

Corollary 3.40. *Let G be a compact Lie group and $f, f_i \in L^2(G)$.*
(1) Then the Parseval–Plancherel *formula holds:*

$$\|f\|_{L^2(G)}^2 = \sum_{[\pi] \in \widehat{G}} \dim E_\pi \, \|\pi(f)\|_{HS}^2.$$

(2) Under the natural inclusion $\text{End}(E_\pi) \hookrightarrow \text{Op}(\widehat{G})$, $\mathcal{I}I_{E_\pi} = (\dim E_\pi)^{\frac{1}{2}} \chi_{\overline{E_\pi}}$ where $I_{E_\pi} \subset \text{End}(E_\pi)$ is the identity operator. Moreover,

$$f = \sum_{[\pi] \in \widehat{G}} (\dim E_\pi) \, f * \chi_{E_\pi}$$

with respect to L^2 convergence.

(3)

$$(f_1, f_2)_{L^2(G)} = \sum_{[\pi] \in \widehat{G}} (\dim E_\pi) \operatorname{tr} \pi(\tilde{f}_2 * f_1).$$

Proof. Part (1) follows immediately from the Plancherel Theorem. Similarly, part (2) will also follow from the Plancherel Theorem once we show $\mathcal{I}I_{E_\pi} = (\dim E_\pi)^{\frac{1}{2}} \chi_{\overline{E_\pi}}$ since $(\dim E_\pi)^{\frac{1}{2}} I_{E_\pi}$ acts on $\operatorname{Op}(\widehat{G})$ by projecting to $\operatorname{End}(E_\pi)$. Although this result is implicitly contained in the proof of Theorem 3.30, it is simple to verify directly. Let $\{x_i\}$ be an orthonormal basis for E_π where $(\cdot, \cdot)$ is a G-invariant inner product. Hence $I_{E_{\bar{\gamma}}} = \sum_i (\cdot, x_i) x_i$. Equation 3.39 shows $\mathcal{I}I_{E_\pi} = (\dim E_\pi)^{\frac{1}{2}} \sum_i f_{(\cdot, x_i) \otimes x_i}$ where $f_{(\cdot, x_i) \otimes x_i}(g) = (g^{-1} x_i, x_i)$. Thus $\mathcal{I}I_{E_\pi} = (\dim E_\pi)^{\frac{1}{2}} \chi_{\overline{E_\pi}}$ by Theorem 3.5.

For part (3), the Plancherel Theorem and Lemma 3.35 imply that

$$(f_1, f_2)_{L^2(G)} = (\mathcal{F}f_1, \mathcal{F}f_2)_{HS} = \sum_{[\pi] \in \widehat{G}} (\dim E_\pi) \operatorname{tr} \left(\pi(f_2)^* \circ \pi(f_1) \right)$$

$$= \sum_{[\pi] \in \widehat{G}} (\dim E_\pi) \operatorname{tr} \pi(\tilde{f}_2 * f_1). \qquad \square$$

Definition 3.41. Let G be a compact Lie group and $f \in L^2(G)$. Define the *scalar valued Fourier transform* by

$$\widehat{f}(\pi) = \operatorname{tr} \pi(f)$$

for $[\pi] \in \widehat{G}$.

Note that $\widehat{f}$ can also be computed by the formula

$$\widehat{f}(\pi) = \int_G f(g) \chi_{E_\pi}(g) \, dg = (f, \chi_{\overline{E_\pi}})_{L^2(G)}$$

since $\widehat{f}(\pi) = \sum_i (\pi(f) v_i, v_i) = \int_G f(g) \sum_i (g v_i, v_i) \, dg$, where $\{v_i\}$ is an orthonormal basis for E_π.

Theorem 3.42 (Scalar Fourier Inversion). *Let G be a compact Lie group and $f \in \operatorname{span} \left(L^2(G) * L^2(G) \right) \subseteq C(G)$. Then*

$$f(e) = \sum_{[\pi] \in \widehat{G}} (\dim E_\pi) \widehat{f}(\pi).$$

Proof. If $f = f_1 * f_2$ for $f_i \in L^2(G)$, then by Corollary 3.40,

$$f(e) = \int_G f_1(g^{-1}) f_2(g) \, dg = \int_G f_2(g) \overline{\tilde{f}_1(g)} \, dg$$

$$= (f_2, \tilde{f}_1)_{L^2(G)} = \sum_{[\pi] \in \widehat{G}} (\dim E_\pi) \operatorname{tr} \pi(f_1 * f_2). \qquad \square$$

As already noted, even for $G = S^1$ the Scalar Fourier Inversion Theorem can fail if f is only assumed to be continuous. However, using Lie algebra techniques and the Plancherel Theorem, it is possible to show that the Scalar Fourier Inversion Theorem holds when f is continuously differentiable. In particular, the Scalar Fourier Inversion Theorem holds for smooth f.

Theorem 3.43. *Let G be a compact Lie group. The map $f \to (\widehat{f}(\pi))_{[\pi] \in \widehat{G}}$ establishes a unitary isomorphism*

$$\{L^2(G) \text{ class functions}\} \cong l^2(\widehat{G}).$$

For $[\gamma] \in \widehat{G}$, the image of χ_{E_γ} under this map is $(\delta_{\pi,\overline{\gamma}})_{[\pi] \in \widehat{G}}$, where $\delta_{\pi,\overline{\gamma}}$ is 1 when $\pi \cong \overline{\gamma}$ and 0 when $\pi \not\cong \overline{\gamma}$.

Proof. This result is implicitly embedded in the proof of Theorem 3.30. However it is trivial to check directly. Observe that

$$\widehat{\chi_{E_\gamma}}(\pi) = \int_G \chi_{E_\gamma}(g) \chi_{E_\pi}(g) \, dg,$$

so that Theorem 3.7 implies χ_{E_π} is mapped to $(\delta_{\pi,\overline{\gamma}})_{[\pi] \in \widehat{G}}$. Since $\{\chi_{E_\pi} \mid [\pi] \in \widehat{G}\}$ is an orthonormal basis for $\{L^2(G)$ class functions$\}$, the result follows. $\square$

3.4.3 Projection Operators and More General Spaces

Let G be a compact Lie group and (γ, V) a unitary representation of G on a Hilbert space. For $[\pi] \in \widehat{G}$, it will turn out that the operator $(\dim E_\pi) \gamma(\chi_{\overline{E_\pi}})$ is the orthogonal G-intertwining projection of V onto $V_{[\pi]}$. In fact, the main part of this result is true in a much more general setting than Hilbert space representations.

Now only assume V is a Hausdorff complete locally convex topological space. The notions of G-finite vector and isotypic component carry over from §3.2.2 and §3.2.3 in the obvious fashion.

Definition 3.44. Let V be a representation of a compact Lie group G on a Hausdorff complete locally convex topological space.
(1) The set of G-*finite vectors*, $V_{G\text{-fin}}$, is the set of $v \in V$ where $\text{span}\{\pi(G)v\}$ is finite dimensional.
(2) For $[\pi] \in \widehat{G}$, let $V_{[\pi]}^0$ be the sum of all irreducible submodules equivalent to E_π.
(3) The closure $V_{[\pi]} = \overline{V_{[\pi]}^0}$ is called the π-*isotypic component* of V.

Theorem 3.45. *Let (γ, V) be a representation of a compact Lie group G on a complete Hausdorff locally convex topological space.*
(1) For $[\pi], [\pi'] \in \widehat{G}$, the operator $(\dim E_\pi) \gamma(\chi_{\overline{E_\pi}})$ is a G-intertwining projection of V onto $V_{[\pi]}$ that acts as the identity on $V_{[\pi]}$ and acts as zero on $V_{[\pi']}$ for $\pi' \not\cong \pi$.
(2) If (γ, V) is a unitary representation on a Hilbert space, then $(\dim E_\pi) \gamma(\chi_{\overline{E_\pi}})$ is also self-adjoint, i.e., orthogonal.

Proof. For part (1), let $g \in G$ and $v \in V$, and observe that

$$\left(g\gamma(\chi_{E\pi})g^{-1}\right) v = \int_G \chi_{E\pi}(h)ghg^{-1}v\,dh = \int_G \chi_{E\pi}(g^{-1}hg)hv\,dh$$

$$= \int_G \chi_{E\pi}(h)hv\,dh = \gamma(\chi_{E\pi})v,$$

so that $\gamma(\chi_{E\pi})$ is a G-map. Applying this to the representation $E_{\pi'}$, Schur's Lemma shows that $\pi'(\chi_{\overline{E\pi}}) = c_{\pi',\pi} I_{E_{\pi'}}$ for $c_{\pi',\pi} \in \mathbb{C}$. Taking traces,

$$(\dim E_{\pi'})c_{\pi',\pi} = \int_G \overline{\chi_{E\pi}(g)}\,\mathrm{tr}\,\pi'(g)\,dg = \int_G \chi_{E\pi'}(g)\overline{\chi_{E\pi}(g)}\,dg.$$

By Theorem 3.7, $c_{\pi',\pi}$ is 0 when $\pi' \not\cong \pi$ and $(\dim E_\pi)^{-1}$ when $\pi' = \pi$. Since any $v \in V_{[\pi']}^0$ lies in a submodule of V that is isomorphic to $E_{\pi'}$, $(\dim E_\pi)\,\pi'(\chi_{\overline{E\pi}})$ acts on $V_{[\pi']}^0$ as the identity when $\pi' = \pi$ and by zero when $\pi' \not\cong \pi$. Continuity finishes part (1).

For part (2), Lemma 3.35 implies that $\gamma(\chi_{E\pi})^* = \gamma(\widetilde{\chi_{E\pi}})$. But Theorem 3.5 shows $\widetilde{\chi_{E\pi}} = \chi_{E\pi}$. $\square$

Theorem 3.46. *Let (γ, V) be a representation of a compact Lie group G on a Hausdorff complete locally convex topological space.*
(1) $V_{G\text{-fin}} = \bigoplus_{[\pi] \in \widehat{G}} V_{[\pi]}^0$.
(2) $V_{G\text{-fin}}$ is dense in V.
(3) If V is irreducible, then V is finite dimensional.

Proof. For part (1), the definitions and Corollary 2.17 imply $V_{G\text{-fin}} = \sum_{[\pi] \in \widehat{G}} V_{[\pi]}^0$, so it only remains to see the sum is direct. However, the existence of the projections in Theorem 3.45 trivially establish this result.

For part (2), suppose $\lambda \in V^*$ vanishes on $V_{G\text{-fin}}$. By the Hahn–Banach Theorem, it suffices to show $\lambda = 0$. For $x \in V$, define $f_x \in C(G)$ by $f_x(g) = \lambda(gx)$. Clearly $\lambda = 0$ if and only if $f_x = 0$ for all x. Looking to use Corollary 3.40, calculate

$$\left(f_x * \chi_{E_\pi}\right)(g) = \int_G \lambda(ghx)\chi_{E_\pi}(h^{-1})\,dh = \lambda\!\left(\int_G \overline{\chi_{E_\pi}(h)}ghx\,dh\right)$$

$$= \lambda(g\pi(\chi_{\overline{E_\pi}})x) = f_{\pi(\chi_{\overline{E_\pi}})x}(g).$$

Since Theorem 3.45 shows that $\pi(\chi_{\overline{E_\pi}})x \in V_{[\pi]}$ and since λ vanishes on each $V_{[\pi]}$ by continuity, $f_x * \chi_{E_\pi} = 0$. Thus $f_x = 0$ and part (2) is finished.

For part (3), observe that part (2) shows V contains a finite-dimensional irreducible submodule W. Since finite-dimensional subspaces are closed, irreducibility implies that $V = W$. $\square$

In particular, notice that even allowing the greater generality of representations on complete locally convex topological spaces still leaves us with the same set of irreducible representations, $\widehat{G}$.

The following corollary will be needed in §7.4.

Corollary 3.47. *Let G be a compact Lie group. Suppose $S \subseteq C(G)$ is a subspace equipped with a topology so that:*
(a) S is dense in $C(G)$,
(b) S is a Hausdorff complete locally convex topological space,
(c) the topology on S is stronger than uniform convergence, i.e., convergence in S implies convergence in $C(G)$, and
(d) S is invariant under l_g and r_g and, with these actions, S is a $G \times G$-module. Then $S_{G\text{-fin}} = C(G)_{G\text{-fin}}$.

Proof. Clearly $S_{[\pi]} \subseteq C(G)_{[\pi]}$ for $[\pi] \in \widehat{G}$. Note $C(G)_{[\pi]} \cong E_\pi^* \otimes E_\pi$ by Theorem 3.24. Arguing by contradiction, suppose $S_{[\pi]} \subsetneq C(G)_{[\pi]}$ for some $[\pi] \in \widehat{G}$. Then there exists a nonzero $f \in C(G)_{[\pi]}$ that is perpendicular to $S_{[\pi]}$ with respect to the L^2 norm. By Corollary 3.26 and Theorem 3.46, it follows that f is perpendicular to all of $S_{G\text{-fin}}$. However, this is a contradiction to the fact that $S_{G\text{-fin}}$ is dense in $L^2(G)$ by (a) and (c).   □

As an example, S could be the set of smooth functions on G or the set of real analytic functions on G. One interpretation of Corollary 3.47 says that $C(G)_{G\text{-fin}}$ is the smallest reasonable class of test functions on G that are usually useful for representation theory. Thus the topological dual $C(G)^*_{G\text{-fin}}$ of distributions is the largest class of useful generalized functions on G.

3.4.4 Exercises

Exercise 3.25 Let G be a compact Lie group. For $f_i \in L^2(G)$ and $g \in G$, examine $\sup_{h \in G} \left| \left(l_g \left(f_1 * f_2 \right) \right)(h) - (f_1 * f_2)(h) \right|$ to show $f_1 * f_2 \in C(G)$.

Exercise 3.26 (a) If V is a (finite-dimensional) vector space and $\|\cdot\|$ is the operator norm on $\text{End}(V)$, show that $\|T \circ S\|_{HS} \leq \|T\| \|S\|_{HS}$ and $\|T\| \leq \|T\|_{HS}$ for $T, S \in \text{End}(V)$.
(b) Let G be a compact Lie group. Show that $((\dim E_\pi)^{-\frac{1}{2}} T_\pi \circ S_\pi)_{[\pi] \in \widehat{G}} \subset \text{Op}(\widehat{G})$ when $(T_\pi)_{[\pi] \in \widehat{G}}, (S_\pi)_{[\pi] \in \widehat{G}} \in \text{Op}(\widehat{G})$. Is the factor $(\dim E_\pi)^{-\frac{1}{2}}$ even needed for this statement?
(c) Show $(g_2 \circ T_\pi \circ g_1^{-1})_{[\pi] \in \widehat{G}} \in \text{Op}(\widehat{G})$ for $g_i \in G$ and that this action defines a representation of $G \times G$ on $\text{Op}(\widehat{G})$.

Exercise 3.27 Let G be a compact Lie group and $f \in \text{span}\left(L^2(G) * L^2(G) \right) \subseteq C(G)$. Show that

$$f(g) = \sum_{[\pi] \in \widehat{G}} (\dim E_\pi) \widehat{\left(r_g f \right)}(\pi).$$

Exercise 3.28 With respect to convolution, show that $C(G)_{G\text{-fin}}$ is an algebra with center spanned by the set of irreducible characters, i.e., by $\{ \chi_{E_\pi} \mid [\pi] \in \widehat{G} \}$.

Exercise 3.29 For this problem, recall Exercise 3.22. Let f be a smooth class function on $SU(2)$. Show that

$$f(I) = \frac{2}{\pi} \sum_{n=0}^{\infty} (n+1) \int_0^{\pi} f(\text{diag}(e^{i\theta}, e^{-i\theta})) \sin\theta \sin(n+1)\theta \, d\theta.$$

Exercise 3.30 Let G be a compact Lie group. Show that G is Abelian if and only if the convolution on $C(G)$ is commutative (c.f. Exercise 3.18).

Exercise 3.31 Let V be a representation of a compact Lie group G on a Hausdorff complete locally convex topological space. For $[\pi] \in \widehat{G}$, show that $V_{[\pi]}^0$ is the largest subspace of V that is a direct sum of irreducible submodules equivalent to E_π.

Exercise 3.32 (a) Let (π, V) be a representation of a compact Lie group G on a Hausdorff complete locally convex topological space and f a continuous class function on G. Show that $\pi(f)$ commutes with $\pi(g)$ for $g \in G$.
(b) Show that $\pi(f)$ acts on $V_{[\pi]}$, $[\pi] \in \widehat{G}$, by $(\dim E_\pi)^{-1}(f, \chi_{\overline{E_\pi}})_{L^2(G)}$.

Exercise 3.33 Let (π, V) be a representation of a compact Lie group G on a Hausdorff complete locally convex topological space, $v \in V_{[\pi]}^0$ for $[\pi] \in \widehat{G}$, and $S = \text{span}\{\pi(G)v\}$. For $\lambda \in S^*$, define $f_\lambda \in C(G)$ by $f_\lambda(g) = \lambda(g^{-1}v)$. Show that $\dim S \leq (\dim E_\pi)^2$.

4

Lie Algebras

By their nature, Lie groups are usually nonlinear objects. However, it turns out there is a way to linearize their study by looking at the tangent space to the identity. The resulting object is called a Lie algebra. Simply by virtue of the fact that vector spaces are simpler than groups, the Lie algebra provides a powerful tool for studying Lie groups and their representations.

4.1 Basic Definitions

4.1.1 Lie Algebras of Linear Lie Groups

Let M be a manifold. Recall that a *vector field* on M is a smooth section of the *tangent bundle*, $T(M) = \cup_{m \in M} T_m(M)$. If G is a Lie group and $g \in G$, then the map $l_g : G \to G$ defined by $l_g h = gh$ for $g \in G$ is a diffeomorphism. A vector field X on G is called *left invariant* if $dl_g X = X$ for all $g \in G$. Since G acts on itself simply transitively under left multiplication, the tangent space of G at e, $T_e(G)$, is clearly in bijection with the space of left invariant vector fields. The correspondence maps $v \in T_e(G)$ to the vector field X where $X_g = dl_g v$, $g \in G$, and conversely maps a left invariant vector field X to $v = X_e \in T_e(G)$.

Elementary differential geometry shows that the set of left invariant vector fields is an algebra under the Lie bracket of vector fields (see [8] or [88]). Using the bijection of left invariant vector fields with $T_e(G)$, it follows that $T_e(G)$ has a natural algebra structure which is called the *Lie algebra* of G.

Since we are interested in compact groups, there is a way to bypass much of this differential geometry. Recall from Theorem 3.28 that a compact group G is a linear group, i.e., G is isomorphic to a closed Lie subgroup of $GL(n, \mathbb{C})$. In the setting of Lie subgroups of $GL(n, \mathbb{C})$, the Lie algebra has an explicit matrix realization which we develop in this chapter. It should be remarked, however, that the theorems in this chapter easily generalize to any Lie group.

Taking our cue from the above discussion, we will define an algebra structure on $T_e(G)$ viewed as a subspace of $T_I(GL(n, \mathbb{C}))$. Since $GL(n, \mathbb{C})$ is an open (dense) set in $M_{n,n}(\mathbb{C}) \cong \mathbb{R}^{2n^2}$, we will identify $T_I(GL(n, \mathbb{C}))$ with $\mathfrak{gl}(n, \mathbb{C})$ where

$$\mathfrak{gl}(n, \mathbb{F}) = M_{n,n}(\mathbb{F}).$$

The identification of $T_I(GL(n, \mathbb{C}))$ with $\mathfrak{gl}(n, \mathbb{C})$ is the standard one for open sets in $\mathbb{R}^{2n^2}$. Namely, to any $X \in T_I(GL(n, \mathbb{C}))$, find a smooth curve $\gamma : (-\epsilon, \epsilon) \to GL(n, \mathbb{C})$, $\epsilon > 0$, so that $\gamma(0) = I$, and so $X(f) = \frac{d}{dt}(f \circ \gamma)|_{t=0}$ for smooth functions f on $GL(n, \mathbb{C})$. The map sending X to $\gamma'(0)$ is a bijection from $T_I(GL(n, \mathbb{C}))$ to $\mathfrak{gl}(n, \mathbb{C})$.

Definition 4.1. Let G be a Lie subgroup of $GL(n, \mathbb{C})$.
(a) The *Lie algebra* of G is

$$\mathfrak{g} = \{\gamma'(0) \mid \gamma(0) = I \text{ and } \gamma : (-\epsilon, \epsilon) \to G, \epsilon > 0, \text{ is smooth}\} \subseteq \mathfrak{gl}(n, \mathbb{C}).$$

(b) The *Lie bracket* on $\mathfrak{g}$ is given by

$$[X, Y] = XY - YX.$$

Given a compact group G, Theorem 3.28 says that there is a faithful representation $\pi : G \to GL(n, \mathbb{C})$. Identifying G with its image under π, G may be viewed as a closed Lie subgroup of $GL(n, \mathbb{C})$. Using this identification, we use Definition 4.1 to define the *Lie algebra* of G. We will see in §4.2.1 that this construction is well defined up to isomorphism.

Theorem 4.2. *Let G be a Lie subgroup of $GL(n, \mathbb{C})$.*
(a) *Then $\mathfrak{g}$ is a real vector space.*
(b) *The Lie bracket is linear in each variable, skew symmetric, i.e., $[X, Y] = -[Y, X]$, and satisfies the* Jacobi identity

$$[[X, Y], Z] + [[Y, Z], X] + [[Z, X], Y] = 0$$

for $X, Y, Z \in \mathfrak{g}$.
(c) *Finally, $\mathfrak{g}$ is closed under the Lie bracket and therefore an algebra.*

Proof. Let $X_i = \gamma_i'(0) \in \mathfrak{g}$. For $r \in \mathbb{R}$, consider the smooth curve γ that maps a neighborhood of $0 \in \mathbb{R}$ to G defined by $\gamma(t) = \gamma_1(rt)\gamma_2(t)$. Then

$$\gamma'(0) = \left(r\gamma_1'(rt)\gamma_2(t) + \gamma_1(rt)\gamma_2'(t)\right)|_{t=0} = rX_1 + X_2$$

so that $\mathfrak{g}$ is a real vector space.

The statements regarding the basic properties of the Lie bracket in part (b) are elementary and left as an exercise (Exercise 4.1). To see that $\mathfrak{g}$ is closed under the bracket, consider the smooth curve σ_s that maps a neighborhood of 0 to G defined by $\sigma_s(t) = \gamma_1(s)\gamma_2(t)(\gamma_1(s))^{-1}$. In particular, $\sigma_s'(0) = \gamma_1(s)X_2(\gamma_1(s))^{-1} \in \mathfrak{g}$. Since the map $s \to \sigma_s'(0)$ is a smooth curve in a finite-dimensional vector space, tangent vectors to this curve also lie in $\mathfrak{g}$. Applying $\frac{d}{ds}|_{s=0}$, we calculate

$$\frac{d}{ds}\left(\gamma_1(s)X_2(\gamma_1(s))^{-1}\right)|_{s=0} = X_1X_2 - X_2X_1 = [X_1, X_2],$$

so that $[X_1, X_2] \in \mathfrak{g}$.   □

4.1.2 Exponential Map

Let G be a Lie subgroup of $GL(n, \mathbb{C})$ and $g \in G$. Since G is a submanifold of $GL(n, \mathbb{C})$, $T_g(G)$ can be identified with

(4.3) $\{\gamma'(0) \mid \gamma(0) = g$ and $\gamma : (-\epsilon, \epsilon) \to G, \epsilon > 0,$ is smooth$\}$

in the usual manner by mapping $\gamma'(0)$ to the element of $T_g(G)$ that acts on a smooth function f by $\frac{d}{dt}(f \circ \gamma)|_{t=0}$. Now if $\gamma(0) = I$ and $\gamma : (-\epsilon, \epsilon) \to G, \epsilon > 0,$ is smooth, then $\sigma(t) = g\gamma(t)$ satisfies $\sigma(0) = g$ and $\sigma'(0) = g\gamma'(0)$. Since left multiplication is a diffeomorphism, Equation 4.3 identifies $T_g(G)$ with the set

$$g\mathfrak{g} = \{gX \mid X \in \mathfrak{g}\}.$$

We make use of this identification without further comment.

Definition 4.4. Let G be a Lie subgroup of $GL(n, \mathbb{C})$ and $X \in \mathfrak{g}$.
(a) Let $\widetilde{X}$ be the vector field on G defined by $\widetilde{X}_g = gX, g \in G$.
(b) Let γ_X be the *integral curve* of $\widetilde{X}$ through I, i.e., γ_X is the unique maximally defined smooth curve in G satisfying

$$\gamma_X(0) = I$$

and

$$\gamma_X'(t) = \widetilde{X}_{\gamma_X(t)} = \gamma_X(t)X.$$

It is well known from the theory of differential equations that integral curves exist and are unique (see [8] or [88]).

Theorem 4.5. *Let G be a Lie subgroup of $GL(n, \mathbb{C})$ and $X \in \mathfrak{g}$.*
(a) Then

$$\gamma_X(t) = \exp(tX) = e^{tX} = \sum_{n=0}^{\infty} \frac{t^n}{n!} X^n.$$

(b) Moreover γ_X is a homomorphism and complete, i.e., it is defined for all $t \in \mathbb{R}$, so that $e^{tX} \in G$ for all $t \in \mathbb{R}$.

Proof. It is a familiar fact that the map $t \to e^{tX}$ is a well-defined smooth homomorphism of $\mathbb{R}$ into $GL(n, \mathbb{C})$ (Exercise 4.3). Hence, first extend $\widetilde{X}$ to a vector field on $GL(n, \mathbb{C})$ by $\widetilde{X}_g = gX, g \in GL(n, \mathbb{C})$. Since $e^{0X} = I$ and $\frac{d}{dt}e^{tX} = e^{tX}X, t \to e^{tX}$ is the unique integral curve for $\widetilde{X}$ passing through I as a vector field on $GL(n, \mathbb{C})$. It is obviously complete. On the other hand, since G is a submanifold of $GL(n, \mathbb{C})$, γ_X may be viewed as a curve in $GL(n, \mathbb{C})$. It is still an integral curve for $\widetilde{X}$ passing through I as a vector field on $GL(n, \mathbb{C})$. By uniqueness, $\gamma_X(t) = e^{tX}$ on the domain of γ_X. In particular, there is an $\epsilon > 0$, so that $\gamma_X(t) = e^{tX}$ for $t \in (-\epsilon, \epsilon)$. Thus $e^{tX} \in G$ for $t \in (-\epsilon, \epsilon)$. But since $e^{ntX} = (e^{tX})^n$ for $n \in \mathbb{N}$, $e^{tX} \in G$ for all $t \in \mathbb{R}$, which finishes the proof. $\square$

Note that Theorem 4.5 shows that the map $t \to e^{tX}$ is actually a smooth map from $\mathbb{R}$ to G for $X \in \mathfrak{g}$.

Theorem 4.6. *Let G be a Lie subgroup of $GL(n, \mathbb{C})$.*
(a) $\mathfrak{g} = \{X \in \mathfrak{gl}(n, \mathbb{C}) \mid e^{tX} \in G \text{ for } t \in \mathbb{R}\}$.
(b) The map $\exp \colon \mathfrak{g} \to G$ is a local diffeomorphism near 0, i.e., there is a neighborhood of 0 in $\mathfrak{g}$ on which $\exp$ restricts to a diffeomorphism onto a neighborhood of I in G.
(c) When G is connected, $\exp \mathfrak{g}$ generates G.

Proof. To see $\mathfrak{g}$ is contained in $\{X \in \mathfrak{gl}(n, \mathbb{C}) \mid e^{tX} \in G \text{ for } t \in \mathbb{R}\}$, use Theorem 4.5. Conversely, if $e^{tX} \in G$ for $t \in \mathbb{R}$ for all $X \in \mathfrak{gl}(n, \mathbb{C})$, apply $\frac{d}{dt}|_{t=0}$ and use the definition to see $X \in \mathfrak{g}$.

For part (b), by the Inverse Mapping theorem, it suffices to show the differential of $\exp \colon \mathfrak{g} \to G$ is invertible at I. In fact, we will see that the differential of $\exp$ at I is the identity map on all of $GL(n, \mathbb{C})$. Let $X \in \mathfrak{gl}(n, \mathbb{C})$. Then, under our tangent space identifications, the differential of $\exp$ maps X to $\frac{d}{dt} e^{tX}|_{t=0} = X$, as claimed. Part (c) follows from Theorem 1.15.   □

Note from the proof of Theorem 4.6 that $X \in \mathfrak{gl}(n, \mathbb{C})$ is an element of $\mathfrak{g}$ if $e^{tX} \in G$ for all t on a neighborhood of 0. However, it is not sufficient to merely verify that $e^X \in G$ (Exercise 4.9). Also in general, $\exp$ need not be onto (Exercise 4.7). However, when G is compact and connected, we will in fact see in §5.1.4 that $G = \exp \mathfrak{g}$.

4.1.3 Lie Algebras for the Compact Classical Lie Groups

We already know that the Lie algebra of $GL(n, \mathbb{F})$ is $\mathfrak{gl}(n, \mathbb{F})$. The Lie algebra of $SL(n, \mathbb{F})$ turns out to be

$$\mathfrak{sl}(n, \mathbb{F}) = \{X \in \mathfrak{gl}(n, \mathbb{F}) \mid \operatorname{tr} X = 0\}.$$

To check this, use Theorem 4.6. Suppose X is in the Lie algebra of $SL(n, \mathbb{F})$. Then $1 = \det e^{tX} = e^{t \operatorname{tr} X}$ for $t \in \mathbb{R}$ (Exercise 4.3). Applying $\frac{d}{dt}|_{t=0}$ implies $0 = \operatorname{tr} X$. On the other hand, if $\operatorname{tr} X = 0$, then $\det e^{tX} = e^{t \operatorname{tr} X} = 1$, so that X is in the Lie algebra of $SL(n, \mathbb{F})$.

It remains to calculate the Lie algebras for the compact classical Lie groups.

4.1.3.1 $SU(n)$ First, we show that the Lie algebra of $U(n)$ is

$$\mathfrak{u}(n) = \{X \in \mathfrak{gl}(n, \mathbb{C}) \mid X^* = -X\}.$$

Again, this follows from Theorem 4.6. Suppose X is in the Lie algebra of $U(n)$. Then $I = e^{tX} \left(e^{tX}\right)^* = e^{tX} e^{tX^*}$ for $t \in \mathbb{R}$ (Exercise 4.3). Applying $\frac{d}{dt}|_{t=0}$ implies that $0 = X + X^*$. On the other hand, if $X^* = -X$, then $e^{tX} e^{tX^*} = e^{tX} e^{-tX} = I$, so that X is in the Lie algebra of $U(n)$.

To calculate the Lie algebra of $SU(n)$, simply toss the determinant condition into the mix. It is handled as in the case of $SL(n, \mathbb{F})$. Thus the Lie algebra of $SU(n)$ is

$$\mathfrak{su}(n) = \{X \in \mathfrak{gl}(n, \mathbb{C}) \mid X^* = -X, \operatorname{tr} X = 0\}.$$

Using the fact that the tangent space has the same dimension as the manifold, we now have a simple way to calculate the dimension of $U(n)$ and $SU(n)$. In particular, since $\dim \mathfrak{u}(n) = 2\frac{n(n-1)}{2} + n$, $\dim U(n) = n^2$ and, since $\dim \mathfrak{su}(n) = 2\frac{n(n-1)}{2} + n - 1$, $\dim SU(n) = n^2 - 1$.

4.1.3.2 $SO(n)$ Working with X^t instead of X^* for $X \in \mathfrak{gl}(n, \mathbb{R})$, $O(n)$ and $SO(n)$ are handled in the same way as $U(n)$ and $SU(n)$. Thus the Lie algebras for $O(n)$ and $SO(n)$ are, respectively,

$$\mathfrak{o}(n) = \{X \in \mathfrak{gl}(n, \mathbb{R}) \mid X^t = -X\}$$
$$\mathfrak{so}(n) = \{X \in \mathfrak{gl}(n, \mathbb{R}) \mid X^t = -X, \operatorname{tr} X = 0\} = \mathfrak{o}(n).$$

Both groups have the same tangent space at I since $SO(n) = O(n)^0$. In particular, both groups also have the same dimension and, since $\dim \mathfrak{o}(n) = \frac{n(n-1)}{2}$, $\dim O(n) = \dim SO(n) = \frac{n(n-1)}{2}$.

4.1.3.3 $Sp(n)$ Recall from §1.1.4.3 that two realizations were given for $Sp(n)$. We give the corresponding Lie algebra for each.

The first realization was $Sp(n) = \{g \in GL(n, \mathbb{H}) \mid g^*g = I\}$. Since $GL(n, \mathbb{H})$ is an open dense set in $\mathfrak{gl}(n, \mathbb{H}) = M_{n,n}(\mathbb{H}) \cong \mathbb{R}^{4n^2}$, $\mathfrak{gl}(n, \mathbb{H})$ can be identified with the tangent space $T_I(GL(n, \mathbb{H}))$. It is therefore clear that Definition 4.1 generalizes in the obvious fashion so as to realize the Lie algebra of $Sp(n)$ inside $\mathfrak{gl}(n, \mathbb{H})$. Working within this scheme and mimicking the case of $U(n)$, it follows that the Lie algebra of this realization of $Sp(n)$ is

$$\mathfrak{sp}(n) = \{X \in \mathfrak{gl}(n, \mathbb{H}) \mid X^* = -X\}.$$

Since $\dim \mathfrak{sp}(n) = 4\frac{n(n-1)}{2} + 3n$, we see that $\dim Sp(n) = 2n^2 + n$.

The second realization of $Sp(n)$ was as $Sp(n) \cong U(2n) \cap Sp(n, \mathbb{C})$, where $Sp(n, \mathbb{C}) = \{g \in GL(2n, \mathbb{C}) \mid g^t J g = J\}$ and

$$J = \begin{pmatrix} 0 & -I_n \\ I_n & 0 \end{pmatrix}.$$

The Lie algebra of this realization of $Sp(n)$ is

$$\mathfrak{sp}(n) \cong \mathfrak{u}(2n) \cap \mathfrak{sp}(n, \mathbb{C}),$$

where the Lie algebra of $Sp(n, \mathbb{C})$ is

$$\mathfrak{sp}(n, \mathbb{C}) = \{X \in \mathfrak{gl}(2n, \mathbb{C}) \mid X^t J = -JX\}.$$

The only statement that needs checking is the identification of $\mathfrak{sp}(n, \mathbb{C})$. As usual, this follows from Theorem 4.6. Suppose X is in the Lie algebra of $Sp(n, \mathbb{C})$. Then $e^{tX^t} J e^{tX} = J$ for $t \in \mathbb{R}$. Applying $\frac{d}{dt}|_{t=0}$ implies $0 = X^t J + JX$. On the other hand, if $X^t J = -JX$, then $JXJ^{-1} = -X^t$, so $e^{tX^t} J e^{tX} J^{-1} = e^{tX^t} e^{tJXJ^{-1}} = e^{tX^t} e^{-tX^t} = I$, so that X is in the Lie algebra of $Sp(n, \mathbb{C})$.

4.1.4 Exercises

Exercise 4.1 Let G be a Lie subgroup of $GL(n, \mathbb{C})$. Show that the Lie bracket is linear in each variable, skew-symmetric, and satisfies the Jacobi identity.

Exercise 4.2 (1) Show that the map $\tilde{\vartheta} : \mathbb{H} \to M_{2,2}(\mathbb{C})$ from Equation 1.13 in §1.1.4.3 induces an isomorphism $Sp(1) \cong SU(2)$ of Lie groups.
(2) Show that

$$\tilde{\vartheta}i = \begin{pmatrix} i & 0 \\ 0 & -i \end{pmatrix}, \quad \tilde{\vartheta}j = \begin{pmatrix} 0 & -1 \\ 1 & 0 \end{pmatrix}, \quad \tilde{\vartheta}k = \begin{pmatrix} 0 & -i \\ -i & 0 \end{pmatrix}.$$

(3) Let $\mathrm{Im}(\mathbb{H}) = \mathrm{span}_{\mathbb{R}}\{i, j, k\}$ and equip $\mathrm{Im}(\mathbb{H})$ with the algebra structure $[u, v] = 2\,\mathrm{Im}(uv) = uv - \overline{uv} = uv - vu$ for $u, v \in \mathrm{Im}(\mathbb{H})$. Show $\tilde{\vartheta}$ induces an isomorphism $\mathrm{Im}(\mathbb{H}) \cong \mathfrak{su}(2)$ as (Lie) algebras.

Exercise 4.3 (1) Let $X, Y \in \mathfrak{gl}(n, \mathbb{C})$. Show that the map $t \to e^{tX}$ is a well-defined smooth homomorphism of $\mathbb{R}$ into $GL(n, \mathbb{C})$.
(2) If X and Y commute, show that $e^{X+Y} = e^X e^Y$. Show by example that this need not be true when X and do not Y commute.
(3) Show that $\det e^X = e^{\mathrm{tr}\, X}$, $\left(e^X\right)^* = e^{X^*}$, $\left(e^X\right)^{-1} = e^{-X}$, $\frac{d}{dt}e^{tX} = e^{tX}X = Xe^{tX}$, and $Ae^X A^{-1} = e^{AXA^{-1}}$ for $A \in GL(n, \mathbb{C})$.

Exercise 4.4 (1) For $x, y \in \mathbb{R}$, show that

$$\exp \begin{pmatrix} x & -y \\ y & x \end{pmatrix} = e^x \begin{pmatrix} \cos y & -\sin y \\ \sin y & \cos y \end{pmatrix}$$

$$\exp \begin{pmatrix} x & y \\ y & x \end{pmatrix} = e^x \begin{pmatrix} \cosh y & \sinh y \\ \sinh y & \cosh y \end{pmatrix}$$

$$\exp \begin{pmatrix} x & 0 \\ y & x \end{pmatrix} = e^x \begin{pmatrix} 1 & 0 \\ y & 1 \end{pmatrix}.$$

(2) Show every matrix in $\mathfrak{gl}(2, \mathbb{R})$ is conjugate to one of the form $\begin{pmatrix} x & -y \\ y & x \end{pmatrix}$, $\begin{pmatrix} x & y \\ y & x \end{pmatrix}$, or $\begin{pmatrix} x & 0 \\ y & x \end{pmatrix}$.

Exercise 4.5 (1) The *Euclidean motion group* on $\mathbb{R}^n$ consists of the set of transformations of $\mathbb{R}^n$ of the form $x \to Ax + b$, where $A \in GL(n, \mathbb{R})$ and $b \in \mathbb{R}^n$ for $x \in \mathbb{R}^n$. Show that the Euclidean motion group can be realized as a linear group of the form

$$\left\{ \begin{pmatrix} A & b \\ 0 & 1 \end{pmatrix} \mid A \in GL(n, \mathbb{R}), b \in \mathbb{R}^n \right\}.$$

(2) Use power series to make sense of $\frac{(e^A - I)}{A}$.
(3) Show that the exponential map is given in this case by

$$\exp \begin{pmatrix} A & b \\ 0 & 1 \end{pmatrix} = \begin{pmatrix} e^A & \frac{(e^A-1)}{A} b \\ 0 & 1 \end{pmatrix}.$$

Exercise 4.6 (1) Let $X \in \mathfrak{sl}(2, \mathbb{C})$ be given by

$$X = \begin{pmatrix} a & b \\ c & -a \end{pmatrix}$$

with $a, b, c \in \mathbb{C}$ and let $\lambda \in \mathbb{C}$ so that $\lambda^2 = a^2 + bc$. Show $e^X = (\cosh \lambda) I + \frac{\sinh \lambda}{\lambda} X$.
(2) Let $X \in \mathfrak{so}(3)$ be given by

$$X = \begin{pmatrix} 0 & a & b \\ -a & 0 & c \\ -b & -c & 0 \end{pmatrix}$$

for $a, b, c \in \mathbb{R}$ and let $\theta = \sqrt{a^2 + b^2 + c^2}$. Show $e^X = I + \frac{\sin \theta}{\theta} X + \frac{1-\cos \theta}{\theta^2} X^2$. Also show that e^X is the rotation about $(c, -b, a)$ through an angle θ.

Exercise 4.7 (1) Show that the map $\exp \colon \mathfrak{sl}(2, \mathbb{R}) \to SL(2, \mathbb{R})$ is not onto by showing that the complement of the image of $\exp$ consists of all $g \in SL(2, \mathbb{R})$ that are conjugate to

$$\begin{pmatrix} -1 & \pm 1 \\ 0 & -1 \end{pmatrix}$$

(i.e., all $g \neq -I$ with both eigenvalues equal to -1).
(2) Calculate the image of $\mathfrak{gl}(2, \mathbb{R})$ under $\exp$.

Exercise 4.8 (1) Use the Jordan canonical form to show $\exp \colon \mathfrak{gl}(n, \mathbb{C}) \to GL(n, \mathbb{C})$ is surjective.
(2) Show that $\exp \colon \mathfrak{u}(n) \to U(n)$ and $\exp \colon \mathfrak{su}(n) \to SU(n)$ are surjective maps.
(3) Show that $\exp \colon \mathfrak{so}(n) \to SO(n)$ is surjective.
(4) Show that $\exp \colon \mathfrak{sp}(n) \to Sp(n)$ is surjective.

Exercise 4.9 Find an $X \in \mathfrak{gl}(2, \mathbb{C})$ so that $e^X \in SL(2, \mathbb{C})$, but $X \notin \mathfrak{sl}(2, \mathbb{C})$.

Exercise 4.10 Let G be a Lie subgroup of $GL(n, \mathbb{C})$. Show $G^0 = \{I\}$ if and only if $\mathfrak{g} = \{0\}$.

Exercise 4.11 Let G be a Lie subgroup of $GL(n, \mathbb{C})$ and $\varphi \colon \mathbb{R} \to G$ a continuous homomorphism.
(1) Show that φ is smooth if and only if φ is smooth at 0.
(2) Let U be a neighborhood of 0 in $\mathfrak{g}$ on which $\exp$ is injective. Show it is possible to linearly reparametrize φ, i.e., replace $\varphi(t)$ by $\varphi(st)$ for some nonzero $s \in \mathbb{R}$, so that $\varphi([-1, 1]) \subseteq \exp U$.
(3) Let $X \in U$ so that $\exp X = 1$. Show that $\varphi(t) = e^{tX}$ for $t \in \mathbb{Q}$.
(4) Show that $\varphi(t) = e^{tX}$ for $t \in \mathbb{R}$ and conclude that φ is actually real analytic and, in particular, smooth.

Exercise 4.12 (1) Let G be a Lie subgroup of $GL(n, \mathbb{C})$. Let $\{X_i\}_{i=1}^n$ be a basis for $\mathfrak{g}$. By calculating the differential on each standard basis vector, show that the map

$$(t_1, \ldots, t_n) \to e^{t_1 X_1} \cdots e^{t_n X_n}$$

is a local diffeomorphism near 0 from $\mathbb{R}^n$ to G. The coordinates $(t_1, \ldots, t_n)$ are called *coordinates of the second kind*.

(2) Show that the map

$$(t_1, \ldots, t_n) \to e^{t_1 X_1 + \cdots + t_n X_n}$$

is a local diffeomorphism near 0 from $\mathbb{R}^n$ to G. The coordinates $(t_1, \ldots, t_n)$ are called *coordinates of the first kind*.

Exercise 4.13 Suppose G and H are Lie subgroups of general linear groups and $\varphi : H \to G$ is a continuous homomorphism. Using Exercises 4.11 and 4.12, show that φ is actually a real analytic and therefore smooth map.

Exercise 4.14 Suppose $B(\cdot, \cdot)$ is a bilinear form on $\mathbb{F}^n$. Let

$$\text{Aut}(B) = \{g \in GL(n, \mathbb{F}) \mid (gv, gw) = (v, w), v, w \in \mathbb{F}^n\}$$
$$\text{Der}(B) = \{X \in \mathfrak{gl}(n, \mathbb{F}) \mid (Xv, w) = -(v, Xw), v, w \in \mathbb{F}^n\}.$$

Show that $\text{Aut}(B)$ is a closed Lie subgroup of $GL(n, \mathbb{F})$ with Lie algebra $\text{Der}(B)$.

Exercise 4.15 Suppose $\mathbb{F}^n$ is equipped with an algebra structure $\cdot$. Let

$$\text{Aut}(\cdot) = \{g \in GL(n, \mathbb{F}) \mid g(v \cdot w) = gv \cdot gw, v, w \in \mathbb{F}^n\}$$
$$\text{Der}(\cdot) = \{X \in \mathfrak{gl}(n, \mathbb{F}) \mid X(v \cdot w) = Xv \cdot w + v \cdot Xw, v, w \in \mathbb{F}^n\}.$$

Show that $\text{Aut}(\cdot)$ is a closed Lie subgroup of $GL(n, \mathbb{F})$ with Lie algebra $\text{Der}(\cdot)$.

Exercise 4.16 Let G be a Lie subgroup of $GL(n, \mathbb{C})$. Use the exponential map to show G has a neighborhood of I that contains no subgroup of G other than $\{e\}$.

Exercise 4.17 For $X, Y \in \mathfrak{gl}(n, \mathbb{C})$, show that $e^{X+Y} = \lim_{n \to \infty} (e^{\frac{X}{n}} e^{\frac{Y}{n}})^n$.

4.2 Further Constructions

4.2.1 Lie Algebra Homomorphisms

Definition 4.7. Suppose $\varphi : H \to G$ is a homomorphism of Lie subgroups of general linear groups. Let the *differential* of φ, $d\varphi : \mathfrak{h} \to \mathfrak{g}$, be given by

$$d\varphi(X) = \frac{d}{dt}\varphi(e^{tX})|_{t=0}.$$

This is well defined by Theorem 4.6 and Definition 4.1. Note by the chain rule that if $\gamma : \mathbb{R} \to G$ is any smooth map with $\gamma'(0) = X$, then $d\varphi$ can be alternately computed as $d\varphi(X) = \frac{d}{dt}\varphi(\gamma(t))|_{t=0}$. If one examines the identifications of Lie algebras with tangent spaces, then it is straightforward to see that the above definition of $d\varphi$ corresponds to the usual differential geometry definition of the differential $d\varphi : T_I(H) \to T_I(G)$. In particular, $d\varphi$ is a linear map and $d(\varphi_1 \circ \varphi_2) = d\varphi_1 \circ d\varphi_2$. Alternatively, this can be verified directly with the chain and product rules (Exercise 4.18).

Theorem 4.8. *Suppose $\varphi, \varphi_i : H \to G$ are homomorphisms of Lie subgroups of general linear groups.*
(a) The following diagram is commutative:

$$\begin{array}{ccc} \mathfrak{h} & \xrightarrow{d\varphi} & \mathfrak{g} \\ \exp \downarrow & & \downarrow \exp \\ H & \xrightarrow{\varphi} & G \end{array}$$

so that $\exp \circ d\varphi = \varphi \circ \exp$, *i.e.,* $e^{d\varphi X} = \varphi(e^X)$ *for $X \in \mathfrak{h}$.*
(b) The differential $d\varphi$ is a homomorphism of Lie algebras, *i.e.,*

$$d\varphi[X, Y] = [d\varphi X, d\varphi Y]$$

for $X, Y \in \mathfrak{h}$.
(c) If H is connected and $d\varphi_1 = d\varphi_2$, then $\varphi_1 = \varphi_2$.

Proof. For part (a), observe that since φ is a homomorphism that

$$\frac{d}{dt}\varphi(e^{tX}) = \frac{d}{ds}\varphi(e^{(t+s)X})|_{s=0} = \varphi(e^{tX})\frac{d}{ds}\varphi(e^{sX})|_{s=0} = \varphi(e^{tX})d\varphi X.$$

Thus $t \to \varphi(e^{tX})$ is the integral curve of $\widetilde{d\varphi X}$ through I. Theorem 4.5 therefore implies $\varphi(e^{tX}) = e^{td\varphi X}$.

For part (b), start with the equality $\varphi(e^{tX} e^{sY} e^{-tX}) = e^{td\varphi X} e^{sd\varphi Y} e^{-td\varphi X}$ that follows from the fact that φ is a homomorphism and part (a). Apply $\frac{\partial}{\partial s}|_{s=0}$ and rewrite $e^{tX} e^{sY} e^{-tX}$ as $e^{se^{tX}Ye^{-tX}}$ (Exercise 4.3) to get

$$d\varphi(e^{tX}Ye^{-tX}) = e^{td\varphi X} d\varphi Y \, e^{-td\varphi X}.$$

Next apply $\frac{d}{dt}|_{t=0}$ to get

$$\frac{d}{dt}\left(d\varphi(e^{tX}Ye^{-tX})\right)|_{t=0} = d\varphi X d\varphi Y - d\varphi Y d\varphi X = [d\varphi X, d\varphi Y]$$

and use the fact that $d\varphi$ is linear to get

$$d\varphi([X, Y]) = d\varphi(XY - YX) = \frac{d}{dt}d\varphi(e^{tX}Ye^{-tX})|_{t=0} = [d\varphi X, d\varphi Y].$$

For part (c), use part (a) to show that φ_1 and φ_2 agree on $\exp \mathfrak{h}$. By Theorem 4.6 and since φ_i is a homomorphism, the proof is finished. $\square$

As a corollary of Theorem 4.8, we can check that the Lie algebra of a compact group is well defined up to isomorphism. To see this, suppose G_i are Lie subgroups of general linear groups with $\varphi : G_1 \to G_2$ an isomorphism. Since $\varphi \circ \varphi^{-1}$ and $\varphi^{-1} \circ \varphi$ are the identity maps, taking differentials shows $d\varphi$ is a Lie algebra isomorphism from $\mathfrak{g}_1$ to $\mathfrak{g}_2$.

A smooth homomorphism of the additive group $\mathbb{R}$ into a Lie group G is called a *one-parameter subgroup*. The next corollary shows that all one-parameter subgroups are of the form $t \to e^{tX}$ for $X \in \mathfrak{g}$.

Corollary 4.9. *Let G be a Lie subgroup of $GL(n, \mathbb{C})$ and let $\gamma : \mathbb{R} \to G$ be a smooth homomorphism, i.e., $\gamma(s + t) = \gamma(s)\gamma(t)$ for $s, t \in \mathbb{R}$. If $\gamma'(0) = X$, then $\gamma(t) = e^{tX}$.*

Proof. View the multiplicative group $\mathbb{R}^+$ as a Lie subgroup of $GL(1, \mathbb{C})$. Let $\widetilde{\gamma}, \sigma : \mathbb{R}^+ \to G$ be the two homomorphisms defined by $\widetilde{\gamma} = \gamma \circ \ln$ and $\sigma(x) = e^{(\ln x)X}$. Then

$$d\widetilde{\gamma}(x) = \frac{d}{dt}\widetilde{\gamma}(e^{tx})|_{t=0} = \frac{d}{dt}\gamma(tx)|_{t=0} = xX$$
$$d\sigma(x) = \frac{d}{dt}\sigma(e^{tx})|_{t=0} = \frac{d}{dt}e^{txX}|_{t=0} = xX.$$

Theorem 4.8 thus shows that $\widetilde{\gamma} = \sigma$ so that $\gamma(t) = e^{tX}$. $\square$

In the definition below any choice of basis can be used to identify $GL(\mathfrak{g})$ and $\text{End}(\mathfrak{g})$ with $GL(\dim \mathfrak{g}, \mathbb{R})$ and $\mathfrak{gl}(\dim \mathfrak{g}, \mathbb{R})$, respectively. Under this identification, exp corresponds to the map exp : $\text{End}(\mathfrak{g}) \to GL(\mathfrak{g})$ with $e^T = \sum_{k=0}^{\infty} \frac{1}{k!}T^k$ for $T \in \text{End}(\mathfrak{g})$ where $T^k X = (T \circ \cdots \circ T)X$ (k copies) for $T \in \text{End}(\mathfrak{g})$ and $X \in \mathfrak{g}$.

Definition 4.10. Let G be a Lie subgroup of $GL(n, \mathbb{C})$.
(a) For $g \in G$, let *conjugation*, $c_g : G \to G$, be the Lie group homomorphism given by $c_g(h) = ghg^{-1}$ for $h \in G$
(b) The *Adjoint representation* of G on $\mathfrak{g}$, $\text{Ad} : G \to GL(\mathfrak{g})$, is given by $\text{Ad}(g) = d\left(c_g\right)$.
(c) The *adjoint representation* of $\mathfrak{g}$ on $\mathfrak{g}$, $\text{ad} : \mathfrak{g} \to \text{End}(\mathfrak{g})$, is given by $\text{ad} = d\,\text{Ad}$, i.e., $(\text{ad}\,X)\,Y = \frac{d}{dt}(\text{Ad}(e^{tX})Y)|_{t=0}$ for $X, Y \in \mathfrak{g}$.

Some notes are in order. Except for the fact that $\mathfrak{g}$ is a real vector space instead of a complex one, Ad is seen to satisfy the key property of a representation,

$$\text{Ad}(g_1 g_2) = \text{Ad}(g_1)\,\text{Ad}(g_2),$$

by taking the differential of the relation $c_{g_1 g_2} = c_{g_1} \circ c_{g_2}$ for $g_i \in G$. More explicitly, however, $dc_g(X) = \frac{d}{dt}(ge^{tX}g^{-1})|_{t=0}$ so that

$$\text{Ad}(g)X = gXg^{-1}.$$

Applying Theorem 4.8, we see that

$$c_g e^X = e^{\mathrm{Ad}(g)X}.$$

Since this is simply the statement $ge^X g^{-1} = e^{gXg^{-1}}$, the equality is already well known from linear algebra.

Secondly, $((d\,\mathrm{Ad})(X))\,Y = \frac{d}{dt}e^{tX}Ye^{-tX}|_{t=0}$ so that

$$(\mathrm{ad}\,X)\,Y = XY - YX = [X, Y].$$

Applying Theorem 4.8, we see that

(4.11)
$$\mathrm{Ad}(e^X) = e^{\mathrm{ad}\,X}.$$

The notion of a representation of a Lie algebra will be developed in §6.1. When that is done, ad will, in fact, be a representation of $\mathfrak{g}$ on itself.

4.2.2 Lie Subgroups and Subalgebras

If M is a manifold and ξ_i are vector fields, the *Lie bracket of vector fields* is defined as $[\xi_1, \xi_2] = \xi_1\xi_2 - \xi_2\xi_1$. For $M = \mathbb{R}^n$, it is easy to see (Exercise 4.19) that the Lie bracket of the vector fields $\xi = \sum_i \xi_i(x)\frac{\partial}{\partial x_i}$ and $\eta = \sum_i \eta_i(x)\frac{\partial}{\partial x_i}$ is given by

(4.12)
$$[\xi, \eta] = \sum_i \sum_j \left(\xi_j \frac{\partial \eta_i}{\partial x_j} - \eta_j \frac{\partial \xi_i}{\partial x_j} \right) \frac{\partial}{\partial x_i}.$$

For $M = GL(n, \mathbb{C})$, recall that $GL(n, \mathbb{C})$ is viewed as an open set in $M_{n,n}(\mathbb{C}) \cong \mathbb{R}^{2n^2} \cong \mathbb{R}^{n^2} \times \mathbb{R}^{n^2}$ by writing $Z \in M_{n,n}(\mathbb{C})$ as $Z = X + iY$, $X, Y \in M_{n,n}(\mathbb{R})$, and mapping Z to (X, Y). For $A \in \mathfrak{gl}(n, \mathbb{C})$, the value of the vector field $\widetilde{A}$ at the point $Z \in GL(n, \mathbb{C})$ is defined as ZA. Unraveling our identifications (see the discussion around Equation 4.3 for the usual identification of $T_g(G)$ with $g\mathfrak{g}$), this means that the vector field $\widetilde{A}$ on $GL(n, \mathbb{C})$ corresponds to the vector field

$$\partial_A = \sum_{i,j} \sum_k \mathrm{Re}(z_{ik}A_{kj})\frac{\partial}{\partial x_{ij}} + \sum_{i,j} \sum_k \mathrm{Im}(z_{ik}A_{kj})\frac{\partial}{\partial y_{ij}}$$

on the open set of $\mathbb{R}^{2n^2}$ cut out by the determinant.

Lemma 4.13. *For* $A, B \in M_n(\mathbb{C})$, $[\partial_A, \partial_B] = \partial_{[A,B]}$.

Proof. For the sake of clarity of exposition, we will verify this lemma for $M_n(\mathbb{R})$ and leave the general case of $M_n(\mathbb{C})$ to the reader. In this setting and with $A \in M_n(\mathbb{R})$, ∂_A is simply $\sum_{i,j}\sum_k x_{ik}A_{kj}\frac{\partial}{\partial x_{ij}}$. Writing $\delta_{i,p}$ for 0 when $i \neq p$ and for 1 when $i = p$, Equation 4.12 shows that

$$[\partial_A, \partial_B] = \sum_{i,j}\sum_{p,q}\left(\sum_k x_{pk}A_{kq}\delta_{i,p}B_{qj} - \sum_k x_{pk}B_{kq}\delta_{i,p}A_{qj} \right)\frac{\partial}{\partial x_{ij}}$$

$$= \sum_{i,j}\sum_{q,k} x_{ik}\left(A_{kq}B_{qj} - B_{kq}A_{qj} \right)\frac{\partial}{\partial x_{ij}}$$

$$= \sum_{i,j}\sum_k x_{ik}[A, B]_{kj}\frac{\partial}{\partial x_{ij}} = \partial_{[A,B]}. \qquad \square$$

For us, the importance of Lemma 4.13 is that if $\mathfrak{h}$ is a k-dimensional subalge-bra of $\mathfrak{gl}(n, \mathbb{C})$, then the vector fields $\{\partial_X \mid X \in \mathfrak{h}\}$ form a subalgebra under the Lie bracket of vector fields. Moreover, on $GL(n, \mathbb{C})$, their value at each point determines a smooth rank k subbundle of the tangent bundle. Thus Frobenius' theorem from dif-ferential geometry (see [8] or [88]) says this subbundle foliates $GL(n, \mathbb{C})$ into *inte-gral submanifolds*. In particular, there is a unique maximal connected k-dimensional submanifold H of $GL(n, \mathbb{C})$ so that $I \in H$ and $T_h(H) = \{(\partial_X)_h \mid X \in \mathfrak{h}\}, h \in H$, where $(\partial_X)_h$ is the value of ∂_X at h. Under our usual identification, this means that the tangent space of H at h corresponds to $h\mathfrak{h}$, i.e., that $\{\gamma'(0) \mid \gamma(0) = h$ and $\gamma : (-\epsilon, \epsilon) \to H, \epsilon > 0$, is smooth$\} = h\mathfrak{h}$. Finally, it is an important fact that integral submanifolds such as H, as was the case for regular submanifolds, satisfy the property that when $f : M \to G$ is a smooth map of manifolds with $f(M) \subseteq H$, then $f : M \to H$ is also a smooth map (see [88]).

Theorem 4.14. *Let G be a Lie subgroup of $GL(n, \mathbb{C})$. There is a bijection between the set of connected Lie subgroups of G and the set of subalgebras of $\mathfrak{g}$. If H is a connected Lie subgroup of G, the correspondence maps H to its Lie algebra $\mathfrak{h}$.*

Proof. Suppose $\mathfrak{h}$ is a subalgebra of $\mathfrak{g}$. Let H be the unique maximal connected submanifold of G so that $I \in H$, and so the tangent space of H at h corresponds to $h\mathfrak{h}$ for $h \in H$. Now the connected submanifold $h_0^{-1}H, h_0 \in H$, contains I. Moreover, since $\frac{d}{dt}(h_0^{-1}\gamma(t))|_{t=0} = h_0^{-1}\gamma'(0)$, the tangent space of $h_0^{-1}H$ at $h_0^{-1}h$ corresponds to $h_0^{-1}h\mathfrak{h}$. Uniqueness of the integral submanifold therefore shows $h_0^{-1}H = H$. A similar argument shows that $h_0H = H$, so that H is a subgroup of G. By the remark above the statement of this theorem, the multiplication and inverse operations are smooth as maps on H, so that H is a Lie subgroup of G. Hence the correspondence is surjective.

To see it is injective, suppose H and H' are connected Lie subgroups of G, so that $\mathfrak{h} = \mathfrak{h}'$. Using the exponential map and Theorem 4.6, H and H' share a neighborhood of I. Since they are both connected, this forces $H = H'$. $\square$

4.2.3 Covering Homomorphisms

Theorem 4.15. *Let H and G be connected Lie subgroups of general linear groups and $\varphi : H \to G$ a homomorphism of Lie groups. Then φ is a covering map if and only if $d\varphi$ is an isomorphism.*

Proof. If φ is a covering, then there is a neighborhood U of I in H and a neighbor-hood V of I in G, so that φ restricts to a diffeomorphism $\varphi : U \to V$. Thus the differential at I, $d\varphi$, is an isomorphism.

Suppose now that $d\varphi$ is an isomorphism. By the Inverse Mapping theorem, there is a neighborhood U_0 of I in H and a neighborhood V_0 of I in G so that φ restricts to a diffeomorphism $\varphi : U_0 \to V_0$. In particular, $\ker \varphi \cap U_0 = \{I\}$. Let V be a connected neighborhood of I in V_0 so that $VV^{-1} \subseteq V_0$ (Exercise 1.4) and let $U = \varphi^{-1}V \cap U_0$ so that U is connected, $UU^{-1} \subseteq U_0$, and $\varphi : U \to V$ is still a diffeomorphism.

As φ is a homomorphism, $\varphi^{-1}V = U\ker\varphi$. To see that φ satisfies the covering condition at $I \in G$, we show that the set of connected components of $\varphi^{-1}V$ is $\{U\gamma \mid \gamma \in \ker\varphi\}$. For this, it suffices to show that $U\gamma_1 \cap U\gamma_2 = \emptyset$ for distinct $\gamma_i \in \ker\varphi$. Suppose $u_1\gamma_1 = u_2\gamma_2$ for $u_i \in U$ and $\gamma_i \in \ker\varphi$. Then $\gamma_2\gamma_1^{-1} = u_2^{-1}u_1$ is in $U_0 \cap \ker\varphi$ and so $\gamma_2\gamma_1^{-1} = I$, as desired.

It remains to see that φ satisfies the covering condition at any $g \in G$. For this, first note that φ is surjective since G is connected, φ is a homomorphism, and the image of φ contains a neighborhood of I (Theorem 1.15). Choose $h \in H$ so that $\varphi(h) = g$. Then gV is a connected neighborhood of g in G and $\varphi^{-1}(gV) = hU\ker\varphi$. The set of connected components of $hU\ker\varphi$ is clearly $\{hU\gamma \mid \gamma \in \ker\varphi\}$. Since φ restricted to $hU\gamma$ is obviously a diffeomorphism to gV, φ is a covering map. $\square$

Theorem 4.16. *Let H and G be connected Lie subgroups of general linear groups with H simply connected. If $\psi : \mathfrak{h} \to \mathfrak{g}$ is a homomorphism of Lie algebras, then there exists a unique homomorphism of Lie groups $\varphi : H \to G$ so that $d\varphi = \psi$.*

Proof. Uniqueness follows from Theorem 4.8. For existence, suppose H is a Lie subgroup of $GL(n, \mathbb{C})$ and G is a subgroup of $GL(m, \mathbb{C})$. Then we may view $H \times G$ as a block diagonal Lie subgroup of $GL(n+m, \mathbb{C})$. When this is done, the Lie algebra of $H \times G$ is clearly the direct sum of $\mathfrak{h}$ and $\mathfrak{g}$ in $\mathfrak{gl}(n + m, \mathbb{C})$. More importantly, note $\mathfrak{h}$ and $\mathfrak{g}$ commute and define

$$\mathfrak{a} = \{X + \psi X \mid X \in \mathfrak{h}\} \subseteq \mathfrak{h} \oplus \mathfrak{g}.$$

Using the fact that ψ is a homomorphism of Lie algebras, if follows that $\mathfrak{a}$ is a subalgebra of $\mathfrak{h} \oplus \mathfrak{g}$ since

$$[X + \psi X, Y + \psi Y] = [X, Y] + [\psi X, \psi Y] = [X, Y] + \psi[X, Y]$$

for $X, Y \in \mathfrak{h}$.

Let A be the connected Lie subgroup of $H \times G$ with Lie algebra $\mathfrak{a}$ (Theorem 4.14) and let π_H and π_G be the Lie group homomorphisms projecting A to H and G, respectively. By the definitions, $d\pi_H(X + \psi X) = X$ and $d\pi_G(X + \psi X) = \psi X$. Then $d\pi_H$ is a Lie algebra isomorphism of $\mathfrak{a}$ and $\mathfrak{h}$, so that Theorem 4.15 implies π_H is a covering map from A to H. Since H is simply connected, this means that $\pi_H : A \to H$ is an isomorphism. Define the Lie group homomorphism $\varphi : H \to G$ by $\varphi = \pi_G \circ \pi_H^{-1}$ to finish the proof. $\square$

Note Theorem 4.16 can easily fail when H is not simply connected (Exercise 4.20).

4.2.4 Exercises

Exercise 4.18 (1) Let $\varphi : H \to G$ be a homomorphism of linear Lie groups. Use the fact that $\frac{d}{dt}\left(e^{trX}e^{tY}\right)|_{t=0} = rX + Y$, $X, Y \in \mathfrak{h}$, to directly show that $d\varphi : \mathfrak{h} \to \mathfrak{g}$ is a linear map.
(2) Let $\varphi' : K \to H$ be a homomorphism of linear Lie groups. Show that $d(\varphi \circ \varphi') = d\varphi \circ d\varphi'$.

Exercise 4.19 Verify that Equation 4.12 holds.

Exercise 4.20 Use the spin representations to show that Theorem 4.16 can fail when H is not simply connected.

Exercise 4.21 (1) Let

$$H = \begin{pmatrix} 1 & 0 \\ 0 & -1 \end{pmatrix}, \quad E = \begin{pmatrix} 0 & 1 \\ 0 & 0 \end{pmatrix}, \text{ and } F = \begin{pmatrix} 0 & 0 \\ 1 & 0 \end{pmatrix}.$$

Show that $[H, E] = 2E$, $[H, F] = -2F$, and $[E, F] = H$.
(2) Up to the Ad action of $SL(2, \mathbb{R})$, find all Lie subalgebras of $\mathfrak{sl}(2, \mathbb{R})$.

Exercise 4.22 (1) Let G be a Lie subgroup of a linear group and $H \subseteq G$. Show that the *centralizer of H in G*,

$$Z_G(H) = \{g \in G \mid gh = hg, h \in H\},$$

is a Lie subgroup of G with Lie algebra the *centralizer of H in $\mathfrak{g}$*,

$$\mathfrak{z}_\mathfrak{g}(H) = \{X \in \mathfrak{g} \mid \mathrm{Ad}(h)X = X, h \in H\}.$$

(2) If $\mathfrak{h} \subseteq \mathfrak{g}$, show that the *centralizer of $\mathfrak{h}$ in G*,

$$Z_G(\mathfrak{h}) = \{g \in G \mid \mathrm{Ad}(g)X = X, X \in \mathfrak{h}\},$$

is a Lie subgroup of G with Lie algebra the *centralizer of $\mathfrak{h}$ in $\mathfrak{g}$*,

$$\mathfrak{z}_\mathfrak{g}(\mathfrak{h}) = \{Y \in \mathfrak{g} \mid [Y, X] = 0, X \in \mathfrak{h}\}.$$

(3) If H is a connected Lie subgroup of G, show $Z_G(H) = Z_G(\mathfrak{h})$ and $\mathfrak{z}_\mathfrak{g}(H) = \mathfrak{z}_\mathfrak{g}(\mathfrak{h})$.

Exercise 4.23 (1) Let G be a Lie subgroup of a linear group and let H be a connected Lie subgroup of G. Show that the *normalizer of H in G*,

$$N_G(H) = \{g \in G \mid gHg^{-1} = H\},$$

is a Lie subgroup of G with Lie algebra the *normalizer of $\mathfrak{h}$ in $\mathfrak{g}$*,

$$\mathfrak{n}_\mathfrak{g}(\mathfrak{h}) = \{Y \in \mathfrak{g} \mid [Y, \mathfrak{h}] \subseteq \mathfrak{h}\}.$$

(2) Show H is normal in G if and only if $\mathfrak{h}$ is an ideal in $\mathfrak{g}$.

Exercise 4.24 (1) Let $\varphi : H \to G$ be a homomorphism of Lie subgroups of linear groups. Show that $\ker \varphi$ is a closed Lie subgroup of H with Lie algebra $\ker d\varphi$.
(2) Show that the Lie subgroup $\varphi(H)$ of G has Lie algebra $d\varphi\mathfrak{h}$.

Exercise 4.25 If G is a Lie subgroup of a linear group satisfying $\mathrm{span}[\mathfrak{g}, \mathfrak{g}] = \mathfrak{g}$, show that $\mathrm{tr}\,(\mathrm{ad}\,X) = 0$ for $X \in \mathfrak{g}$.

Exercise 4.26 (1) For $X, Y \in \mathfrak{gl}(n, \mathbb{C})$, show that $e^{tX}e^{tY} = e^{t(X+Y)+\frac{1}{2}t^2[X,Y]+O(t^3)}$ for t near 0.
(2) Show that $e^{tX}e^{tY}e^{-tX} = e^{tY+t^2[X,Y]+O(t^3)}$ for t near 0.

Exercise 4.27 For $X, Y \in \mathfrak{gl}(n, \mathbb{C})$, show that $e^{[X,Y]} = \lim_{n \to \infty} \left(e^{\frac{X}{n}}e^{\frac{Y}{n}}e^{-\frac{X}{n}}e^{-\frac{Y}{n}}\right)^{n^2}$.

Exercise 4.28 This exercise gives a proof of Theorem 1.6. Recall the well-known fact that an n-dimensional submanifold N of an m-dimensional manifold M is regular if and only each $n \in N$ lies in an open set U of M with the property that there is a chart $\varphi : U \to \mathbb{R}^m$ of M so that $N \cap U = \varphi^{-1}(\mathbb{R}^n)$, where $\mathbb{R}^n$ is viewed as sitting in $\mathbb{R}^m$ in the usual manner ([8]). Such a chart is called cubical. Let $G \subseteq GL(n, \mathbb{C})$ be a Lie subgroup and $H \subseteq G$ be a subgroup (with no manifold assumption).
(1) Assume first that H is a regular submanifold of G and $h_i \to h$ with $h_i \in H$ and $h \in G$. Show that there is a cubical chart U of G around e and open sets $V \subseteq W \subseteq U$, so that $V^{-1}V \subseteq \overline{W} \subseteq U$. Noting that $h_i^{-1}h_j \in V^{-1}V$ for big i, j, use the definitions to show that H is closed.
(2) For the remainder, only assume H is closed. Let $\mathfrak{h} = \{X \in \mathfrak{g} \mid e^{tX} \in H, t \in \mathbb{R}\}$. Show that $e^{tX}e^{tY} = e^{t(X+Y)+O(t^2)}$, $X, Y \in \mathfrak{g}$, and use induction to see that

$$e^{t(X+Y)} = \lim_{n \to \infty} \left(e^{\frac{t}{n}X}e^{\frac{t}{n}Y}\right)^n.$$

Conclude that $\mathfrak{h}$ is a subspace and choose a complementary subspace $\mathfrak{s} \subseteq \mathfrak{g}$, so that $\mathfrak{s} \oplus \mathfrak{h} = \mathfrak{g}$.
(3) Temporarily, assume there are no neighborhoods V of 0 in $\mathfrak{g}$ with $\exp(\mathfrak{h} \cap V) = H \cap \exp V$. Using this assumption and the fact that the map $(Y, Z) \to e^Y e^Z$ is a local diffeomorphism at $(0, 0)$ from $\mathfrak{s} \oplus \mathfrak{h}$ to G, construct a nonzero sequence $Y_n \in \mathfrak{s}$, so that $Y_n \to 0$ and $e^{Y_n} \in H$. Show that you can pass to a subsequence and further assume $Y_n / \|Y_n\| \to Y$ for some nonzero $Y \in \mathfrak{s}$.
(4) For any $t \in \mathbb{R}$, show that there is $k_n \in \mathbb{Z}$ so that $k_n \|Y_n\| \to t$. Conclude that $\left(e^{Y_n}\right)^{k_n} \to e^{tY}$.
(5) Obtain a contradiction to the assumption in part (3) by showing that $Y \in \mathfrak{h}$. Conclude that there is a neighborhood V of 0 in $\mathfrak{g}$, so that $\exp$ is a diffeomorphism from V to its image in G and $\exp(\mathfrak{h} \cap V) = H \cap \exp V$.
(6) Given any $h \in H$, consider the neighborhood $U = h \exp V$ of h in G and the chart $\varphi = \exp^{-1} \circ l_h^{-1} : G \to \mathfrak{g}$ of G. Show that $\varphi^{-1}(\mathfrak{h}) = H \cap U$, so that H is a regular submanifold, as desired.

5

Abelian Lie Subgroups and Structure

Since a compact Lie group, G, can be thought of as a Lie subgroup of $U(n)$, Theorems 3.28 and 2.15, it is possible to diagonalize each $g \in G$ using conjugation in $U(n)$. In fact, the main theorem of this chapter shows it is possible to diagonalize each $g \in G$ using conjugation in G. This result will have far-reaching consequences, including various structure theorems.

5.1 Abelian Subgroups and Subalgebras

5.1.1 Maximal Tori and Cartan Subalgebras

If G is a Lie group, recall G is called *Abelian* if $g_1 g_2 = g_2 g_1$ for all $g_i \in G$. Similarly, if $\mathfrak{a}$ is a subalgebra of $\mathfrak{gl}(n, \mathbb{C})$, $\mathfrak{a}$ is called *Abelian* if $[X, Y] = 0$ for all $X, Y \in \mathfrak{a}$.

Theorem 5.1. *Let G be a Lie subgroup of $GL(n, \mathbb{C})$.*
(a) For $X, Y \in \mathfrak{g}$, $[X, Y] = 0$ if and only if e^{tX} and e^{sY} commute for $s, t \in \mathbb{R}$. In this case, $e^{X+Y} = e^X e^Y$.
(b) If A is a connected Lie subgroup of G with Lie algebra $\mathfrak{a}$, then A is Abelian if and only if $\mathfrak{a}$ is Abelian.

Proof. Since part (b) follows from part (a) and Theorems 1.15 and 4.6, it suffices to prove part (a). It is a familiar fact (Exercise 4.3) that when X and Y commute, i.e., $[X, Y] = 0$, that $e^{tX+sY} = e^{tX}e^{sY}$. Since $e^{tX+sY} = e^{sY+tX}$, it follows that e^{tX} and e^{sY} commute. Conversely, if e^{tX} and e^{sY} commute, then $e^{tX}e^{sY}e^{-tX} = e^{sY}$. Applying $\frac{\partial}{\partial s}|_{s=0}$ and then $\frac{d}{dt}|_{t=0}$ yields $XY - YX = 0$, as desired. $\qquad\square$

It is well known (Exercise 5.1) that the discrete (additive) subgroups of $\mathbb{R}^n$, up to application of an invertible linear transformation, are of the form

$$\Gamma_k = \{x = (x_1, \ldots, x_n) \in \mathbb{R}^n \mid x_1, \ldots, x_k \in \mathbb{Z} \text{ and } x_{k+1} = \cdots x_n = 0\}.$$

In the next theorem, recall that a *torus* is a Lie group of the form $T^k = \left(S^1\right)^k \cong \mathbb{R}^k / \Gamma_k = \mathbb{R}^k / \mathbb{Z}^k$.

Theorem 5.2. *(a) The most general compact Abelian Lie group is isomorphic to* $T^k \times F$, *where* F *is a finite Abelian group. In particular, the most general compact connected Abelian Lie group is a torus.*
(b) If G *is a compact Abelian Lie group, then* exp *is a surjective map to* G^0, *the connected component of* G.

Proof. Let G be a compact Abelian group. By Theorem 5.1, $\exp : \mathfrak{g} \to G^0$ is a homomorphism. By Theorems 1.15 and 4.6 it follows that exp is surjective, so $G^0 \cong \mathfrak{g}/\ker(\exp)$. Since $\mathfrak{g} \cong \mathbb{R}^{\dim \mathfrak{g}}$ as a vector space and since Theorem 4.6 shows that $\ker(\exp)$ is discrete, $\ker(\exp) \cong \Gamma_k$ for some $k \leq \dim \mathfrak{g}$. But as G is compact with $G \cong \mathbb{R}^{\dim \mathfrak{g}}/\Gamma_k \cong T^k \times \mathbb{R}^{\dim \mathfrak{g}-k}$, k must be $\dim \mathfrak{g}$, so that $G^0 \cong T^{\dim \mathfrak{g}}$.

Next, G/G^0 is a finite Abelian group by compactness. It is well known that a finite Abelian group is isomorphic to a direct product of (additive) groups of the form $\mathbb{Z}/(n_i\mathbb{Z})$, $n_i \in \mathbb{N}$. For each such product, pick $g_i \in G$ whose image in G/G^0 corresponds to $1+n_i\mathbb{Z}$ in $\mathbb{Z}/(n_i\mathbb{Z})$. Then $g_i^{n_i} \in G^0$. Choose $X_i \in \mathfrak{g}$ so that $e^{n_iX_i} = g_i^{n_i}$ and let $h_i = g_ie^{-X_i}$. Then h_i is in the same connected component as is g_i, but now $h_i^{n_i} = I$. It follows easily that the map $G^0 \times \prod_i \mathbb{Z}/(n_i\mathbb{Z}) \to G$ given by mapping $(g, (m_i + n_i\mathbb{Z})) \to g \prod_i h_i^{m_i}$ is an isomorphism.   □

Definition 5.3. (a) Let G be a compact Lie group with Lie algebra $\mathfrak{g}$. A *maximal torus* of G is a maximal connected Abelian subgroup of G.
(b) A maximal Abelian subalgebra of $\mathfrak{g}$ is called a *Cartan subalgebra* of $\mathfrak{g}$.

By Theorem 5.2, a maximal torus T of a compact Lie group G is indeed isomorphic to a torus T^k. It should also be noted that the definition of Cartan subalgebra needs to be tweaked when working outside the category of compact Lie groups.

Theorem 5.4. *Let* G *be a compact Lie group and* T *a connected Lie subgroup of* G. *Then* T *is a maximal torus if and only if* $\mathfrak{t}$ *is a Cartan subalgebra. In particular, maximal tori and Cartan subalgebras exist.*

Proof. Theorems 4.14 and 5.1 show that T is a maximal torus of G if and only if $\mathfrak{t}$ is a Cartan subalgebra of $\mathfrak{g}$. Since maximal Abelian subalgebras clearly exist, this also shows that maximal tori exist.   □

5.1.2 Examples

Recall that the Lie algebras for the compact classical Lie groups were computed in §4.1.3.

5.1.2.1 $SU(n)$ For $U(n)$ with $\mathfrak{u}(n) = \{X \in \mathfrak{gl}(n, \mathbb{C}) \mid X^* = -X\}$, let

$$(5.5) \qquad T = \{\mathrm{diag}(e^{i\theta_1}, \dots, e^{i\theta_n}) \mid \theta_i \in \mathbb{R}\}$$
$$\mathfrak{t} = \{\mathrm{diag}(i\theta_1, \dots, i\theta_n) \mid \theta_i \in \mathbb{R}\}$$

Clearly $\mathfrak{t}$ is the Lie algebra of the connected Lie subgroup T. Since it is easy to see $\mathfrak{t}$ is a maximal Abelian subalgebra of $\mathfrak{u}(n)$ (Exercise 5.2), it follows that T is a maximal torus and $\mathfrak{t}$ is its corresponding Cartan subalgebra.

For $SU(n)$ with $\mathfrak{su}(n) = \{X \in \mathfrak{gl}(n, \mathbb{C}) \mid X^* = -X, \operatorname{tr} X = 0\}$, a similar construction yields a maximal torus and Cartan subalgebra. Simply use the definition for T and $\mathfrak{t}$ as in Equation 5.5 coupled with the additional requirement that $\sum_{i=1}^{n} \theta_i = 0$.

5.1.2.2 $Sp(n)$ For the first realization of $Sp(n)$ as

$$Sp(n) = \{g \in GL(n, \mathbb{H}) \mid g^* g = I\}$$

with $\mathfrak{sp}(n) = \{X \in \mathfrak{gl}(n, \mathbb{H}) \mid X^* = -X\}$, use the definition for T and $\mathfrak{t}$ as in Equation 5.5 to construct a maximal torus and Cartan subalgebra. It is straightforward to verify that $\mathfrak{t}$ is a Cartan subalgebra (Exercise 5.2).

For the second realization of $Sp(n)$ as

$$Sp(n) \cong U(2n) \cap Sp(n, \mathbb{C})$$

with $\mathfrak{sp}(n) \cong \mathfrak{u}(2n) \cap \mathfrak{sp}(n, \mathbb{C})$, let

$$T = \{\operatorname{diag}(e^{i\theta_1}, \ldots, e^{i\theta_n}, e^{-i\theta_1}, \ldots, e^{-i\theta_n}) \mid \theta_i \in \mathbb{R}\}$$
$$\mathfrak{t} = \{\operatorname{diag}(i\theta_1, \ldots, i\theta_n, -i\theta_1, \ldots, -i\theta_n) \mid \theta_i \in \mathbb{R}\}$$

for $\theta_i \in \mathbb{R}$. Then T is a maximal torus and $\mathfrak{t}$ is its corresponding Cartan subalgebra (Exercise 5.2).

5.1.2.3 $SO(2n)$ For $SO(2n)$ with $\mathfrak{so}(2n) = \{X \in \mathfrak{gl}(2n, \mathbb{R}) \mid X^t = -X, \operatorname{tr} X = 0\}$, define the set of block diagonal matrices

$$T = \left\{ \begin{pmatrix} \cos\theta_1 & \sin\theta_1 & & & \\ -\sin\theta_1 & \cos\theta_1 & & & \\ & & \ddots & & \\ & & & \cos\theta_n & \sin\theta_n \\ & & & -\sin\theta_n & \cos\theta_n \end{pmatrix} \mid \theta_i \in \mathbb{R} \right\}$$

$$\mathfrak{t} = \left\{ \begin{pmatrix} 0 & \theta_1 & & & \\ -\theta_1 & 0 & & & \\ & & \ddots & & \\ & & & 0 & \theta_n \\ & & & -\theta_n & 0 \end{pmatrix} \mid \theta_i \in \mathbb{R} \right\}.$$

Then T is a maximal torus and $\mathfrak{t}$ is its corresponding Cartan subalgebra (Exercise 5.2).

5.1.2.4 $SO(2n+1)$ For $SO(2n)$ with $\mathfrak{so}(2n) = \{X \in \mathfrak{gl}(2n+1, \mathbb{R}) \mid X^t = -X, \operatorname{tr} X = 0\}$, define the set of block diagonal matrices

$$T = \left\{ \begin{pmatrix} \cos\theta_1 & \sin\theta_1 & & & & \\ -\sin\theta_1 & \cos\theta_1 & & & & \\ & & \ddots & & & \\ & & & \cos\theta_n & \sin\theta_n & \\ & & & -\sin\theta_n & \cos\theta_n & \\ & & & & & 1 \end{pmatrix} \,\middle|\, \theta_i \in \mathbb{R} \right\}$$

$$\mathfrak{t} = \left\{ \begin{pmatrix} 0 & \theta_1 & & & & \\ -\theta_1 & 0 & & & & \\ & & \ddots & & & \\ & & & 0 & \theta_n & \\ & & & -\theta_n & 0 & \\ & & & & & 0 \end{pmatrix} \,\middle|\, \theta_i \in \mathbb{R} \right\}.$$

Then T is a maximal torus and $\mathfrak{t}$ is its corresponding Cartan subalgebra (Exercise 5.2).

5.1.3 Conjugacy of Cartan Subalgebras

Lemma 5.6. *Let G be a compact Lie group and (π, V) a finite-dimensional representation of G.*
(a) There exists a G-invariant inner product, $(\cdot, \cdot)$, on V and for any such G-invariant inner product on V, $d\pi X$ is skew-Hermitian, i.e., $(d\pi(X)v, w) = -(v, d\pi(X)w)$ for $X \in \mathfrak{g}$ and $v, w \in V$;
(b) There exists an Ad-invariant inner product, $(\cdot, \cdot)$, on $\mathfrak{g}$, that is, $(\mathrm{Ad}(g)Y_1, \mathrm{Ad}(g)Y_2) = (Y_1, Y_2)$ for $g \in G$ and $Y_i \in \mathfrak{g}$. For any such Ad-invariant inner product on $\mathfrak{g}$, ad is skew-symmetric, i.e., $(\mathrm{ad}(X)Y_1, Y_2) = -(Y_1, \mathrm{ad}(X)Y_2)$.

Proof. Part (b) is simply a special case of part (a), where π is the Adjoint representation on $\mathfrak{g}$. To prove part (a), recall that Theorem 2.15 provides the existence of a G-invariant inner product on V. If $(\cdot, \cdot)$ is a G-invariant inner product on V, apply $\frac{d}{dt}|_{t=0}$ to $(\pi(e^{tX})Y_1, \pi(e^{tX})Y_2) = (Y_1, Y_2)$ to get $(d\pi(X)Y_1, Y_2) + (Y_1, d\pi(X)Y_2) = 0$. $\square$

Lemma 5.7. *Let G be a compact Lie group and $\mathfrak{t}$ a Cartan subalgebra of $\mathfrak{g}$. There exists $X \in \mathfrak{t}$, so that $\mathfrak{t} = \mathfrak{z}_\mathfrak{g}(X)$ where $\mathfrak{z}_\mathfrak{g}(X) = \{Y \in \mathfrak{g} \mid [Y, X] = 0\}$.*

Proof. By choosing a basis for $\mathfrak{t}$ and using the fact that $\mathfrak{t}$ is maximal Abelian in $\mathfrak{g}$, there exist independent $\{X_i\}_{i=1}^n$, $X_i \in \mathfrak{t}$, so that $\mathfrak{t} = \cap_i \ker(\mathrm{ad}\, X_i)$. Below we show that there exists $t \in \mathbb{R}$, so that $\ker(\mathrm{ad}(X_1 + tX_2)) = \ker(\mathrm{ad}\, X_1) \cap \ker(\mathrm{ad}\, X_2)$. Once this result is established, it is clear that an inductive argument finishes the proof.

Let $(\cdot, \cdot)$ be an invariant inner product on $\mathfrak{g}$ for which ad is skew-symmetric. Let $\mathfrak{k}_X = \ker(\mathrm{ad}\, X)$ and $\mathfrak{r}_X = (\ker(\mathrm{ad}\, X))^\perp$. It follows that $\mathfrak{r}_X$ is an $\mathrm{ad}\, X$-invariant subspace. Since $\mathfrak{g} = \mathfrak{k}_X \oplus \mathfrak{r}_X$, it is easy to see $\mathfrak{r}_X$ is the range of $\mathrm{ad}\, X$ acting on $\mathfrak{g}$.

If non-central $X, Y \in \mathfrak{t}$, then the fact that $\mathrm{ad}\, X$ and $\mathrm{ad}\, Y$ commute, Theorem 4.8, implies that $\mathrm{ad}\, Y$ preserves the subspaces $\mathfrak{k}_X$ and $\mathfrak{r}_X$. In particular,

$$\mathfrak{g} = (\mathfrak{k}_X \cap \mathfrak{k}_Y) \oplus (\mathfrak{k}_X \cap \mathfrak{r}_Y) \oplus (\mathfrak{r}_X \cap \mathfrak{k}_Y) \oplus (\mathfrak{r}_X \cap \mathfrak{r}_Y).$$

If $\mathfrak{r}_X \cap \mathfrak{r}_Y = \{0\}$, then $\mathfrak{k}_X \cap \mathfrak{k}_Y = \ker(\mathrm{ad}(X + Y))$. Otherwise, restrict $\mathrm{ad}(X + tY)$ to $\mathfrak{r}_X \cap \mathfrak{r}_Y$ and take the determinant. The resulting polynomial in t is nonzero since it is nonzero when $t = 0$. Thus there is a $t_0 \neq 0$, so $\mathrm{ad}(X + t_0 Y)$ is invertible on $\mathfrak{r}_X \cap \mathfrak{r}_Y$. Hence, in either case, there exists a $t_0 \in \mathbb{R}$, so that $\ker(\mathrm{ad}(X + t_0 Y)) = \mathfrak{k}_X \cap \mathfrak{k}_Y$. □

Definition 5.8. Let G be a compact Lie group and $X \in \mathfrak{g}$. If $\mathfrak{z}_\mathfrak{g}(X)$ is a Cartan subalgebra, then X is called a *regular* element of $\mathfrak{g}$.

We will see in §7.2.1 that the set of regular elements is an open dense set in $\mathfrak{g}$.

Theorem 5.9. *Let G be a compact Lie group and $\mathfrak{t}$ a Cartan subalgebra. For $X \in \mathfrak{g}$, there exists $g \in G$ so that $\mathrm{Ad}(g)X \in \mathfrak{t}$.*

Proof. Let $(\cdot, \cdot)$ be an Ad-invariant inner product on $\mathfrak{g}$. Using Lemma 5.7, write $\mathfrak{t} = \mathfrak{z}_\mathfrak{g}(Y)$ for some $Y \in \mathfrak{g}$. It is necessary to find $g_0 \in G$ so that $[\mathrm{Ad}(g_0)X, Y] = 0$. For this, it suffices to show that $([\mathrm{Ad}(g_0)X, Y], Z) = 0$ for all $Z \in \mathfrak{g}$. Using the skew-symmetry of ad, Lemma 5.6, this is equivalent to showing $(Y, [Z, \mathrm{Ad}(g_0)X]) = 0$.

Consider the continuous function f on G defined by $f(g) = (Y, \mathrm{Ad}(g)X)$. Since G is compact, choose $g_0 \in G$ so that f has a maximum at g_0. Then the function $t \to (Y, \mathrm{Ad}(e^{tZ})\,\mathrm{Ad}(g)X)$ has a maximum at $t = 0$. Applying $\frac{d}{dt}|_{t=0}$ therefore yields $0 = (Y, [Z, \mathrm{Ad}(g_0)X])$, as desired. □

The corresponding theorem is true on the group level as well, although much harder to prove (see §5.1.4).

Corollary 5.10. *(a) Let G be a compact Lie group with Lie algebra $\mathfrak{g}$. Then $\mathrm{Ad}(G)$ acts transitively on the set of Cartan subalgebras of G.*
(b) Via conjugation, G acts transitively on the set of maximal tori of G.

Proof. For part (a), let $\mathfrak{t}_i = \mathfrak{z}_\mathfrak{g}(X_i)$, $X_i \in \mathfrak{g}$, be Cartan subalgebras. Using Theorem 5.9, there is a $g \in G$ so that $\mathrm{Ad}(g)X_1 \in \mathfrak{t}_2$. Using the fact that $\mathrm{Ad}(g)$ is a Lie algebra homomorphism, Theorem 4.8, it follows that

$$\begin{aligned}
\mathrm{Ad}(g)\mathfrak{t}_1 &= \{\mathrm{Ad}(g)Y \in \mathfrak{g} \mid [Y, X_1] = 0\} \\
&= \{Y' \in \mathfrak{g} \mid [\mathrm{Ad}(g)^{-1}Y', X_1] = 0\} \\
&= \{Y' \in \mathfrak{g} \mid [Y', \mathrm{Ad}(g)X_1] = 0\} = \mathfrak{z}_\mathfrak{g}(\mathrm{Ad}(g)X_1).
\end{aligned}$$

Since $\mathrm{Ad}(g)X_1 \in \mathfrak{t}_2$ and $\mathfrak{t}_2$ is Abelian, $\mathrm{Ad}(g)\mathfrak{t}_1 \supseteq \mathfrak{t}_2$. Since $\mathrm{Ad}(g)$ is a Lie algebra homomorphism, $\mathrm{Ad}(g)\mathfrak{t}_1$ is still Abelian. By maximality of Cartan subalgebras, $\mathrm{Ad}(g)\mathfrak{t}_1 = \mathfrak{t}_2$.

For part (b), let T_i be the maximal torus of G corresponding to $\mathfrak{t}_i$. Use Theorem 5.2 to write $T_i = \exp \mathfrak{t}_i$. Then Theorem 4.8 shows that

$$c_g T_1 = c_g \exp \mathfrak{t}_1 = \exp(\mathrm{Ad}(g)\mathfrak{t}_1) = \exp \mathfrak{t}_2 = T_2. \qquad \square$$

5.1.4 Maximal Torus Theorem

Lemma 5.11. *Let G be a compact connected Lie group. The kernel of the Adjoint map is the center of G, i.e., $\mathrm{Ad}(g) = I$ if and only if $g \in Z(G)$, where $Z(G) = \{h \in G \mid gh = hg\}$.*

Proof. If $g \in Z(G)$, then c_g is the identity, so that its differential, $\mathrm{Ad}(g)$, is trivial as well. On the other hand, if $\mathrm{Ad}(g) = I$, then $c_g e^X = e^{\mathrm{Ad}(g)X} = e^X$ for $X \in \mathfrak{g}$. Thus c_g is the identity on a neighborhood of I in G. Since G is connected and c_g is a homomorphism, Theorem 1.15 shows that c_g is the identity on all of G.   □

Theorem 5.12 (Maximal Torus Theorem). *Let G be a compact connected Lie group, T a maximal torus of G, and $g_0 \in G$.*
(a) There exists $g \in G$ so that $g g_0 g^{-1} \in T$.
(b) The exponential map is surjective, i.e., $G = \exp \mathfrak{g}$.

Proof. Use Theorems 5.2, 4.8, and 5.9 to observe that

$$\bigcup_{g \in G} c_g T = \bigcup_{g \in G} c_g \exp \mathfrak{t} = \bigcup_{g \in G} \exp\left(\mathrm{Ad}(g)\mathfrak{t}\right) = \exp \mathfrak{g}.$$

Thus $\bigcup_{g \in G} c_g T = G$ if and only if $\exp \mathfrak{g} = G$. In other words parts (a) and (b) are equivalent.

We will prove part (b) by induction on $\dim \mathfrak{g}$. If $\dim \mathfrak{g} = 1$, then $\mathfrak{g}$ is Abelian, and so Theorem 5.2 shows that $\exp \mathfrak{g} = G$. For $\dim \mathfrak{g} > 1$, we will use the inductive hypothesis to show that $\exp \mathfrak{g}$ is open and closed to finish the proof. Since $\bigcup_{g \in G} c_g T$ is the continuous image of the compact set $G \times T$, $\exp \mathfrak{g}$ is compact and therefore closed. Thus it remains to show that $\exp \mathfrak{g}$ is open.

Fix $X_0 \in \mathfrak{g}$ and write $g_0 = \exp X_0$. It is necessary to show that there is a neighborhood of g_0 contained in $\exp \mathfrak{g}$. By Theorem 4.6, assume $X_0 \neq 0$. Using Lemma 5.6, let $(\cdot, \cdot)$ be an Ad-invariant inner product on $\mathfrak{g}$ so that $\mathrm{Ad}(g_0)$ is unitary. Define

$$\mathfrak{a} = \mathfrak{z}_\mathfrak{g}(g_0) = \{Y \in \mathfrak{g} \mid \mathrm{Ad}(g_0)Y = Y\}$$
$$\mathfrak{b} = \mathfrak{a}^\perp,$$

so $\mathfrak{g} = \mathfrak{a} \oplus \mathfrak{b}$ with $\mathrm{Ad}(g_0) - I$ an invertible endomorphism of $\mathfrak{b}$. Note that $X_0 \in \mathfrak{a}$ since $\mathrm{Ad}(\exp X_0)X_0 = e^{\mathrm{ad}(X_0)}X_0 = X_0$.

Consider the smooth map $\varphi : \mathfrak{a} \oplus \mathfrak{b} \to G$ given by $\varphi(X, Y) = g_0^{-1} e^Y g_0 e^X e^{-Y}$. Under the usual tangent space identifications, the differential of φ can be computed at 0 by

$$d\varphi(X, 0) = \frac{d}{dt}\varphi(tX, 0)|_{t=0} = \frac{d}{dt}e^{tX}|_{t=0} = X$$

$$d\varphi(0, Y) = \frac{d}{dt}\varphi(0, tY)|_{t=0} = \frac{d}{dt}\left(g_0^{-1} e^{tY} g_0 e^{-tY}\right)|_{t=0} = (\mathrm{Ad}(g_0) - I)\,Y.$$

Thus $d\varphi$ is an isomorphism, so that $\{g_0^{-1} e^Y g_0 e^X e^{-Y} \mid X \in \mathfrak{a},\, Y \in \mathfrak{b}\}$ contains a neighborhood of I in G. Since $l_{g_0^{-1}}$ is a diffeomorphism,

$$\{e^Y g_0 e^X e^{-Y} \mid X \in \mathfrak{a}, Y \in \mathfrak{b}\}$$

contains a neighborhood of g_0 in G.

Let $A = Z_G(g_0)^0 = \{g \in G \mid g g_0 = g_0 g\}^0$, a closed and therefore compact Lie subgroup of G. Rewriting the condition $g g_0 = g_0 g$ as $c_{g_0} g = g$, the usual argument using Theorem 4.6 shows that the Lie algebra of A is $\mathfrak{a}$ (Exercise 4.22). In particular, $e^{\mathfrak{a}} \subseteq A$, so that $g_0 e^{\mathfrak{a}} \subseteq A$ since $g_0 \in A$. Thus

$$\{e^Y g_0 e^X e^{-Y} \mid X \in \mathfrak{a}, Y \in \mathfrak{b}\} \subseteq \{e^Y A e^{-Y} \mid Y \in \mathfrak{b}\},$$

and so $\bigcup_{g \in G} g^{-1} A g$ certainly contains a neighborhood of g_0 in G.

Note that $\dim \mathfrak{a} \geq 1$ as $X_0 \in \mathfrak{g}$. If $\dim \mathfrak{a} < \dim \mathfrak{g}$, the inductive hypothesis shows that $A = \exp \mathfrak{a}$, so that $\bigcup_{g \in G} g^{-1} A g = \bigcup_{g \in G} \exp(\mathrm{Ad}(g)\mathfrak{a}) \subseteq \exp \mathfrak{g}$. Thus $\exp \mathfrak{g}$ contains a neighborhood of g_0, as desired.

On the other hand, if $\dim \mathfrak{a} = \dim \mathfrak{g}$, then $\mathrm{Ad}(g_0) = I$ so that Lemma 5.11 shows $g_0 \in Z(G)$. Let $\mathfrak{t}$ be a Cartan subalgebra containing X_0. By Theorem 5.9, $\mathfrak{g} = \bigcup_{g \in G} \mathrm{Ad}(g)\mathfrak{t}$. For any $X \in \mathfrak{t}$, use the facts that $g_0 = e^{X_0} \in Z(G)$ and $[X_0, X] = 0$ to compute

$$g_0 \exp(\mathrm{Ad}(g)X) = g_0 c_g e^X = c_g \left(e^{X_0} e^X\right) = c_g e^{X_0 + X} = \exp(\mathrm{Ad}(g)(X_0 + X))$$

for $g \in G$. Since $X_0 + X \in \mathfrak{t}$, $g_0 \exp \mathfrak{g} \subseteq \exp \mathfrak{g}$. However, Theorem 4.6 shows $\exp \mathfrak{g}$ contains a neighborhood of I so that $g_0 \exp \mathfrak{g}$ contains a neighborhood of g_0. The inclusion $g_0 \exp \mathfrak{g} \subseteq \exp \mathfrak{g}$ finishes the proof. $\square$

Corollary 5.13. *Let G be a compact connected Lie group with maximal torus T.*
(a) Then $Z_G(T) = T$, where $Z_G(T) = \{g \in G \mid gt = tg \text{ for } t \in T\}$. In particular, T is maximal Abelian.
(b) The center of G is contained in T, i.e., $Z(G) \subseteq T$.

Proof. Part (b) clearly follows from part (a). For part (a), obviously $T \subseteq Z_G(T)$. Conversely, let $g_0 \in Z_G(T)$ and consider the closed, therefore compact, connected Lie subgroup $Z_G(g_0)^0$. Using the Maximal Torus Theorem, write $g_0 = e^{X_0}$. Looking at the path $t \to e^{tX_0} \in Z_G(g_0)$, it follows that $g_0 \in Z_G(g_0)^0$. Moreover, since T is connected and contains I, clearly $T \subseteq Z_G(g_0)^0$, so that T is a maximal torus in $Z_G(g_0)^0$. By the Maximal Torus theorem applied to $Z_G(g_0)^0$, there is $h \in Z_G(g_0)^0$, so that $c_h g_0 \in T$. But by construction, $c_h g_0 = g_0$, so $g_0 \in T$, as desired. $\square$

Note that there exist maximal Abelian subgroups that are not maximal tori (Exercise 5.6).

5.1.5 Exercises

Exercise 5.1 Let Γ be a discrete subgroup of $\mathbb{R}^n$. Pick an indivisible element $e_1 \in \Gamma$ and show that $\Gamma / \mathbb{Z} e_1$ is a discrete subgroup in $\mathbb{R}^n / \mathbb{R} e_1$. Use induction to show that Γ is isomorphic to Γ_k.

Exercise 5.2 For each compact classical Lie group in §5.1.2, show that the given subgroup T is a maximal torus and that the given subalgebra $\mathfrak{t}$ is its corresponding Cartan subalgebra.

Exercise 5.3 Show that the most general connected Abelian Lie subgroup of a general linear Lie group is isomorphic to $T^k \times \mathbb{R}^n$.

Exercise 5.4 Classify the irreducible representations of $T^k \times (\mathbb{Z}/n_1\mathbb{Z}) \times \cdots \times (\mathbb{Z}/n_k\mathbb{Z})$.

Exercise 5.* *(Kronecker's Theorem).* View $T^k \cong \mathbb{R}^k/\mathbb{Z}^2$ and let $x = (x_i) \in \mathbb{R}^k$. Show that the following statements are equivalent.
(a) The set $\{1, x_1, \ldots, x_n\}$ is linearly dependent over $\mathbb{Q}$.
(b) There is a nonzero $n = (n_i) \in \mathbb{Z}^k$, so $n \cdot x \in \mathbb{Z}$.
(c) There is a nontrivial homomorphism $\pi : \mathbb{R}^k/\mathbb{Z}^k \to S^1$ with $x + \mathbb{Z}^k \in \ker \pi$.
(d) The set $\overline{\mathbb{Z}x + \mathbb{Z}^k} \neq \mathbb{R}^k/\mathbb{Z}^k$.

Exercise 5.5 Working in $\mathrm{Spin}_{2n}(\mathbb{R})$ or $\mathrm{Spin}_{2n+1}(\mathbb{R})$, let

$$T = \{(\cos t_1 + e_1 e_2 \sin t_1) \cdots (\cos t_n + e_1 e_2 \sin t_n) \mid t_k \in \mathbb{R}\}.$$

Show that T is a maximal torus (c.f. Exercise 1.33).

Exercise 5.6 Find a maximal Abelian subgroup of $SO(3)$ that is isomorphic to $(\mathbb{Z}/2\mathbb{Z})^2$ and therefore not a maximal torus.

Exercise 5.7 If H is a closed connected Lie subgroup of a compact Lie group G, show that $\exp \mathfrak{h} = H$.

Exercise 5.8 Let G be a connected Lie subgroup of a general linear group. Show that the center of G, $Z(G)$, is a closed Lie subgroup of G with Lie algebra $\mathfrak{z}(\mathfrak{g})$ (c.f. Exercise 4.22).

Exercise 5.9 Let G be a compact connected Lie group. Show that the center of G, $Z(G)$, is the intersection of all maximal tori in G.

Exercise 5.10 Let G be a compact connected Lie group. For $g \in G$ and positive $n \in \mathbb{N}$, show that there exists $h \in G$ so that $h^n = g$.

Exercise 5.11 Let T be the maximal torus of $SO(3)$ given by

$$T = \left\{ \begin{pmatrix} \cos \theta & \sin \theta & 0 \\ -\sin \theta & \cos \theta & 0 \\ 0 & 0 & 1 \end{pmatrix} \mid \theta \in \mathbb{R} \right\}.$$

Find $g \in SO(3)$, so that $Z_G(g)^0 = T$ but with $Z_G(g)$ disconnected.

Exercise 5.12 Let G be a compact connected Lie group. Suppose S is a connected Abelian Lie subgroup of G.
(1) If $g \in Z_G(S)$, show that there exists a maximal torus T of G, so that $g \in T$ and $S \subseteq T$.
(2) Show that $Z_G(S)$ is the union of all maximal tori containing S and therefore is connected.
(3) For $g \in G$, show that $Z_G(g)^0$ is the union of all maximal tori containing g.

Exercise 5.13 Let G be a compact connected Lie group. Suppose that G is also a complex manifold whose group operations are holomorphic. Then the map $g \to \mathrm{Ad}(g)$, $g \in G$, is holomorphic. Show that G is Abelian and isomorphic to $\mathbb{C}^n / \Gamma$ for some discrete subgroup Γ of $\mathbb{C}^n$.

5.2 Structure

5.2.1 Exponential Map Revisited

5.2.1.1 Local Diffeomorphism Let G be a Lie subgroup of $GL(n, \mathbb{C})$. We already know from Theorem 4.6 that $\exp: \mathfrak{g} \to G$ is a local diffeomorphism near 0. In fact, more is true. Before beginning, use power series to define

$$\frac{I - e^{-\mathrm{ad}\,X}}{\mathrm{ad}\,X} = \sum_{n=0}^{\infty} \frac{(-1)^n}{(n+1)!} (\mathrm{ad}\,X)^n$$

for $X \in \mathfrak{g}$.

Theorem 5.14. (a) *Let G be a Lie subgroup of $GL(n, \mathbb{C})$ and $\gamma : \mathbb{R} \to \mathfrak{g}$ a smooth curve. Then*

$$\frac{d}{dt} e^{\gamma(t)} = e^{\gamma(t)} \left(\frac{I - e^{-\mathrm{ad}\,\gamma(t)}}{\mathrm{ad}\,\gamma(t)} \right) (\gamma'(t))$$

$$= \left[\left(\frac{e^{\mathrm{ad}\,\gamma(t)} - I}{\mathrm{ad}\,\gamma(t)} \right) (\gamma'(t)) \right] e^{\gamma(t)}.$$

(b) *For $X \in \mathfrak{g}$, the map $\exp: \mathfrak{g} \to G$ is a local diffeomorphism near X if and only if the eigenvalues of $\mathrm{ad}\,X$ on $\mathfrak{g}$ are disjoint from $2\pi i \mathbb{Z} \setminus \{0\}$.*

Proof. In part (a), consider the special case of, say, $\gamma(t) = X + tY$ for $Y \in \mathfrak{g}$. Using the usual tangent space identifications at $t = 0$, the first part of (a) calculates the differential at X of the map $\exp : \mathfrak{g} \to G$ evaluated on Y. If $(\mathrm{ad}\,X)\,Y = \lambda Y$ for $\lambda \in \mathbb{C}$, then

$$\left(\frac{I - e^{-\mathrm{ad}\,X}}{\mathrm{ad}\,X} \right) (Y) = \begin{cases} \frac{1 - e^{-\lambda}}{\lambda} Y & \text{if } \lambda \in 0 \\ Y & \text{if } \lambda = 0 \end{cases}$$

which is zero if and only if $\lambda \in 2\pi i \mathbb{Z} \backslash \{0\}$. Since Lemma 5.6 shows that ad X is normal and therefore diagonalizable, part (b) follows from the Inverse Mapping Theorem and part (a).

With sufficient patience, the proof of part (a) can be accomplished by explicit power series calculations. As is common in mathematics, we instead resort to a trick. Define $\varphi : \mathbb{R}^2 \to G$ by

$$\varphi(s, t) = e^{-s\gamma(t)} \frac{\partial}{\partial t} e^{s\gamma(t)}.$$

To prove the first part of (a), it is necessary to show $\varphi(1, t) = \left(\frac{I - e^{-\operatorname{ad}\gamma(t)}}{\operatorname{ad}\gamma(t)} \right) (\gamma'(t))$.

Begin by observing that $\varphi(0, t) = 0$, and so $\varphi(1, t) = \int_0^1 \frac{\partial}{\partial s} \varphi(s, t) \, ds$. However,

$$\frac{\partial}{\partial s} \varphi(s, t) = -\gamma(t) e^{-s\gamma(t)} \frac{\partial}{\partial t} e^{s\gamma(t)} + e^{-s\gamma(t)} \frac{\partial}{\partial t} \left[\gamma(t) e^{s\gamma(t)} \right]$$

$$= -e^{-s\gamma(t)} \gamma(t) \frac{\partial}{\partial t} e^{s\gamma(t)} + e^{-s\gamma(t)} \gamma'(t) e^{s\gamma(t)} + e^{-s\gamma(t)} \gamma(t) \frac{\partial}{\partial t} e^{s\gamma(t)}$$

$$= e^{-s\gamma(t)} \gamma'(t) e^{s\gamma(t)},$$

so that $\frac{\partial}{\partial s} \varphi(s, t) = \operatorname{Ad}(e^{-s\gamma(t)}) \gamma'(t) = e^{-s \operatorname{ad}\gamma(t)} \gamma'(t)$ by Equation 4.11. Thus

$$\varphi(1, t) = \int_0^1 e^{-s \operatorname{ad}\gamma(t)} \gamma'(t) \, ds = \int_0^1 \sum_{n=0}^{\infty} \frac{(-s)^n}{n!} (\operatorname{ad}\gamma(t))^n \, \gamma'(t) \, ds$$

$$= \left(\sum_{n=0}^{\infty} \frac{(-1)^n s^{n+1}}{(n+1)!} (\operatorname{ad}\gamma(t))^n \, \gamma'(t) \right) |_{s=0}^{s=1} = \left(\frac{I - e^{-\operatorname{ad}\gamma(t)}}{\operatorname{ad}\gamma(t)} \right) \gamma'(t),$$

as desired. To show the second part of (a), use the relation $l_{e^{\gamma(t)}} = r_{e^{\gamma(t)}} \circ \operatorname{Ad}(e^{\gamma(t)}) = r_{e^{\gamma(t)}} \circ e^{\operatorname{ad}\gamma(t)}$, where $l_{e^{\gamma(t)}}$ and $r_{e^{\gamma(t)}}$ stand for left and right multiplication by $e^{\gamma(t)}$.   □

5.2.1.2 Dynkin's Formula Let G be a Lie subgroup of $GL(n, \mathbb{C})$. For $X_i \in \mathfrak{g}$, write $[X_n, \ldots, X_3, X_2, X_1]$ for the *iterated Lie bracket*

$$\left[X_n, \ldots, \left[X_3, \left[X_2, X_1 \right] \right], \ldots \right]$$

and write $[X_n^{(i_n)}, \ldots, X_1^{(i_1)}]$ for the iterated Lie bracket

$$[\overbrace{X_n, \ldots, X_n}^{i_n \text{ copies}}, \ldots, \overbrace{X_1, \ldots, X_1}^{i_1 \text{ copies}}].$$

Although now known as the *Campbell–Baker–Hausdorff Series* ([21], [5], and [49]), the following explicit formula is actually due to Dynkin ([35]). In the proof we use the well-known fact that $\ln(X)$ inverts e^X on a neighborhood of I, where $\ln(I + X) = \sum_{n=1}^{\infty} \frac{(-1)^{n+1}}{n} X^n$ converges absolutely on a neighborhood 0 (Exercise 5.15).

Theorem 5.15 (Dynkin's Formula). *Let G be a Lie subgroup of $GL(n, \mathbb{C})$. For $X, Y \in \mathfrak{g}$ in a sufficiently small neighborhood of 0,*

$$e^X e^Y = e^Z,$$

where Z is given by the formula

$$Z = \sum \frac{(-1)^{n+1}}{n} \frac{1}{(i_1 + j_1) + \cdots + (i_n + j_n)} \frac{[X^{(i_1)}, Y^{(j_1)}, \ldots, X^{(i_n)}, Y^{(j_n)}]}{i_1! j_1! \cdots i_n! j_n!},$$

where the sum is taken over all $2n$-tuples $(i_1, \ldots, i_n, j_1, \ldots, j_n) \in \mathbb{N}^{2n}$ satisfying $i_k + j_k \geq 1$ for positive $n \in \mathbb{N}$.

Proof. The approach of this proof follows [34]. Using Theorem 4.6, choose a neighborhood U_0 of 0 in $\mathfrak{g}$ on which $\exp$ is a local diffeomorphism and where $\ln$ is well defined on $\exp U$. Let $U \subseteq U_0$ be an open ball about of 0 in $\mathfrak{g}$, so that $(\exp U)^2 (\exp U)^{-2} \subseteq \exp U_0$ (by continuity of the group structure as in Exercise 1.4). For $X, Y \in U$, define $\gamma(t) = e^{tX} e^{tY}$ mapping a neighborhood of $[0, 1]$ to $\exp U$. Therefore there is a unique smooth curve $Z(t) \in U_0$, so that $e^{Z(t)} = e^{tX} e^{tY}$. Apply $\frac{d}{dt}$ to this equation and use Theorem 5.14 to see that

$$\left[\left(\frac{e^{\operatorname{ad} Z(t)} - I}{\operatorname{ad} Z(t)} \right) (Z'(t)) \right] e^{Z(t)} = X e^{Z(t)} + e^{Z(t)} Y.$$

Since $Z(t) \in U_0$, $\exp$ is a local diffeomorphism near $Z(t)$. Thus the proof of Theorem 5.14 shows that $\left(\frac{I - e^{-\operatorname{ad} Z(t)}}{\operatorname{ad} Z(t)} \right)$ is an invertible map on $\mathfrak{g}$. As $e^{Z(t)} = e^{tX} e^{tY}$, $\operatorname{Ad}(e^{Z(t)}) = \operatorname{Ad}(e^{tX}) \operatorname{Ad}(e^{tY})$, so that $e^{\operatorname{ad} Z(t)} = e^{t \operatorname{ad} X} e^{t \operatorname{ad} Y}$ by Equation 4.11. Thus

$$Z'(t) = \left(\frac{\operatorname{ad} Z(t)}{e^{\operatorname{ad} Z(t)} - I} \right) (X + \operatorname{Ad}(e^{Z(t)}) Y) = \left(\frac{\operatorname{ad} Z(t)}{e^{\operatorname{ad} Z(t)} - I} \right) (X + e^{\operatorname{ad} Z(t)} Y)$$

$$= \left(\frac{\operatorname{ad} Z(t)}{e^{\operatorname{ad} Z(t)} - I} \right) (X + e^{t \operatorname{ad} X} e^{t \operatorname{ad} Y} Y) = \left(\frac{\operatorname{ad} Z(t)}{e^{\operatorname{ad} Z(t)} - I} \right) (X + e^{t \operatorname{ad} X} Y).$$

Using the relation $A = \ln(I + (e^A - I)) = \sum_{n=1}^{\infty} \frac{(-1)^{n-1}}{n} (e^A - I)^n$ for $A = \operatorname{ad} Z(t)$ and $e^A = e^{t \operatorname{ad} X} e^{t \operatorname{ad} Y}$, we get

$$\frac{\operatorname{ad} Z(t)}{e^{\operatorname{ad} Z(t)} - I} = \sum_{n=1}^{\infty} \frac{(-1)^{n-1}}{n} (e^{t \operatorname{ad} X} e^{t \operatorname{ad} Y} - I)^{n-1}.$$

Hence

$$Z'(t) = \sum_{n=1}^{\infty} \frac{(-1)^{n-1}}{n} (e^{t \operatorname{ad} X} e^{t \operatorname{ad} Y} - I)^{n-1} (X + e^{t \operatorname{ad} X} Y)$$

$$= \sum_{n=1}^{\infty} \frac{(-1)^{n-1}}{n} \left[\sum_{i,j=0, \, (i,j) \neq (0,0)}^{\infty} \frac{t^{i+j}}{i! j!} (\operatorname{ad} X)^i (\operatorname{ad} Y)^j \right]^{n-1} \left(X + \left(\sum_{i=0}^{\infty} \frac{t^i}{i!} (\operatorname{ad} X)^i \right) Y \right)$$

$$= \sum_{n=1}^{\infty} \frac{(-1)^{n-1}}{n} \left[\sum \frac{t^{i_1+j_1+\cdots i_{n-1}+j_{n-1}}}{i_1! j_1! \cdots i_{n-1}! j_{n-1}!} [(X)^{i_1}, (Y)^{j_1}, \ldots, (X)^{i_{k-1}}, (Y)^{j_{k-1}}, X] \right.$$

$$\left. + \sum \frac{t^{i_1+j_1+\cdots i_{n-1}+j_{n-1}+i_n}}{i_1! j_1! \cdots i_{n-1}! j_{n-1}! i_n!} [(X)^{i_1}, (Y)^{j_1}, \ldots, (X)^{i_{k-1}}, (Y)^{j_{k-1}}, (X)^{i_n}, Y] \right]$$

where the second and third sum are taken over all $i_k, j_k \in \mathbb{N}$ with $i_k + j_k \geq 1$. Since $Z(0) = 0$, $Z(1) = \int_0^1 \frac{d}{dt} Z(t)\, dt$. Integrating the above displayed equation finishes the proof. □

The explicit formula for Z in Dynkin's Formula is actually not important. In practice it is much too difficult to use. However, what *is* important is the fact that such a formula exists using only Lie brackets.

Corollary 5.16. *Let N be a connected Lie subgroup of $GL(n, \mathbb{C})$ whose Lie algebra $\mathfrak{n}$ lies in the set of strictly upper triangular matrices, i.e., if $X \in \mathfrak{n}$, then $X_{i,j} = 0$ when $i \geq j$. Then the map $\exp: \mathfrak{n} \to N$ is surjective, i.e., $N = \exp \mathfrak{n}$.*

Proof. It is a simple exercise to see that $[X_n, \ldots, X_3, X_2, X_1] = 0$ for any strictly upper triangular $X, X_i \in \mathfrak{gl}(n, \mathbb{C})$ and that e^X is polynomial in X (Exercise 5.18). In particular, for $X, Y \in \mathfrak{n}$ near 0, Dynkin's Formula gives a polynomial expression for $Z \in \mathfrak{n}$ solving $e^X e^Y = e^Z$. Since both sides of this expression are polynomials in X and Y that agree on a neighborhood, they agree everywhere. Because the formula for Z involves only the algebra structure of $\mathfrak{n}$, Z remains in $\mathfrak{n}$ for $X, Y \in \mathfrak{n}$. In other words, $(\exp \mathfrak{n})^2 \subseteq \exp \mathfrak{n}$. Since $\exp \mathfrak{n}$ generates N by Theorem 1.15, this shows that $\exp \mathfrak{n} = N$. □

5.2.2 Lie Algebra Structure

If G_i are Lie subgroups of a linear group, then, as in the proof of Theorem 4.16, recall that the *direct sum* of $\mathfrak{g}_1$ and $\mathfrak{g}_2$, $\mathfrak{g}_1 \oplus \mathfrak{g}_2$, may be viewed as the Lie algebra of $G_1 \times G_2$ with $[X_1 + X_2, Y_1 + Y_2] = [X_1, X_2] + [Y_1, Y_2]$ for $X_i, Y_i \in \mathfrak{g}_i$.

Definition 5.17. (a) Let $\mathfrak{g}$ be the Lie algebra of a Lie subgroup of a linear group. Then $\mathfrak{g}$ is called *simple* if $\mathfrak{g}$ has no proper ideals and if $\dim \mathfrak{g} > 1$, i.e., if the only ideals of $\mathfrak{g}$ are $\{0\}$ and $\mathfrak{g}$ and $\mathfrak{g}$ is non-Abelian.
(b) The Lie algebra $\mathfrak{g}$ is called *semisimple* if $\mathfrak{g}$ is a direct sum of simple Lie algebras.
(c) The Lie algebra $\mathfrak{g}$ is called *reductive* if $\mathfrak{g}$ is a direct sum of a semisimple Lie algebra and an Abelian Lie algebra.
(d) Let $\mathfrak{g}'$ be the ideal of $\mathfrak{g}$ spanned by $[\mathfrak{g}, \mathfrak{g}]$.

Theorem 5.18. *Let G be a compact Lie group with Lie algebra $\mathfrak{g}$. Then $\mathfrak{g}$ is reductive. If $\mathfrak{z}(\mathfrak{g})$ is the center of $\mathfrak{g}$, i.e., $\mathfrak{z}(\mathfrak{g}) = \{X \in \mathfrak{g} \mid [X, \mathfrak{g}] = 0\}$, then*

$$\mathfrak{g} = \mathfrak{g}' \oplus \mathfrak{z}(\mathfrak{g}),$$

$\mathfrak{g}'$ is semisimple, and $\mathfrak{z}(\mathfrak{g})$ is Abelian. Moreover, there are simple ideals $\mathfrak{s}_i$ of $\mathfrak{g}'$, so that

$$g' = \bigoplus_{i=1}^{k} \mathfrak{s}_i$$

with $[\mathfrak{s}_i, \mathfrak{s}_j] = 0$ for $i \neq j$ and span$[\mathfrak{s}_i, \mathfrak{s}_i] = \mathfrak{s}_i$.

Proof. Using Lemma 5.6, let $(\cdot, \cdot)$ be an Ad-invariant inner product on $\mathfrak{g}$, so that ad X, $X \in \mathfrak{g}$, is skew-Hermitian. If $\mathfrak{a}$ is an ideal of $\mathfrak{g}$, then $\mathfrak{a}^{\perp}$ is also an ideal. It follows that $\mathfrak{g}$ can be written as a direct sum of minimal ideals

(5.19) $$\mathfrak{g} = \mathfrak{s}_1 \oplus \cdots \oplus \mathfrak{s}_k \oplus \mathfrak{z}_1 \oplus \cdots \oplus \mathfrak{z}_n,$$

where dim $\mathfrak{s}_i > 1$ and dim $\mathfrak{z}_j = 1$. Since $\mathfrak{s}_i$ is an ideal, $[\mathfrak{s}_i, \mathfrak{s}_j] \subseteq \mathfrak{s}_i \cap \mathfrak{s}_j$. Thus $[\mathfrak{s}_i, \mathfrak{s}_j] = 0$ for $i \neq j$ and $[\mathfrak{s}_i, \mathfrak{s}_i] \subseteq \mathfrak{s}_i$. Similarly, $[\mathfrak{s}_i, \mathfrak{z}_j] = 0$ and $[\mathfrak{z}_i, \mathfrak{z}_j] = 0$ for $i \neq j$. Moreover, $[\mathfrak{z}_i, \mathfrak{z}_i] = 0$ since dim $\mathfrak{z}_i = 1$ and $[\cdot, \cdot]$ is skew-symmetric.

In particular, $\mathfrak{z}_1 \oplus \cdots \oplus \mathfrak{z}_n \subseteq \mathfrak{z}(\mathfrak{g})$. On the other hand, if $Z \in \mathfrak{z}(\mathfrak{g})$ is decomposed according to Equation 5.19 as $Z = \sum_i S_i + \sum_j Z_j$, then $0 = [Z, \mathfrak{s}_i] = [S_i, \mathfrak{s}_i]$. This suffices to show that $S_i \in \mathfrak{z}(\mathfrak{g})$ which, by construction of $\mathfrak{s}_i$ as a minimal ideal with dim $\mathfrak{s}_i > 1$, implies that $S_i = 0$. Thus $\mathfrak{z}(\mathfrak{g}) = \mathfrak{z}_1 \oplus \cdots \oplus \mathfrak{z}_n$. The remainder of the proof follows by showing that span$[\mathfrak{s}_i, \mathfrak{s}_i] = \mathfrak{s}_i$. However, this too follows from the construction of $\mathfrak{s}_i$ as a minimal ideal. Since $\mathfrak{s}_i$ is not central, dim(span$[\mathfrak{s}_i, \mathfrak{s}_i]) \geq 1$. As a result, dim(span$[\mathfrak{s}_i, \mathfrak{s}_i])$ cannot be less than dim $\mathfrak{s}_i$ either, or else span$[\mathfrak{s}_i, \mathfrak{s}_i]$ would be a proper ideal. □

It is an important theorem from the study of Lie algebras (see [56], [61], or [70]) that the simple Lie algebras are classified. It is rather remarkable that there are, relatively speaking, so few of them. In §6.1.2 we will discuss the complexification of our Lie algebras. In that setting, there are four infinite families of simple complex Lie algebras. They arise from the compact classical Lie groups $SU(n)$, $SO(2n+1)$, $Sp(n)$, and $SO(2n)$. Beside these families, there are only five other simple complex Lie algebras. They are called exceptional and go by the names G_2, F_4, E_6, E_7, and E_8. They have dimensions are 14, 52, 78, 133, and 248, respectively.

5.2.3 Commutator Theorem

Definition 5.20. Let G be a Lie subgroup of a linear group. The *commutator subgroup*, G', is the normal subgroup of G generated by

$$\{g_1 g_2 g_1^{-1} g_2^{-1} \mid g_i \in G\}.$$

In a more general setting, G' need not be closed, however this nuisance does not arise for compact Lie groups.

Theorem 5.21. *Let G be a compact connected Lie group. Then G' is a connected closed normal Lie subgroup of G with Lie algebra $\mathfrak{g}'$.*

Proof. As usual, Theorem 3.28, assume G is a closed Lie subgroup of $U(n)$. With respect to the standard representation on $\mathbb{C}^n$, decompose $\mathbb{C}^n$ into its irreducible summands under the action of G, $\mathbb{C}^n \cong \mathbb{C}^{n_1} \oplus \cdots \oplus \mathbb{C}^{n_k}$ with $n_1 + \cdots + n_k = n$ and $n_i \geq 1$. Thus G can be viewed as a closed Lie subgroup of $U(n_1) \times \cdots \times U(n_k)$, so that the induced projection $\pi_i : G \to U(n_i)$ yields an irreducible representation of G.

Let $\varphi : G \to S^1 \times \cdots \times S^1$ (k copies) be the homomorphism induced by taking the determinant of each $\pi_i(g)$. Define H to be the closed Lie subgroup of G given by $H = \ker \varphi$. We will show that $\mathfrak{h} = \mathfrak{g}'$ and that $H^0 = G'$, which will finish the proof.

Recall that it follows easily from Theorem 4.6 that $\mathfrak{h} = \ker d\varphi$ (Exercise 4.24) and that the Lie algebra of $Z(G)$ is $\mathfrak{z}(\mathfrak{g})$ (c.f. Exercises 4.22 or 5.8). Now if $Z \in \mathfrak{z}(\mathfrak{g})$, then $e^{tZ} \in Z(G)$, so that Schur's Lemma implies $\pi_i e^{tZ} = c_i(t) I$ for some scalar $c_i(t)$ with $c_i(0) = 1$. Evaluating at $\frac{d}{dt}|_0$, this means that

$$Z = \operatorname{diag}(\overbrace{c_1'(0), \ldots, c_1'(0)}^{n_1 \text{ copies}}, \ldots, \overbrace{c_k'(0), \ldots, c_k'(0)}^{n_k \text{ copies}}).$$

Hence $d\varphi(Z) = (n_1 c_1'(0), \ldots, n_k c_k'(0))$ and $Z \in \ker \varphi$ if and only if $Z = 0$. On the other hand, since $\mathfrak{g}'$ is spanned by $[\mathfrak{g}, \mathfrak{g}]$, clearly $\operatorname{tr}(d\pi_i X) = 0$ for $X \in \mathfrak{g}'$ (Exercise 5.20). Thus $\det \pi_i e^{tX} = 1$ (Exercise 4.3), so that $\mathfrak{g}' \subseteq \ker d\varphi$. Combined with the decomposition from Theorem 5.18, it follows that $\mathfrak{h} = \mathfrak{g}'$.

Turning to G', let $U = \{g_1 g_2 g_1^{-1} g_2^{-1} \mid g_i \in G\}$. Since U is the continuous image of $G \times G$ under the obvious map, U is connected. As $I \in U$, $I \in U^j$ and since $G' = \cup_j U^j$, G' is therefore connected.

Next, by the multiplicative nature of determinants and the definition of the commutator, it follows that $\pi_i G' \subseteq SU(n_i)$ so that $G' \subseteq H$. It only remains to see $H^0 \subseteq G'$ since G' is connected. For this, it suffices to show that G' contains a neighborhood of I in H by Theorem 1.15.

To this end, for $X, Y \in \mathfrak{h}$, define $c_{X,Y}(t) \in H \cap G', t \in \mathbb{R}$, by

$$c_{X,Y}(t) = \begin{cases} e^{\sqrt{t}X} e^{\sqrt{t}Y} e^{-\sqrt{t}X} e^{-\sqrt{t}Y} & t \geq 0 \\ e^{\sqrt{|t|}X} e^{-\sqrt{|t|}Y} e^{-\sqrt{|t|}X} e^{\sqrt{|t|}Y} & t < 0. \end{cases}$$

Using either Dynkin's Formula (Exercise 5.21, c.f. Exercises 4.26 and 5.16) or elementary power series calculations, it easily follows that $c_{X,Y}(t) = e^{t[X,Y] + O(|t|^{\frac{3}{2}})}$ for t near 0, so that $c_{X,Y}$ is continuously differentiable with $c_{X,Y}'(0) = [X, Y]$.

Let $\{[X_i, Y_i]\}_{i=1}^p$ be a basis for $\mathfrak{g}'$ and consider the map $c : \mathbb{R}^p \to H$ given by $c(t_1, \ldots, t_p) = \prod_i c_{X_i, Y_i}(t)$. As $c_{X,Y}'(0) = [X, Y]$, the differential of c at 0 is an isomorphism to $\mathfrak{h}$ (c.f. Exercise 4.12). Thus the image of c contains a neighborhood of I in H and, since $c(t)$ is also in G', the proof is finished. $\square$

5.2.4 Compact Lie Group Structure

Theorem 5.22. *(a) Let G be a compact connected Lie group. Then $G = G' Z(G)^0$, $Z(G') = G' \cap Z(G)$ is a finite Abelian group, $Z(G)^0$ is a torus, and*

$$G \cong [G' \times Z(G)^0]/F,$$

where the finite Abelian group $F = G' \cap Z(G)^0$ is embedded in $G' \times Z(G)^0$ as $\{(f, f^{-1}) \mid f \in F\}$.

(b) Decompose $\mathfrak{g}' = \mathfrak{s}_1 \oplus \cdots \oplus \mathfrak{s}_k$ into simple ideals as in Theorem 5.18 and let $S_i = \exp \mathfrak{s}_i$. Then S_i is a connected closed normal Lie subgroup of G' with Lie algebra $\mathfrak{s}_i$. The only proper closed normal Lie subgroups of S_i are discrete, finite, and central in G. Moreover, the map $(s_1, \ldots, s_k) \rightarrow s_i \cdots s_k$ from $S_1 \times \cdots \times S_k$ to G' is a surjective homomorphism with finite central kernel, F', so that

$$G' \cong [S_1 \times \cdots \times S_k]/F'.$$

Proof. For part (a), first note that G' is closed and therefore compact. It follows from the Maximal Torus Theorem that $G' = \exp \mathfrak{g}'$. Using the decomposition $\mathfrak{g} = \mathfrak{g}' \oplus \mathfrak{z}(\mathfrak{g})$ and the fact that the Lie algebra of $Z(G)$ is $\mathfrak{z}(\mathfrak{g})$, Theorems 5.1 and 5.2 show $\exp \mathfrak{g} = G'Z(G)^0$. Thus $G = G'Z(G)^0$. This relation also shows that $Z(G') = G' \cap Z(G)$.

Using the machinery from the proof Theorem 5.21, any $Z \in Z(G)$ must be of the form $Z = \text{diag}(c_1, \ldots, c_1, \ldots, c_k, \ldots, c_k)$. If also $Z \in G'$, then $c_1^{n_1} = \ldots, c_k^{n_k} = 1$, so that c_i is an n_i^{th}-root of unity. In particular, $G' \cap Z(G)$ is a finite Abelian group. Finally, consider the surjective homomorphism mapping $(g, z) \rightarrow gz$ from $G' \times Z(G)^0 \rightarrow G$. Clearly the kernel is $\{(f, f^{-1}) \mid f \in F\}$, as desired.

For part (b), the fact that $\mathfrak{s}_i$ and $\mathfrak{s}_j$, $i \neq j$, commute and $G' = \exp \mathfrak{g}'$ show $G' = S_1 \cdots S_k$ with S_i and S_j, $i \neq j$, commuting. It is necessary to verify that S_i is a closed Lie subgroup. To this end, write $\text{Ad}(g)|_{\mathfrak{s}_j}$ for $\text{Ad}(g)$ restricted to $\mathfrak{s}_j$ and let $K_i = \{g \in G' \mid \text{Ad}(g)|_{\mathfrak{s}_j} = I, j \neq i\}^0$. Obviously K_i is a connected closed Lie subgroup of G'. We will show that $K_i = S_i$.

Now $X \in \mathfrak{k}_i$ if and only if $e^{tX} \in K_i$, $t \in \mathbb{R}$, if and only if $\text{Ad}(e^{tX})|_{\mathfrak{s}_j} = e^{t \, \text{ad}(X)}|_{\mathfrak{s}_j} = I$ for $j \neq i$. Using $\frac{d}{dt}|_{t=0}$, $X \in \mathfrak{k}_i$ if and only if $[X, \mathfrak{s}_j] = 0$ for $j \neq i$. Since $\mathfrak{s}_j$ is an ideal with $\text{span}[\mathfrak{s}_j, \mathfrak{s}_j] = \mathfrak{s}_j$ and $[\mathfrak{s}_i, \mathfrak{s}_j] = 0$, decomposing X according to $\mathfrak{s}_1 \oplus \cdots \oplus \mathfrak{s}_k$ shows $\mathfrak{k}_i = \mathfrak{s}_i$. Thus $K_i = \exp \mathfrak{k}_i = S_i$ and in particular, S_i is a closed connected normal Lie subgroup with Lie algebra $\mathfrak{s}_i$.

If N is a normal Lie subgroup of S_i, then $c_s N = N$ for $s \in S_i$. Since $\text{Ad}(s)$ is the differential of c_s, $\text{Ad}(s)\mathfrak{n} = \mathfrak{n}$. Since ad is the differential of Ad, $\text{ad}(X)\mathfrak{n} \subseteq \mathfrak{n}$, $X \in \mathfrak{s}_i$, so that $\mathfrak{n}$ is an ideal (c.f. Exercise 4.23). By construction, this forces $\mathfrak{n}$ to be $\mathfrak{s}_i$ or $\{0\}$, so that $N = S_i$ or $N^0 = I$. Assuming further that N is proper and closed, therefore compact, N must be discrete and finite. Lemma 1.21 shows that N is central. Finally, the differential of the map $(s_1, \ldots, s_k) \rightarrow s_i \cdots s_k$ is obviously the identity map, so that F' is discrete and normal. As above, this shows that F' is finite and central as well. $\square$

The effect of Theorem 5.22 is to reduce the study of connected compact groups to the study of connected compact groups with simple Lie algebras.

5.2.5 Exercises

Exercise 5.14 Let $X \in \mathfrak{gl}(n, \mathbb{C})$ be diagonalizable with eigenvalues $\{\lambda_i\}_{i=1}^n$. Show that $\text{ad}(X)$ has eigenvalues $\{\lambda_i - \lambda_j\}_{i,j=1}^n$.

Exercise 5.15 Show that $\ln(I + X) = \sum_{n=1}^{\infty} \frac{(-1)^{n+1}}{n} X^n$ is well defined for X in a neighborhood of I in $GL(n, \mathbb{C})$. On that neighborhood, show that $\ln X$ is the inverse function to e^X.

Exercise 5.16 Let G be a Lie subgroup of $GL(n, \mathbb{C})$. For $X, Y \in \mathfrak{g}$ in a sufficiently small neighborhood of 0, show that

$$e^X e^Y = e^{X+Y+\frac{1}{2}[X,Y]+\frac{1}{12}[X,[X,Y]]+\frac{1}{12}[Y,[Y,X]]+\frac{1}{24}[Y,[X,[Y,X]]]+\cdots}.$$

Exercise 5.17 Let G be a Lie subgroup of $GL(n, \mathbb{C})$. For $X, Y \in \mathfrak{g}$ in a sufficiently small neighborhood of 0, write $e^X e^Y = e^Z$. Modify the proof of Dynkin's Formula to show Z is also given by the formula

$$Z = \sum \frac{(-1)^n}{n+1} \frac{1}{i_1 + \cdots + i_n + 1} \frac{[X^{(i_1)}, Y^{(j_1)}, \ldots, X^{(i_n)}, Y^{(j_n)}, X]}{i_1! j_1! \cdots i_n! j_n!}$$

by starting with $e^{Z(t)} = e^{tX} e^Y$.

Exercise 5.18 Let $X, X_i \in \mathfrak{gl}(n, \mathbb{C})$ be strictly upper triangular. Show that $X_n \cdots X_2 X_1 = 0$, so that $[X_n, \ldots [X_3, [X_2, X_1]] \ldots] = 0$ and e^X is a polynomial in X.

Exercise 5.19 Let N_i be connected Lie subgroups of $GL(n, \mathbb{C})$ whose Lie algebras $\mathfrak{n}_i$ lie in the set of strictly upper triangular matrices. Suppose $\psi : \mathfrak{n}_1 \to \mathfrak{n}_2$ is a linear map. Show ψ descends to a well-defined homomorphism of groups $\varphi : N_1 \to N_2$ by $\varphi(e^X) = e^{\psi X}$ if and only if ψ is a Lie algebra homomorphism.

Exercise 5.20 For $X, Y \in \mathfrak{gl}(n, \mathbb{C})$, show that $\operatorname{tr} XY = \operatorname{tr} YX$.

Exercise 5.21 In the proof of Theorem 5.21, verify that $c_{X,Y}(t) = e^{t[X,Y]+O(|t|^{\frac{3}{2}})}$.

Exercise 5.22 (1) Take advantage of diagonalization to show directly that $U(n)' = SU(n)$.
(2) Show that $GL(n, \mathbb{F})' = SL(n, \mathbb{F})$.

Exercise 5.23 For a Lie subgroup $G \subseteq U(n)$, show that the differential of the determinant $\det : G \to S^1$ is the trace.

Exercise 5.24 (1) Let G be a Lie subgroup of $GL(n, \mathbb{C})$. Show that G' is the smallest normal subgroup subgroup of G whose quotient group in G is commutative.
(2) Show that $\mathfrak{g}'$ is the smallest ideal of $\mathfrak{g}$ whose quotient algebra in $\mathfrak{g}$ is commutative.

Exercise 5.25 Let G be a compact connected Lie group and write $G = S_1 \cdots S_k Z(G)^0$ as in Theorem 5.22. Show that any closed normal Lie subgroup of G is a product of some of the S_i with a central subgroup.

6

Roots and Associated Structures

By examining the joint eigenvalues of a Cartan subalgebra under the ad-action, a great deal of information about a Lie group and its Lie algebra may be encoded. For instance, the fundamental group can be read off from this data (§6.3.3). Moreover, this encoding is a key step in the classification of irreducible representations (§7).

6.1 Root Theory

6.1.1 Representations of Lie Algebras

Definition 6.1. (a) Let $\mathfrak{g}$ be the Lie algebra of a Lie subgroup of $GL(n, \mathbb{C})$. A *representation* of $\mathfrak{g}$ is a pair (ψ, V), where V is a finite-dimensional complex vector space and ψ is a linear map $\psi : \mathfrak{g} \to \text{End}(V)$, satisfying $\psi([X, Y]) = \psi(X) \circ \psi(Y) - \psi(Y) \circ \psi(X)$ for $X, Y \in \mathfrak{g}$.
(b) The representation (ψ, V) is said to be *irreducible* if there are no proper $\psi(\mathfrak{g})$-invariant subspaces, i.e., the only $\psi(\mathfrak{g})$-invariant subspaces are $\{0\}$ and V. Otherwise (ψ, V) is called *reducible*.

As with group representations, a Lie algebra representation (ψ, V) may simply be written as ψ or as V when no ambiguity can arise. Also similar to the group case, $\psi(X)v, v \in V$, may be denoted by $X \cdot v$ or by Xv.

It should be noted that if V is m-dimensional, a choice of basis allows us to view a representation of $\mathfrak{g}$ as a homomorphism $\psi : \mathfrak{g} \to \mathfrak{gl}(m, \mathbb{C})$, i.e., ψ is linear and satisfies $\psi[X, Y] = [\psi X, \psi Y]$. We will often make this identification without comment.

Theorem 6.2. (a) Let G be a Lie subgroup of $GL(n, \mathbb{C})$ and (π, V) a finite-dimensional representation of G. Then $(d\pi, V)$ is a representation of $\mathfrak{g}$ satisfying $e^{d\pi X} = \pi(e^X)$, where the differential of π is given by $d\pi(X) = \frac{d}{dt}\pi(e^{tX})|_{t=0}$ for $X \in \mathfrak{g}$. If G is connected, π is completely determined by $d\pi$.
(b) For connected G, a subspace $W \subseteq V$ is $\pi(G)$-invariant if and only if it is $d\pi(\mathfrak{g})$-invariant. In particular, V is irreducible under G if and only if it is irreducible under $\mathfrak{g}$.

(c) For connected compact G, V is irreducible if and only if the only endomorphisms of V commuting with all the operators $d\pi(\mathfrak{g})$ are scalar multiples of the identity map.

Proof. Part (a) follows immediately from Theorem 4.8 by looking at the homomorphism $\pi : G \to GL(V)$ and choosing a basis for V. Part (b) follows from the relation $e^{d\pi X} = \pi(e^X)$, the definition of $d\pi$, and the fact that $\exp \mathfrak{g}$ generates G. For part (c), let $T \in \mathrm{End}(V)$ and embed G in $GL(n, \mathbb{C})$. Observe that $[T, d\pi X] = 0$ if and only if $e^{t\,\mathrm{ad}(d\pi X)}T = T$, $t \in \mathbb{R}$, if and only if $\mathrm{Ad}(e^{t d\pi X})T = T$ if and only if T commutes with $e^{t d\pi X} = \pi(e^{tX})$. Using the fact that G is connected, part (c) follows from part (b) and Schur's Lemma.   □

As an example, let G be a Lie subgroup of $GL(n, \mathbb{C})$ and let $(\pi, \mathbb{C})$ be the trivial representation of G. Then $d\pi = 0$. This representation of $\mathfrak{g}$ is called the *trivial representation*.

As a second example, let G be a Lie subgroup of $GL(n, \mathbb{C})$ and let $(\pi, \mathbb{C}^n)$ be the standard representation. Then $d\pi(X)v = Xv$ for $v \in \mathbb{C}^n$. This representation is called the *standard representation*. In the cases of G equal to $GL(n, \mathbb{F})$, $SL(n, \mathbb{F})$, $U(n)$, $SU(n)$, or $SO(n)$, the standard representation is known to be irreducible on the Lie group level (§2.2.2), so that each is irreducible on the Lie algebra level.

As a last example, consider the representation $V_n(\mathbb{C}^2)$ of $SU(2)$ from §2.1.2.2 given by

$$\begin{pmatrix} a & -\overline{b} \\ b & \overline{a} \end{pmatrix} \cdot z_1^k z_2^{n-k} = (\overline{a}z_1 + \overline{b}z_2)^k (-bz_1 + az_2)^{n-k}.$$

From §4.1.3, $\mathfrak{su}(2) = \{X = \begin{pmatrix} ix & -\overline{w} \\ w & -ix \end{pmatrix} \mid x \in \mathbb{R}, w \in \mathbb{C}\}$. Using either power series calculations or Corollary 4.9, $\exp tX = (\cos \lambda t) I + \left(\frac{1}{\lambda} \sin \lambda t\right) X$ where $\lambda = \sqrt{\det X}$ (Exercise 6.2). It follows that the Lie algebra acts by

$$X \cdot (z_1^k z_2^{n-k}) = \frac{d}{dt}\left(\begin{pmatrix} \cos \lambda t + \frac{ix}{\lambda} \sin \lambda t & -\frac{\overline{w}}{\lambda} \sin \lambda t \\ \frac{w}{\lambda} \sin \lambda t & \cos \lambda t - \frac{ix}{\lambda} \sin \lambda t \end{pmatrix} \cdot z_1^k z_2^{n-k}\right)|_{t=0}$$

$$= k\,(-ixz_1 + \overline{w}z_2)\, z_1^{k-1} z_2^{n-k} + (n-k)\,(-wz_1 + ixz_2)\, z_1^k z_2^{n-k-1}$$

(6.3) $$= k\overline{w}\, z_1^{k-1} z_2^{n-k+1} + i\,(n-2k)\, x \cdot z_1^k z_2^{n-k} + (k-n)w\, z_1^{k+1} z_2^{n-k-1}.$$

It is easy to use Equation 6.3 and Theorem 6.2 to show that $V_n(\mathbb{C}^2)$ is irreducible. In fact, this is the idea underpinning the argument given in §2.1.2.2.

As in the case of representations of Lie groups, new Lie algebra representations can be created using linear algebra. It is straightforward to verify (Exercise 6.1) that differentials of the Lie group representations listed in Definition 2.10 yield the following Lie algebra representations.

Definition 6.4. Let V and W be representations of a Lie algebra $\mathfrak{g}$ of a Lie subgroup of $GL(n, \mathbb{C})$.

(1) $\mathfrak{g}$ acts on $V \oplus W$ by $X(v, w) = (Xv, Xw)$.

(2) $\mathfrak{g}$ acts on $V \otimes W$ by $X \sum v_i \otimes w_j = \sum X v_i \otimes w_j + \sum v_i \otimes X w_j$.

(3) $\mathfrak{g}$ acts on $\mathrm{Hom}(V, W)$ by $(XT)(v) = XT(v) - T(Xv)$.

(4) $\mathfrak{g}$ acts on $\bigotimes^k V$ by $X \sum v_{i_1} \otimes \cdots v_{i_k} = \sum (X v_{i_1}) \otimes \cdots v_{i_k} + \cdots \sum v_{i_1} \otimes \cdots (X v_{i_k})$.

(5) $\mathfrak{g}$ acts on $\bigwedge^k V$ by $X \sum v_{i_1} \wedge \cdots v_{i_k} = \sum (X v_{i_1}) \wedge \cdots v_{i_k} + \cdots \sum v_{i_1} \wedge \cdots (X v_{i_k})$.

(6) $\mathfrak{g}$ acts on $S^k(V)$ by $X \sum v_{i_1} \cdots v_{i_k} = \sum (X v_{i_1}) \cdots v_{i_k} + \cdots \sum v_{i_1} \cdots (X v_{i_k})$.

(7) $\mathfrak{g}$ acts on V^* by $(XT)(v) = -T(Xv)$.

(8) $\mathfrak{g}$ acts on $\overline{V}$ by the same action as it does on V.

6.1.2 Complexification of Lie Algebras

Definition 6.5. **(a)** Let $\mathfrak{g}$ be the Lie algebra of a Lie subgroup of $GL(n, \mathbb{C})$. The *complexification* of $\mathfrak{g}$, $\mathfrak{g}_\mathbb{C}$, is defined as $\mathfrak{g}_\mathbb{C} = \mathfrak{g} \otimes_\mathbb{R} \mathbb{C}$. The Lie bracket on $\mathfrak{g}$ is extended to $\mathfrak{g}_\mathbb{C}$ by $\mathbb{C}$-linearity.

(b) If (ψ, V) is a representation of $\mathfrak{g}$, extend the domain of ψ to $\mathfrak{g}_\mathbb{C}$ by $\mathbb{C}$-linearity. Then (ψ, V) is said to be *irreducible* under $\mathfrak{g}_\mathbb{C}$ if there are no proper $\psi(\mathfrak{g}_\mathbb{C})$-invariant subspaces.

Writing a matrix in terms of its skew-Hermitian and Hermitian parts, observe that $\mathfrak{gl}(n, \mathbb{C}) = \mathfrak{u}(n) \oplus i\mathfrak{u}(n)$. It follows that if $\mathfrak{g}$ is the Lie algebra of a compact Lie group G realized with $G \subseteq U(n)$, $\mathfrak{g}_\mathbb{C}$ may be identified with $\mathfrak{g} \oplus i\mathfrak{g}$ equipped with the standard Lie bracket inherited from $\mathfrak{gl}(n, \mathbb{C})$ (Exercise 6.3). We will often make this identification without comment. In particular, $\mathfrak{u}(n)_\mathbb{C} = \mathfrak{gl}(n, \mathbb{C})$. Similarly, $\mathfrak{su}(n)_\mathbb{C} = \mathfrak{sl}(n, \mathbb{C})$, $\mathfrak{so}(n)_\mathbb{C}$ is realized by

$$\mathfrak{so}(n, \mathbb{C}) = \{X \in \mathfrak{sl}(n, \mathbb{C}) \mid X^t = -X\},$$

and, realizing $\mathfrak{sp}(n)$ as $\mathfrak{u}(2n) \cap \mathfrak{sp}(n, \mathbb{C})$ as in §4.1.3, $\mathfrak{sp}(n)_\mathbb{C}$ is realized by $\mathfrak{sp}(n, \mathbb{C})$ (Exercise 6.3).

Lemma 6.6. *Let $\mathfrak{g}$ be the Lie algebra of a Lie subgroup of $GL(n, \mathbb{C})$ and let (ψ, V) be a representation of $\mathfrak{g}$. Then V is irreducible under $\mathfrak{g}$ if and only if it is irreducible under $\mathfrak{g}_\mathbb{C}$.*

Proof. Simply observe that since a subspace $W \subseteq V$ is a complex subspace, W is $\psi(\mathfrak{g})$-invariant if and only if it is $\psi(\mathfrak{g}_\mathbb{C})$-invariant. □

For example, $\mathfrak{su}(2)_\mathbb{C} = \mathfrak{sl}(2, \mathbb{C})$ is equipped with the *standard basis*

$$E = \begin{pmatrix} 0 & 1 \\ 0 & 0 \end{pmatrix}, \quad H = \begin{pmatrix} 1 & 0 \\ 0 & -1 \end{pmatrix}, \quad F = \begin{pmatrix} 0 & 0 \\ 1 & 0 \end{pmatrix}$$

(c.f. Exercise 4.21). Since $E = \frac{1}{2} \begin{pmatrix} 0 & 1 \\ -1 & 0 \end{pmatrix} - \frac{i}{2} \begin{pmatrix} 0 & i \\ i & 0 \end{pmatrix}$, Equation 6.3 shows that the resulting action of E on $V_n(\mathbb{C}^2)$ is given by

$$E \cdot (z_1^k z_2^{n-k}) = \frac{1}{2} \left[-k \, z_1^{k-1} z_2^{n-k+1} - (k-n) \, z_1^{k+1} z_2^{n-k-1} \right]$$
$$- \frac{i}{2} \left[-ik \, z_1^{k-1} z_2^{n-k+1} + i(k-n) \, z_1^{k+1} z_2^{n-k-1} \right]$$
$$= -k \, z_1^{k-1} z_2^{n-k+1}.$$

Similarly (Exercise 6.4), the action of H and F on $V_n(\mathbb{C}^2)$ is given by

(6.7)
$$H \cdot (z_1^k z_2^{n-k}) = (n-2k) \, z_1^k z_2^{n-k}$$
$$F \cdot (z_1^k z_2^{n-k}) = (k-n) \, z_1^{k+1} z_2^{n-k-1}.$$

Irreducibility of $V_n(\mathbb{C}^2)$ is immediately apparent from these formulas (Exercise 6.7).

6.1.3 Weights

Let G be a compact Lie group and (π, V) a finite-dimensional representation of G. Fix a Cartan subalgebra $\mathfrak{t}$ of $\mathfrak{g}$ and write $\mathfrak{t}_\mathbb{C}$ for its complexification. By Theorem 5.6, there exists an inner product, $(\cdot, \cdot)$, on V that is G-invariant and for which $d\pi$ is skew-Hermitian on $\mathfrak{g}$ and is Hermitian on $i\mathfrak{g}$. Thus $\mathfrak{t}_\mathbb{C}$ acts on V as a family of commuting normal operators and so V is simultaneously diagonalizable under the action of $\mathfrak{t}_\mathbb{C}$. In particular, the following definition is well defined.

Definition 6.8. Let G be a compact Lie group, (π, V) a finite-dimensional representation of G, and $\mathfrak{t}$ a Cartan subalgebra of $\mathfrak{g}$. There is a finite set $\Delta(V) = \Delta(V, \mathfrak{t}_\mathbb{C}) \subseteq \mathfrak{t}_\mathbb{C}^*$, called the *weights* of V, so that

$$V = \bigoplus_{\alpha \in \Delta(V)} V_\alpha,$$

where

$$V_\alpha = \{v \in V \mid d\pi(H)v = \alpha(H)v, \ H \in \mathfrak{t}_\mathbb{C}\}$$

is nonzero. The above displayed equation is called the *weight space decomposition* of V with respect to $\mathfrak{t}_\mathbb{C}$.

As an example, take $G = SU(2)$, $V = V_n(\mathbb{C}^2)$, and $\mathfrak{t}$ to be the diagonal matrices in $\mathfrak{su}(2)$. Define $\alpha_m \in \mathfrak{t}_\mathbb{C}^*$ by requiring $\alpha_m(H) = m$. Then Equation 6.7 shows that the weight space decomposition for $V_n(\mathbb{C}^2)$ is $V_n(\mathbb{C}^2) = \bigoplus_{k=0}^n V_n(\mathbb{C}^2)_{\alpha_{n-2k}}$, where $V_n(\mathbb{C}^2)_{\alpha_{n-2k}} = \mathbb{C} z_1^k z_2^{n-k}$.

Theorem 6.9. *(a) Let G be a compact Lie group, (π, V) a finite-dimensional representation of G, T a maximal torus of G, and $V = \bigoplus_{\alpha \in \Delta(V, \mathfrak{t}_\mathbb{C})} V_\alpha$ the weight space decomposition. For each weight $\alpha \in \Delta(V)$, α is purely imaginary on $\mathfrak{t}$ and is real valued on $i\mathfrak{t}$.*
(b) For $t \in T$, choose $H \in \mathfrak{t}$ so that $e^H = t$. Then $t v_\alpha = e^{\alpha(H)} v_\alpha$ for $v_\alpha \in V_\alpha$.

Proof. Part (a) follows from the facts that $d\pi$ is skew-Hermitian on t and is Hermitian on it. Part (b) follows from the fact that $\exp t = T$ and the relation $e^{d\pi H} = \pi(e^H)$. □

By $\mathbb{C}$-linearity, $\alpha \in \Delta(V)$ is completely determined by its restriction to either t or it. Thus we permit ourselves to interchangeably view α as an element of any of the dual spaces $t_{\mathbb{C}}^*$, $(it)^*$ (real valued), or t^* (purely imaginary valued). In alternate notation (not used in this text), it is sometimes written $t_{\mathbb{C}}(\mathbb{R})$.

6.1.4 Roots

Let G be a compact Lie group. For $g \in G$, extend the domain of $\mathrm{Ad}(g)$ from $\mathfrak{g}$ to $\mathfrak{g}_{\mathbb{C}}$ by $\mathbb{C}$-linearity. Then $(\mathrm{Ad}, \mathfrak{g}_{\mathbb{C}})$ is a representation of G with differential given by ad (extended by $\mathbb{C}$-linearity). It has a weight space decomposition

$$\mathfrak{g}_{\mathbb{C}} = \bigoplus_{\alpha \in \Delta(\mathfrak{g}_{\mathbb{C}}, t_{\mathbb{C}})} \mathfrak{g}_\alpha$$

that is important enough to warrant its own name. Notice the zero weight space is $\mathfrak{g}_0 = \{Z \in \mathfrak{g}_{\mathbb{C}} \mid [H, Z] = 0, H \in t_{\mathbb{C}}\}$. Thus

$$\mathfrak{g}_0 = t_{\mathbb{C}}$$

since t is a maximal Abelian subspace of $\mathfrak{g}$. In the definition below, it turns out to be advantageous to separate this zero weight space from the remaining nonzero weight spaces.

Definition 6.10. Let G be a compact Lie group and t a Cartan subalgebra of $\mathfrak{g}$. There is a finite set of *nonzero* elements $\Delta(\mathfrak{g}_{\mathbb{C}}) = \Delta(\mathfrak{g}_{\mathbb{C}}, t_{\mathbb{C}}) \subseteq t_{\mathbb{C}}^*$, called the *roots* of $\mathfrak{g}_{\mathbb{C}}$, so that

$$\mathfrak{g}_{\mathbb{C}} = t_{\mathbb{C}} \oplus \bigoplus_{\alpha \in \Delta(\mathfrak{g}_{\mathbb{C}})} \mathfrak{g}_\alpha,$$

where $\mathfrak{g}_\alpha = \{Z \in \mathfrak{g}_{\mathbb{C}} \mid [H, Z] = \alpha(H)Z, H \in t_{\mathbb{C}}\}$ is nonzero. The above displayed equation is called the *root space decomposition* of $\mathfrak{g}_{\mathbb{C}}$ with respect to $t_{\mathbb{C}}$.

Theorem 6.11. *(a) Let G be a compact Lie group, (π, V) a finite-dimensional representation of G, and t a Cartan subalgebra of $\mathfrak{g}$. For $\alpha \in \Delta(\mathfrak{g}_{\mathbb{C}})$ and $\beta \in \Delta(V)$, $d\pi(\mathfrak{g}_\alpha)V_\beta \subseteq V_{\alpha+\beta}$.*
(b) In particular for $\alpha, \beta \in \Delta(\mathfrak{g}_{\mathbb{C}}) \cup \{0\}$, $[\mathfrak{g}_\alpha, \mathfrak{g}_\beta] \subseteq \mathfrak{g}_{\alpha+\beta}$.
(c) Let $(\cdot, \cdot)$ be an $\mathrm{Ad}(G)$-invariant inner product on $\mathfrak{g}_{\mathbb{C}}$. For $\alpha, \beta \in \Delta(\mathfrak{g}_{\mathbb{C}}) \cup \{0\}$, $(\mathfrak{g}_\alpha, \mathfrak{g}_\beta) = 0$ when $\alpha + \beta \neq 0$.
(d) If $\mathfrak{g}$ has trivial center (i.e., if $\mathfrak{g}$ is semisimple), then $\Delta(\mathfrak{g}_{\mathbb{C}})$ spans $t_{\mathbb{C}}^*$.

Proof. For part (a), let $H \in t_{\mathbb{C}}$, $X_\alpha \in \mathfrak{g}_\alpha$, and $v_\beta \in V_\beta$ and calculate

$$d\pi(H)d\pi(X_\alpha)v_\beta = (d\pi(X_\alpha)d\pi(H) + [d\pi(H), d\pi(X_\alpha)]) \, v_\beta$$
$$= (d\pi(X_\alpha)d\pi(H) + d\pi\,[H, X_\alpha]) \, v_\beta$$
$$= (d\pi(X_\alpha)d\pi(H) + \alpha(H)d\pi\,(X_\alpha)) \, v_\beta$$
$$= (\beta(H) + \alpha(H)) \, d\pi(X_\alpha)v_\beta,$$

so that $d\pi(X_\alpha)v_\beta \in V_{\alpha+\beta}$ as desired. Part (b) clearly follows from part (a).

For part (c), recall that Lemma 5.6 shows that ad is skew-Hermitian. Thus

$$\alpha(H)(X_\alpha, X_\beta) = ([H, X_\alpha], X_\beta) = -(X_\alpha, [H, X_\beta]) = -\beta(H)(X_\alpha, X_\beta).$$

For part (d), suppose $H \in \mathfrak{t}_\mathbb{C}$ satisfies $\alpha(H) = 0$ for all $\alpha \in \Delta(\mathfrak{g}_\mathbb{C})$. It suffices to show that $H = 0$. However, the condition $\alpha(H) = 0$ for all $\alpha \in \Delta(\mathfrak{g}_\mathbb{C})$ is equivalent to saying that H is central in $\mathfrak{g}_\mathbb{C}$. Since it is easy to see that $\mathfrak{z}(\mathfrak{g}_\mathbb{C}) = \mathfrak{z}(\mathfrak{g})_\mathbb{C}$ (Exercise 6.5), it follows from semisimplicity and Theorem 5.18 that $H = 0$.   □

In §6.2.3 we will further see that $\dim \mathfrak{g}_\alpha = 1$ for $\alpha \in \Delta(\mathfrak{g}_\mathbb{C})$ and that the only multiples of α in $\Delta(\mathfrak{g}_\mathbb{C})$ are $\pm\alpha$.

6.1.5 Compact Classical Lie Group Examples

The root space decomposition for the complexification of the Lie algebra of each compact classical Lie group is given below. The details are straightforward to verify (Exercise 6.10).

6.1.5.1 $\mathfrak{su}(n)$ For $G = U(n)$ with $\mathfrak{t} = \{\mathrm{diag}(i\theta_1, \ldots, i\theta_n) \mid \theta_i \in \mathbb{R}\}$, $\mathfrak{g}_\mathbb{C} = \mathfrak{gl}(n, \mathbb{C})$ and $\mathfrak{t}_\mathbb{C} = \{\mathrm{diag}(z_1, \ldots, z_n) \mid z_i \in \mathbb{C}\}$. For $G = SU(n)$ with $\mathfrak{t} = \{\mathrm{diag}(i\theta_1, \ldots, i\theta_n) \mid \theta_i \in \mathbb{R}, \sum_i \theta_i = 0\}$, $\mathfrak{g}_\mathbb{C} = \mathfrak{sl}(n, \mathbb{C})$ and $\mathfrak{t}_\mathbb{C} = \{\mathrm{diag}(z_1, \ldots, z_n) \mid z_i \in \mathbb{C}, \sum_i z_i = 0\}$. In either case, it is straightforward to check that the set of roots is given by

$$\Delta(\mathfrak{g}_\mathbb{C}) = \{\pm(\epsilon_i - \epsilon_j) \mid 1 \le i < j \le n\},$$

where $\epsilon_i(\mathrm{diag}(z_1, \ldots, z_n)) = z_i$. In the theory of Lie algebras, this root system is called A_{n-1}. The corresponding root space is

$$\mathfrak{g}_{\epsilon_i - \epsilon_j} = \mathbb{C}E_{i,j},$$

where $\{E_{i,j}\}$ is the standard basis for $n \times n$ matrices.

6.1.5.2 $\mathfrak{sp}(n)$ For $G = Sp(n)$ realized as $Sp(n) \cong U(2n) \cap Sp(n, \mathbb{C})$ with $\mathfrak{t} = \{\mathrm{diag}(i\theta_1, \ldots, i\theta_n, -i\theta_1, \ldots, -i\theta_n) \mid \theta_i \in \mathbb{R}\}$, $\mathfrak{g}_\mathbb{C} = \mathfrak{sp}(n, \mathbb{C})$ and $\mathfrak{t}_\mathbb{C} = \{\mathrm{diag}(z_1, \ldots, z_n, -z_1, \ldots, -z_n) \mid z_i \in \mathbb{C}\}$. Then

$$\Delta(\mathfrak{g}_\mathbb{C}) = \{\pm(\epsilon_i - \epsilon_j) \mid 1 \le i < j \le n\} \cup \{\pm(\epsilon_i + \epsilon_j) \mid 1 \le i \le j \le n\},$$

where $\epsilon_i(\mathrm{diag}(z_1, \ldots, z_n, -z_1, \ldots, -z_n)) = z_i$. In the theory of Lie algebras, this root system is called C_n. The corresponding root spaces are

$$\mathfrak{g}_{\epsilon_i - \epsilon_j} = \mathbb{C}\left(E_{i,j} - E_{j+n,i+n}\right)$$

$$\mathfrak{g}_{\epsilon_i + \epsilon_j} = \mathbb{C}\left(E_{i,j+n} + E_{j,i+n}\right), \quad \mathfrak{g}_{-\epsilon_i - \epsilon_j} = \mathbb{C}\left(E_{i+n,j} + E_{j+n,i}\right)$$

$$\mathfrak{g}_{2\epsilon_i} = \mathbb{C}E_{i,i+n}, \quad \mathfrak{g}_{-2\epsilon_i} = \mathbb{C}E_{i+n,i}.$$

6.1.5.3 $\mathfrak{so}(E_n)$ For $SO(n)$, it turns out that the root space decomposition is a bit messy (see Exercise 6.14 for details). The results are much cleaner if we diagonalize by making a change of variables. In other words, we will examine an isomorphic copy of $SO(n)$ instead of $SO(n)$ itself. Define

$$T_{2m} = \frac{1}{\sqrt{2}} \begin{pmatrix} I_m & I_m \\ iI_m & -iI_m \end{pmatrix}, \quad E_{2m} = \begin{pmatrix} 0 & I_m \\ I_m & 0 \end{pmatrix},$$

$$T_{2m+1} = \begin{pmatrix} T_{2m} & 0 \\ 0 & 1 \end{pmatrix}, \quad E_{2m+1} = \begin{pmatrix} E_{2m} & 0 \\ 0 & 1 \end{pmatrix},$$

$$SO(E_n) = \{g \in SL(n, \mathbb{C}) \mid \overline{g} = E_n g E_n, \ g^t E_n g = E_n\},$$

$$\mathfrak{so}(E_n) = \{X \in \mathfrak{gl}(n, \mathbb{C}) \mid \overline{X} = E_n X E_n, \ X^t E_n + E_n X = 0\},$$

$$\mathfrak{so}(E_n, \mathbb{C}) = \{X \in \mathfrak{gl}(n, \mathbb{C}) \mid X^t E_n + E_n X = 0\}.$$

Notice that $E_n = T_n^t T_n$ and $\overline{T}_n = T_n^{-1, t}$. The following lemma is straightforward and left as an exercise (Exercise 6.12).

Lemma 6.12. *(a) $SO(E_n)$ is a compact Lie subgroup of $SU(n)$ with Lie algebra $\mathfrak{so}(E_n)$ and with complexified Lie algebra $\mathfrak{so}(E_n, \mathbb{C})$.*
(b) The map $g \to T_n^{-1} g T_n$ induces an isomorphism of Lie groups $SO(n) \cong SO(E_n)$.
(c) The map $X \to T_n^{-1} X T_n$ induces an isomorphism of Lie algebras $\mathfrak{so}(n) \cong \mathfrak{so}(E_n)$ and $\mathfrak{so}(n, \mathbb{C}) \cong \mathfrak{so}(E_n, \mathbb{C})$.
(d) For $n = 2m$, a maximal torus is given by

$$T = \{\mathrm{diag}(e^{i\theta_1}, \ldots, e^{i\theta_m}, e^{-i\theta_1}, \ldots, e^{-i\theta_m}) \mid \theta_i \in \mathbb{R}\}$$

with corresponding Cartan subalgebra

$$\mathfrak{t} = \{\mathrm{diag}(i\theta_1, \ldots, i\theta_m, -i\theta_1, \ldots, -i\theta_m) \mid \theta_i \subset \mathbb{R}\}$$

and complexification

$$\mathfrak{t}_{\mathbb{C}} = \{\mathrm{diag}(z_1, \ldots, z_m, -z_1, \ldots, -z_m) \mid z_i \in \mathbb{C}\}.$$

(e) For $n = 2m + 1$, a maximal torus is given by

$$T = \{\mathrm{diag}(e^{i\theta_1}, \ldots, e^{i\theta_m}, e^{-i\theta_1}, \ldots, e^{-i\theta_m}, 1) \mid \theta_i \in \mathbb{R}\}$$

with corresponding Cartan subalgebra

$$\mathfrak{t} = \{\mathrm{diag}(i\theta_1, \ldots, i\theta_m, -i\theta_1, \ldots, -i\theta_m, 0) \mid \theta_i \in \mathbb{R}\}$$

and complexification

$$\mathfrak{t}_{\mathbb{C}} = \{\mathrm{diag}(z_1, \ldots, z_m, -z_1, \ldots, -z_m, 0) \mid z_i \in \mathbb{C}\}.$$

6.1.5.4 $\mathfrak{so}(2n)$ Working with $G = SO(E_{2n})$ and the Cartan subalgebra from Lemma 6.12, the set of roots is

$$\Delta(\mathfrak{g}_{\mathbb{C}}) = \{\pm(\epsilon_i \pm \epsilon_j) \mid 1 \leq i < j \leq n\},$$

where $\epsilon_i(\mathrm{diag}(z_1, \ldots, z_n, -z_1, \ldots, -z_n)) = z_i$. In the theory of Lie algebras, this root system is called D_n. The corresponding root spaces are

$$\mathfrak{g}_{\epsilon_i-\epsilon_j} = \mathbb{C}\left(E_{i,j} - E_{j+n,i+n}\right), \quad \mathfrak{g}_{-\epsilon_i+\epsilon_j} = \mathbb{C}\left(E_{j,i} - E_{i+n,j+n}\right)$$

$$\mathfrak{g}_{\epsilon_i+\epsilon_j} = \mathbb{C}\left(E_{i,j+n} - E_{j,i+n}\right), \quad \mathfrak{g}_{-\epsilon_i-\epsilon_j} = \mathbb{C}\left(E_{i+n,j} - E_{j+n,i}\right).$$

6.1.5.5 $\mathfrak{so}(2n+1)$ Working with $G = SO(E_{2n+1})$ and with the Cartan subalgebra from Lemma 6.12, the set of roots is

$$\Delta(\mathfrak{g}_{\mathbb{C}}) = \{\pm(\epsilon_i \pm \epsilon_j) \mid 1 \leq i < j \leq n\} \cup \{\pm\epsilon_i \mid 1 \leq i \leq n\},$$

where $\epsilon_i(\mathrm{diag}(z_1, \ldots, z_n, -z_1, \ldots, -z_n, 0)) = z_i$. In the theory of Lie algebras, this root system is called B_n. The corresponding root spaces are

$$\mathfrak{g}_{\epsilon_i-\epsilon_j} = \mathbb{C}\left(E_{i,j} - E_{j+n,i+n}\right), \quad \mathfrak{g}_{-\epsilon_i+\epsilon_j} = \mathbb{C}\left(E_{j,i} - E_{i+n,j+n}\right)$$

$$\mathfrak{g}_{\epsilon_i+\epsilon_j} = \mathbb{C}\left(E_{i,j+n} - E_{j,i+n}\right), \quad \mathfrak{g}_{-\epsilon_i-\epsilon_j} = \mathbb{C}\left(E_{i+n,j} - E_{j+n,i}\right).$$

$$\mathfrak{g}_{\epsilon_i} = \mathbb{C}\left(E_{i,2n+1} - E_{2n+1,i+n}\right), \quad \mathfrak{g}_{-\epsilon_i} = \mathbb{C}\left(E_{i+n,2n+1} - E_{2n+1,i}\right).$$

6.1.6 Exercises

Exercise 6.1 Verify that the differentials of the actions given in Definition 2.10 give rise to the actions given in Definition 6.4.

Exercise 6.2 For $X = \begin{pmatrix} ix & z \\ -\bar{z} & -ix \end{pmatrix}$, $x \in \mathbb{R}$ and $z \in \mathbb{C}$, let $\lambda = \sqrt{x^2 + |z|^2}$. Show that $\exp X = (\cos \lambda) I + \frac{\sin \lambda}{\lambda} X$.

Exercise 6.3 (1) Show that $\mathfrak{gl}(n, \mathbb{C}) = \mathfrak{u}(n) \oplus i\mathfrak{u}(n)$.
(2) Suppose $\mathfrak{g}$ is the Lie algebra of a compact Lie group G with G a Lie subgroup of $U(n)$. Show that there is an isomorphism of algebras $\mathfrak{g} \otimes_{\mathbb{R}} \mathbb{C} \cong \mathfrak{g} \oplus i\mathfrak{g}$ induced by mapping $X \otimes (a + ib)$ to $aX + ibX$ for $X \in \mathfrak{g}$ and $a, b \in \mathbb{R}$.
(3) Show that $\mathfrak{su}(n)_{\mathbb{C}} = \mathfrak{sl}(n, \mathbb{C})$ and that $\mathfrak{so}(n)_{\mathbb{C}} = \mathfrak{so}(n, \mathbb{C})$.
(4) Show that $\mathfrak{sp}(n)_{\mathbb{C}} \cong \mathfrak{sp}(n, \mathbb{C})$ and that

$$\mathfrak{sp}(n, \mathbb{C}) = \{\begin{pmatrix} X & Y \\ Z & -X^t \end{pmatrix} \mid X, Y, Z \in \mathfrak{gl}(n, \mathbb{C}), Y^t = Y, Z^t = Z\}.$$

(5) Show that $\mathfrak{so}(E_{2n})_{\mathbb{C}} = \mathfrak{so}(E_{2n}, \mathbb{C})$ and that

$$\mathfrak{so}(E_{2n}, \mathbb{C}) = \{ \begin{pmatrix} X & Y \\ Z & -X^t \end{pmatrix} \mid X, Y, Z \in \mathfrak{gl}(n, \mathbb{C}), Y^t = -Y, Z^t = -Z \}.$$

(6) Show that $\mathfrak{so}(E_{2n+1})_{\mathbb{C}} = \mathfrak{so}(E_{2n+1}, \mathbb{C})$ and that

$$\mathfrak{so}(E_{2n+1}, \mathbb{C})$$
$$= \{ \begin{pmatrix} X & Y & u \\ Z & -X^t & v \\ -v^t & -u^t & 0 \end{pmatrix} \mid X, Y, Z \in \mathfrak{gl}(n, \mathbb{C}), Y^t = -Y, Z^t = -Z, u, v \in \mathbb{C}^n \}.$$

Exercise 6.4 Verify Equation 6.7.

Exercise 6.5 Let G be a compact Lie group. Show that $\mathfrak{z}(\mathfrak{g}_{\mathbb{C}}) = \mathfrak{z}(\mathfrak{g})_{\mathbb{C}}$.

Exercise 6.6 (1) Let G be a Lie subgroup of $GL(n, \mathbb{C})$ and assume $\mathfrak{g}$ is semisimple. Show that any one-dimensional representation of $\mathfrak{g}$ is trivial, i.e., $\mathfrak{g}$ acts by 0.
(2) Show that any one-dimensional representation of G is trivial.

Exercise 6.7 Use Equation 6.7 to verify that $V_n(\mathbb{C}^2)$ is an irreducible representation of $SU(2)$.

Exercise 6.8 This exercise gives an algebraic proof of the classification of irreducible representations of $SU(2)$ (c.f. Theorem 3.32).
(1) Given any irreducible representation V of $SU(2)$, show that there is a nonzero $v_0 \in V$, so that $H v_0 = \lambda v_0$, $\lambda \in \mathbb{C}$, and so that $E v_0 = 0$.
(2) Let $v_i = F^i v_0$. Show that $H v_i = (\lambda - 2i) v_i$ and $E v_i = i(\lambda - i + 1) v_{i-1}$.
(3) Let m be the smallest natural number satisfying $v_{m+1} = 0$. Show that $\{v_i\}_{i=0}^m$ is a basis for V.
(4) Show that the trace of the H action on V is zero.
(5) Show that $\lambda = m$ and use this to show that $V \cong V_m(\mathbb{C}^2)$.

Exercise 6.9 (1) Find the weight space decomposition for the standard representation of $SU(n)$ on $\mathbb{C}^n$.
(2) Find the weight space decomposition for the standard representation of $SO(n)$ on $\mathbb{C}^n$.

Exercise 6.10 Verify that the roots and root spaces listed in §6.1.5 are correct (c.f., Exercise 6.3).

Exercise 6.11 Let G be a compact Lie group and $\mathfrak{t}$ a Cartan subalgebra of $\mathfrak{g}$. Use root theory to show directly that there exists $X \in \mathfrak{t}$, so that $\mathfrak{t} = \mathfrak{z}_\mathfrak{g}(X)$ (c.f. Lemma 5.7).

Exercise 6.12 Prove Lemma 6.12.

Exercise 6.13 (1) Let $\mathfrak{g}$ be the Lie algebra of a Lie subgroup of a linear group. Then $\mathfrak{g}_{\mathbb{C}}$ is called *simple* if $\mathfrak{g}_{\mathbb{C}}$ has no (complex) proper ideals and if $\dim_{\mathbb{C}} \mathfrak{g} > 1$, i.e., if the only ideals of $\mathfrak{g}_{\mathbb{C}}$ are $\{0\}$ and $\mathfrak{g}_{\mathbb{C}}$ and if $\mathfrak{g}_{\mathbb{C}}$ is non-Abelian. Show that $\mathfrak{g}$ is simple if and only if $\mathfrak{g}_{\mathbb{C}}$ is simple.

(2) Use the root decomposition to show that $\mathfrak{sl}(n, \mathbb{C})$ is simple, $n \geq 2$.

(3) Show that $\mathfrak{sp}(n, \mathbb{C})$ is simple, $n \geq 1$.

(4) Show that $\mathfrak{so}(2n, \mathbb{C})$ is simple for $n \geq 3$, but that $\mathfrak{so}(4, \mathbb{C}) \cong \mathfrak{sl}(2, \mathbb{C}) \oplus \mathfrak{sl}(2, \mathbb{C})$.

(5) Show that $\mathfrak{so}(2n + 1, \mathbb{C})$ is simple, $n \geq 1$.

Exercise 6.14 (1) For $G = SO(2n)$ and

$$\mathfrak{t} = \{\text{blockdiag}\left(\begin{pmatrix} 0 & \theta_1 \\ -\theta_1 & 0 \end{pmatrix}, \ldots, \begin{pmatrix} 0 & \theta_n \\ -\theta_n & 0 \end{pmatrix}\right) \mid \theta_i \in \mathbb{R}\}$$

as in §5.1.2, $\mathfrak{g}_\mathbb{C} = \mathfrak{so}(2n, \mathbb{C})$ and

$$\mathfrak{t}_\mathbb{C} = \{\text{blockdiag}\left(\begin{pmatrix} 0 & z_1 \\ -z_1 & 0 \end{pmatrix}, \ldots, \begin{pmatrix} 0 & z_n \\ -z_n & 0 \end{pmatrix}\right) \mid z_i \in \mathbb{C}\}.$$

Show that

$$\Delta(\mathfrak{g}_\mathbb{C}, \mathfrak{t}_\mathbb{C}) = \{\pm(\epsilon_i - \epsilon_j), \ \pm(\epsilon_i + \epsilon_j) \mid 1 \leq i < j \leq n\},$$

where $\epsilon_j \left(\text{blockdiag}\left(\begin{pmatrix} 0 & z_1 \\ -z_1 & 0 \end{pmatrix}, \ldots, \begin{pmatrix} 0 & z_n \\ -z_n & 0 \end{pmatrix}\right)\right) = -iz_j$. Partition each $2n \times 2n$ matrix into n^2 blocks of size 2×2. For $\alpha = \pm\epsilon_i \pm \epsilon_j$, show that the root space is $\mathfrak{g}_\alpha = \mathbb{C}E_\alpha$, where E_α is 0 on all 2×2 blocks except for the ij^{th} block and the ji^{th} block. Show that E_α is given by the matrix X_α on the ij^{th} block and by $-X_\alpha^t$ on the ji^{th} block, where

$$X_{\epsilon_i - \epsilon_j} = \begin{pmatrix} 1 & i \\ -i & 1 \end{pmatrix}, \ X_{-\epsilon_i + \epsilon_j} = \begin{pmatrix} 1 & -i \\ i & 1 \end{pmatrix}$$

$$X_{\epsilon_i + \epsilon_j} = \begin{pmatrix} 1 & -i \\ -i & -1 \end{pmatrix}, \ X_{-\epsilon_i - \epsilon_j} = \begin{pmatrix} 1 & i \\ i & -1 \end{pmatrix}.$$

(2) For $G = SO(2n + 1)$ and

$$\mathfrak{t} = \{\text{blockdiag}\left(\begin{pmatrix} 0 & \theta_1 \\ -\theta_1 & 0 \end{pmatrix}, \ldots, \begin{pmatrix} 0 & \theta_n \\ -\theta_n & 0 \end{pmatrix}, 0\right) \mid \theta_i \in \mathbb{R}\}$$

as in §5.1.2, $\mathfrak{g}_\mathbb{C} = \mathfrak{so}(2n + 1, \mathbb{C})$ and

$$\mathfrak{t}_\mathbb{C} = \{\text{blockdiag}\left(\begin{pmatrix} 0 & z_1 \\ -z_1 & 0 \end{pmatrix}, \ldots, \begin{pmatrix} 0 & z_n \\ -z_n & 0 \end{pmatrix}, 0\right) \mid z_i \in \mathbb{C}\}.$$

Show that

$$\Delta(\mathfrak{g}_\mathbb{C}, \mathfrak{t}_\mathbb{C}) = \{\pm(\epsilon_i \pm \epsilon_j) \mid 1 \leq i < j \leq n\} \cup \{\pm\epsilon_i \mid 1 \leq i \leq n\},$$

where

$$\epsilon_j \left(\text{blockdiag}\left(\begin{pmatrix} 0 & z_1 \\ -z_1 & 0 \end{pmatrix}, \ldots, \begin{pmatrix} 0 & z_n \\ -z_n & 0 \end{pmatrix}, 0\right)\right) = -iz_j.$$

For $\alpha = \pm \left(\epsilon_i \pm \epsilon_j \right)$, show that the root space is obtained by embedding the corresponding root space from $\mathfrak{so}(2n, \mathbb{C})$ into $\mathfrak{so}(2n + 1, \mathbb{C})$ via the map $X \to \begin{pmatrix} X & 0 \\ 0 & 0 \end{pmatrix}$. For $\alpha = \pm \epsilon_j$, show that the root space is $\mathfrak{g}_\alpha = \mathbb{C} E_\alpha$, where E_α is 0 except on the last column and last row. Writing $v \in \mathbb{C}^{2n+1}$ for the last column, show the last row of E_α is given by $-v^t$, where v is given in terms of the standard basis vectors by $v = e_{2j-1} \mp i e_{2j}$.

6.2 The Standard $\mathfrak{sl}(2, \mathbb{C})$ Triple

6.2.1 Cartan Involution

Definition 6.13. Let G be a compact Lie group. The *Cartan involution*, θ, of $\mathfrak{g}_\mathbb{C}$ with respect to $\mathfrak{g}$ is the Lie algebra involution of $\mathfrak{g}_\mathbb{C}$ given by $\theta(X \otimes z) = X \otimes \bar{z}$ for $X \in \mathfrak{g}$ and $z \in \mathbb{C}$. In other words, if $Z \in \mathfrak{g}_\mathbb{C}$ is uniquely written as $Z = X + iY$ for $X, Y \in \mathfrak{g} \otimes 1$, then $\theta Z = X - iY$.

It must be verified that θ is a Lie algebra involution, but this follows from a simple calculation (Exercise 6.15). Under the natural embedding of $\mathfrak{g}$ in $\mathfrak{g}_\mathbb{C}$, notice that the $+1$ eigenspace of θ on $\mathfrak{g}_\mathbb{C}$ is $\mathfrak{g}$ and that the -1 eigenspace is $i\mathfrak{g}$. Notice also that when $\mathfrak{g} \subseteq \mathfrak{u}(n)$, then $\theta Z = -Z^*$ for $Z \in \mathfrak{g}_\mathbb{C}$ since $X^* = -X$ for $X \in \mathfrak{u}(n)$. In particular,

$$\theta Z = -Z^*$$

when $\mathfrak{g}$ is $\mathfrak{u}(n)$, $\mathfrak{su}(n)$, $\mathfrak{sp}(n)$, $\mathfrak{so}(n)$, or $\mathfrak{so}(E_n)$.

Lemma 6.14. *Let G be a compact Lie group and $\mathfrak{t}$ be a Cartan subalgebra of $\mathfrak{g}$.*
(a) If $\alpha \in \Delta(\mathfrak{g}_\mathbb{C})$, then $-\alpha \in \Delta(\mathfrak{g}_\mathbb{C})$ and $\mathfrak{g}_{-\alpha} = \theta \mathfrak{g}_\alpha$.
(b) $\theta \mathfrak{t}_\mathbb{C} = \mathfrak{t}_\mathbb{C}$.

Proof. Let $\alpha \in \Delta(\mathfrak{g}_\mathbb{C}) \cup \{0\}$. Recalling that θ is an involution, it suffices to show $\theta \mathfrak{g}_\alpha \subseteq \mathfrak{g}_{-\alpha}$. Write $Z \in \mathfrak{g}_\alpha$ uniquely as $Z = X + iY$ for $X, Y \in \mathfrak{g} \otimes 1$. Then for $H \in \mathfrak{t}$,

$$\alpha(H)(X + iY) = [H, X + iY] = [H, X] + i[H, Y].$$

Since $\alpha(H) \in i\mathbb{R}$ by Theorem 6.9 and since $[H, X], [H, Y] \in \mathfrak{g} \otimes 1$,

$$\alpha(H)X = i[H, Y] \text{ and } \alpha(H)Y = -i[H, X].$$

Thus

$$[H, \theta Z] = [H, X] - i[H, Y] = -\alpha(H)(X - iY) = -\alpha(H)(\theta Z),$$

so that $\theta Z \in \mathfrak{g}_{-\alpha}$, as desired. $\square$

In particular, notice that $\mathfrak{g}$ is spanned by elements of the form $Z + \theta Z$ for $Z \in \mathfrak{g}_\alpha$, $\alpha \in \Delta(\mathfrak{g}_\mathbb{C}) \cup \{0\}$.

6.2.2 Killing Form

Definition 6.15. Let $\mathfrak{g}$ be the Lie algebra of a Lie subgroup of $GL(n, \mathbb{C})$. For $X, Y \in \mathfrak{g}_\mathbb{C}$, the symmetric complex bilinear form $B(X, Y) = \text{tr}(\text{ad } X \circ \text{ad } Y)$ on $\mathfrak{g}_\mathbb{C}$ is called the *Killing form*.

Theorem 6.16. *Let $\mathfrak{g}$ be the Lie algebra of a compact Lie group G.*
(a) For $X, Y \in \mathfrak{g}$, $B(X, Y) = \text{tr}(\text{ad } X \circ \text{ad } Y)$ on $\mathfrak{g}$.
(b) B is Ad-invariant, i.e., $B(X, Y) = B(\text{Ad}(g)X, \text{Ad}(g)Y)$ for $g \in G$ and $X, Y \in \mathfrak{g}_\mathbb{C}$.
(c) B is skew ad-invariant, i.e., $B(\text{ad}(Z)X, Y) = -B(X, \text{ad}(Z)Y)$ for $Z, X, Y \in \mathfrak{g}_\mathbb{C}$.
(d) B restricted to $\mathfrak{g}' \times \mathfrak{g}'$ is negative definite.
(e) B restricted to $\mathfrak{g}_\alpha \times \mathfrak{g}_\beta$ is zero when $\alpha + \beta \neq 0$ for $\alpha, \beta \in \Delta(\mathfrak{g}_\mathbb{C}) \cup \{0\}$.
(f) B is nonsingular on $\mathfrak{g}_\alpha \times \mathfrak{g}_{-\alpha}$. If $\mathfrak{g}$ is semisimple with a Cartan subalgebra $\mathfrak{t}$, then B is also nonsingular on $\mathfrak{t}_\mathbb{C} \times \mathfrak{t}_\mathbb{C}$.
(g) The radical of B, $\text{rad } B = \{X \in \mathfrak{g}_\mathbb{C} \mid B(X, \mathfrak{g}_\mathbb{C}) = 0\}$, is the center of $\mathfrak{g}_\mathbb{C}$, $\mathfrak{z}(\mathfrak{g}_\mathbb{C})$.
(h) If $\mathfrak{g}$ is semisimple, the form $(X, Y) = -B(X, \theta Y)$, $X, Y \in \mathfrak{g}_\mathbb{C}$, is an Ad-invariant inner product on $\mathfrak{g}_\mathbb{C}$.
(i) Let $\mathfrak{g}$ be simple and choose a linear realization of G, so that $\mathfrak{g} \subseteq \mathfrak{u}(n)$. Then there exists a positive $c \in \mathbb{R}$, so that $B(X, Y) = c \text{ tr}(XY)$ for $X, Y \in \mathfrak{g}_\mathbb{C}$.

Proof. Part (a) is elementary. For part (b), recall that $\text{Ad } g$ preserves the Lie bracket by Theorem 4.8. Thus $\text{ad}(\text{Ad}(g)X) = \text{Ad}(g)\text{ad}(X)\text{Ad}(g^{-1})$ and part (b) follows. As usual, part (c) follows from part (b) by examining the case of $g = \exp tZ$ and applying $\frac{d}{dt}|_{t=0}$ when $Z \in \mathfrak{g}$. For $Z \in \mathfrak{g}_\mathbb{C}$, use the fact that B is complex bilinear.

For part (d), let $X \in \mathfrak{g}$. Using Theorem 5.9, choose a Cartan subalgebra $\mathfrak{t}$ containing X. Then the root space decomposition shows $B(X, X) = \sum_{\alpha \in \Delta(\mathfrak{g}_\mathbb{C})} \alpha^2(X)$. Since G is compact, $\alpha(X) \in i\mathbb{R}$ by Theorem 6.9. Thus B is negative semidefinite on $\mathfrak{g}$. Moreover, $B(X, X) = 0$ if and only if $\alpha(X) = 0$ for all $\alpha \in \Delta(\mathfrak{g}_\mathbb{C})$, i.e., if and only if $X \in \mathfrak{z}(\mathfrak{g})$. Thus the decomposition $\mathfrak{g} = \mathfrak{z}(\mathfrak{g}) \oplus \mathfrak{g}'$ from Theorem 5.18 finishes part (d).

For part (e), let $X_\alpha \in \mathfrak{g}_\alpha$ and $H \in \mathfrak{t}$. Use part (c) to see that

$$0 = B(\text{ad}(H)X_\alpha, X_\beta) + B(X_\alpha, \text{ad}(H)X_\beta) = [(\alpha + \beta)(H)](X_\alpha, X_\beta).$$

In particular (e) follows.

For part (f), recall that $\mathfrak{g}_{-\alpha} = \theta\mathfrak{g}_\alpha$. Thus if $X_\alpha = U_\alpha + iV_\alpha$ with $U_\alpha, V_\alpha \in \mathfrak{g}$, then $U_\alpha - iV_\alpha \in \mathfrak{g}_{-\alpha}$ and

(6.17) $$B(U_\alpha + iV_\alpha, U_\alpha - iV_\alpha) = B(U_\alpha, U_\alpha) + B(V_\alpha, V_\alpha).$$

In light of part (d), the above expression is zero if and only if $X_\alpha \in \mathbb{C}\mathfrak{z}(\mathfrak{g}) = \mathfrak{z}(\mathfrak{g}_\mathbb{C})$ (Exercise 6.5). Since $\mathfrak{g}_\alpha \subseteq (\mathfrak{g}')_\mathbb{C}$ for $\alpha \neq 0$, part (f) is complete.

For part (g), first observe that $\mathfrak{z}(\mathfrak{g}_\mathbb{C}) \subseteq \text{rad } B$ since $\text{ad } Z = 0$ for $Z \in \mathfrak{z}(\mathfrak{g}_\mathbb{C})$. On the other hand, since $\mathfrak{g}_\mathbb{C} = \mathfrak{z}(\mathfrak{g}_\mathbb{C}) \oplus (\mathfrak{g}')_\mathbb{C}$, the root space decomposition and part (f) finishes part (g).

Except for verifying positive definiteness, the assertion in part (h) follows from the definitions. To check positive definiteness, use the root space decomposition, the relation $\mathfrak{g}_{-\alpha} = \theta\mathfrak{g}_\alpha$, parts (d), (e) and (f), and Equation 6.17.

For part (i), first note that the trace form mapping $X, Y \in \mathfrak{g}_{\mathbb{C}}$ to $\mathrm{tr}(XY)$ is Ad-invariant since $\mathrm{Ad}(g)X = gXg^{-1}$. For $X \in \mathfrak{u}(n)$, X is diagonalizable with eigenvalues in $i\mathbb{R}$. In particular, the trace form is negative definite on $\mathfrak{g}$. Arguing as in Equation 6.17, this shows the trace form is nondegenerate on $\mathfrak{g}_{\mathbb{C}}$. In particular, both $-B(X, \theta Y)$ and $-\mathrm{tr}(X\theta Y)$ are Ad-invariant inner products on $\mathfrak{g}_{\mathbb{C}}$. However, since $\mathfrak{g}$ is simple, $\mathfrak{g}_{\mathbb{C}}$ is an irreducible representation of $\mathfrak{g}$ under ad (Exercise 6.17) and therefore an irreducible representation of G under Ad by Lemma 6.6 and Theorem 6.2. Corollary 2.20 finishes the argument. $\square$

6.2.3 The Standard $\mathfrak{sl}(2, \mathbb{C})$ and $\mathfrak{su}(2)$ Triples

Let G be a compact Lie algebra and $\mathfrak{t}$ a Cartan subalgebra of $\mathfrak{g}$. When $\mathfrak{g}$ is semisimple, recall that B is negative definite on $\mathfrak{t}$ by Theorem 6.16. It follows that B restricts to a real inner product on the real vector space $i\mathfrak{t}$. Continuing to write $(i\mathfrak{t})^*$ for the set of $\mathbb{R}$-linear functionals on $i\mathfrak{t}$, B induces an isomorphism between $i\mathfrak{t}$ and $(i\mathfrak{t})^*$ as follows.

Definition 6.18. Let G be a compact Lie group with semisimple Lie algebra, $\mathfrak{t}$ a Cartan subalgebra of $\mathfrak{g}$, and $\alpha \in (i\mathfrak{t})^*$. Let $u_\alpha \in i\mathfrak{t}$ be uniquely determined by the equation

$$\alpha(H) = B(H, u_\alpha)$$

for all $H \in i\mathfrak{t}$ and, when $\alpha \neq 0$, let

$$h_\alpha = \frac{2u_\alpha}{B(u_\alpha, u_\alpha)}.$$

In case $\mathfrak{g}$ is not semisimple, define $u_\alpha \in i\mathfrak{t}' \subseteq i\mathfrak{t} \subseteq \mathfrak{t}$ by first restricting B to $i\mathfrak{t}'$. For $\alpha \in \Delta(\mathfrak{g}_{\mathbb{C}})$, recall that α is determined by its restriction to $i\mathfrak{t}$. On $i\mathfrak{t}$, α is a real-valued linear functional by Theorem 6.9. Viewing α as an element of $(i\mathfrak{t})^*$, define u_α and h_α via Definition 6.18. Note that the equation $\alpha(H) = B(H, u_\alpha)$ now holds for all $H \in \mathfrak{t}_{\mathbb{C}}$ by $\mathbb{C}$-linear extension. An alternate notation for h_α is $\alpha^\vee$, and so we write

$$\Delta(\mathfrak{g}_{\mathbb{C}})^\vee = \{h_\alpha \mid \alpha \in \Delta(\mathfrak{g}_{\mathbb{C}})\}.$$

When $\mathfrak{g} \subseteq \mathfrak{u}(n)$ is simple, Theorem 6.16 shows that there exists a positive $c \in \mathbb{R}$, so that $B(X, Y) = c\,\mathrm{tr}(XY)$ for $X, Y \in \mathfrak{g}_{\mathbb{C}}$. Thus if $\alpha \in (i\mathfrak{t})^*$ and $u'_\alpha, h'_\alpha \in i\mathfrak{t}$ are determined by the equations $\alpha(H) = \mathrm{tr}(Hu'_\alpha)$ and $h'_\alpha = \frac{2u'_\alpha}{\mathrm{tr}(u'_\alpha u'_\alpha)}$, it follows that $u'_\alpha = cu_\alpha$ but that $h'_\alpha = h_\alpha$. In particular, h_α can be computed with respect to the trace form instead of the Killing form.

For the classical compact groups, this calculation is straightforward (see §6.1.5 and Exercise 6.21). Notice also that $h_{-\alpha} = -h_\alpha$.

For $SU(n)$ with $\mathfrak{t} = \{\operatorname{diag}(i\theta_1, \ldots, i\theta_n) \mid \theta_i \in \mathbb{R}, \sum_i \theta_i = 0\}$, that is the A_{n-1} root system,

$$h_{\epsilon_i - \epsilon_j} = E_i - E_j,$$

where $E_i = \operatorname{diag}(0, \ldots, 0, 1, 0, \ldots, 0)$ with the 1 in the ith position

For $Sp(n)$ realized as $Sp(n) \cong U(2n) \cap Sp(n, \mathbb{C})$ with

$$\mathfrak{t} = \{\operatorname{diag}(i\theta_1, \ldots, i\theta_n, -i\theta_1, \ldots, -i\theta_n) \mid \theta_i \in \mathbb{R}\},$$

that is the C_n root system,

$$
\begin{aligned}
h_{\epsilon_i - \epsilon_j} &= \left(E_i - E_j\right) - \left(E_{i+n} - E_{j+n}\right) \\
h_{\epsilon_i + \epsilon_j} &= \left(E_i + E_j\right) - \left(E_{i+n} + E_{j+n}\right) \\
h_{2\epsilon_i} &= E_i - E_{i+n}.
\end{aligned}
$$

For $SO(E_{2n})$ with $\mathfrak{t} = \{\operatorname{diag}(i\theta_1, \ldots, i\theta_n, -i\theta_1, \ldots, -i\theta_n) \mid \theta_i \in \mathbb{R}\}$, that is the D_n root system,

$$
\begin{aligned}
h_{\epsilon_i - \epsilon_j} &= \left(E_i - E_j\right) - \left(E_{i+n} - E_{j+n}\right) \\
h_{\epsilon_i + \epsilon_j} &= \left(E_i + E_j\right) - \left(E_{i+n} + E_{j+n}\right).
\end{aligned}
$$

For $SO(E_{2n+1})$ with $\mathfrak{t} = \{\operatorname{diag}(i\theta_1, \ldots, i\theta_n, -i\theta_1, \ldots, -i\theta_n, 0) \mid \theta_i \in \mathbb{R}\}$, that is the B_n root system,

$$
\begin{aligned}
h_{\epsilon_i - \epsilon_j} &= \left(E_i - E_j\right) - \left(E_{i+n} - E_{j+n}\right) \\
h_{\epsilon_i + \epsilon_j} &= \left(E_i + E_j\right) - \left(E_{i+n} + E_{j+n}\right) \\
h_{\epsilon_i} &= 2E_i - 2E_{i+n}.
\end{aligned}
$$

Lemma 6.19. *Let G be a compact Lie group, $\mathfrak{t}$ a Cartan subalgebra of $\mathfrak{g}$, and $\alpha \in \Delta(\mathfrak{g}_{\mathbb{C}})$.*
(a) Then $\alpha(h_\alpha) = 2$.
(b) For $E \in \mathfrak{g}_\alpha$ and $F \in \mathfrak{g}_{-\alpha}$,

$$[E, F] = B(E, F)u_\alpha = \frac{1}{2}B(E, F)B(u_\alpha, u_\alpha)h_\alpha.$$

(c) Given a nonzero $E \in \mathfrak{g}_\alpha$, E may be rescaled by an element of $\mathbb{R}$, so that $[E, F] = h_\alpha$, where $F = -\theta E$.

Proof. For part (a) simply use the definitions

$$\alpha(h_\alpha) = \frac{2\alpha(u_\alpha)}{B(u_\alpha, u_\alpha)} = \frac{2B(u_\alpha, u_\alpha)}{B(u_\alpha, u_\alpha)} = 2.$$

For part (b), first note that $[E, F] \subseteq \mathfrak{t}_{\mathbb{C}}$ by Theorem 6.11. Given any $H \in \mathfrak{t}_{\mathbb{C}}$, calculate

$$B([E, F], H) = B(E, [F, H]) = \alpha(H)B(E, F) = B(u_\alpha, H)B(E, F)$$
$$= B(B(E, F)u_\alpha, H).$$

Since B is nonsingular on $\mathfrak{t}_\mathbb{C}$ by Theorem 6.16, part (b) is finished. For part (c), replace E by cE, where

$$c^2 = \frac{2}{-B(E, \theta E)B(u_\alpha, u_\alpha)},$$

and use Theorem 6.16 to check that $-B(E, \theta E) > 0$ and $B(u_\alpha, u_\alpha) > 0$.   □

For the next theorem, recall that B is nonsingular on $\mathfrak{g}_\alpha \times \mathfrak{g}_{-\alpha}$ by Theorem 6.16 and that $-B(X, \theta Y)$ is an Ad-invariant inner product on $\mathfrak{g}_\mathbb{C}'$ for $X, Y \in \mathfrak{g}_\mathbb{C}$.

Theorem 6.20. *Let G be a compact Lie group, $\mathfrak{t}$ a Cartan subalgebra of $\mathfrak{g}$, and $\alpha \in \Delta(\mathfrak{g}_\mathbb{C})$. Fix a nonzero $E_\alpha \in \mathfrak{g}_\alpha$ and let $F_\alpha = -\theta E_\alpha$. Using Lemma 6.19, rescale E_α (and therefore F_α), so that $[E_\alpha, F_\alpha] = H_\alpha$ where $H_\alpha = h_\alpha$.*
(a) Then $\mathfrak{sl}(2, \mathbb{C}) \cong \mathrm{span}_\mathbb{C}\{E_\alpha, H_\alpha, F_\alpha\}$ with $\{E_\alpha, H_\alpha, F_\alpha\}$ corresponding to the standard basis

$$E = \begin{pmatrix} 0 & 1 \\ 0 & 0 \end{pmatrix}, \quad H = \begin{pmatrix} 1 & 0 \\ 0 & -1 \end{pmatrix}, \quad F = \begin{pmatrix} 0 & 0 \\ 1 & 0 \end{pmatrix}$$

of $\mathfrak{sl}(2, \mathbb{C})$.
(b) Let $\mathcal{I}_\alpha = iH_\alpha$, $\mathcal{J}_\alpha = -E_\alpha + F_\alpha$, and $\mathcal{K}_\alpha = -i(E_\alpha + F_\alpha)$. Then $\mathcal{I}_\alpha, \mathcal{J}_\alpha, \mathcal{K}_\alpha \in \mathfrak{g}$ and $\mathfrak{su}(2) \cong \mathrm{span}_\mathbb{R}\{\mathcal{I}_\alpha, \mathcal{J}_\alpha, \mathcal{K}_\alpha\}$ with $\{\mathcal{I}_\alpha, \mathcal{J}_\alpha, \mathcal{K}_\alpha\}$ corresponding to the basis

$$\begin{pmatrix} i & 0 \\ 0 & -i \end{pmatrix}, \quad \begin{pmatrix} 0 & -1 \\ 1 & 0 \end{pmatrix}, \quad \begin{pmatrix} 0 & -i \\ -i & 0 \end{pmatrix}$$

of $\mathfrak{su}(2)$ (c.f. Exercise 4.2 for the isomorphism $\mathrm{Im}(\mathbb{H}) \cong \mathfrak{su}(2)$).
(c) There exists a Lie algebra homomorphism $\varphi_\alpha : SU(2) \to G$, so that $d\varphi : \mathfrak{su}(2) \to \mathfrak{g}$ implements the isomorphism in part (b) and whose complexification $d\varphi : \mathfrak{sl}(2, \mathbb{C}) \to \mathfrak{g}_\mathbb{C}$ implements the isomorphism in part (a).
(d) The image of φ_α in G is a Lie subgroup of G isomorphic to either $SU(2)$ or $SO(3)$ depending on whether the kernel of φ_α is $\{I\}$ or $\{\pm I\}$.

Proof. For part (a), Lemma 6.19 and the definitions show that $[H_\alpha, E_\alpha] = 2E_\alpha$, $[H_\alpha, F_\alpha] = -2F_\alpha$, and $[E_\alpha, F_\alpha] = H_\alpha$. Since these are the bracket relations for the standard basis of $\mathfrak{sl}(2, \mathbb{C})$, part (a) is finished (c.f. Exercise 4.21). For part (b), observe that θ fixes $\mathcal{I}_\alpha, \mathcal{J}_\alpha$, and $\mathcal{K}_\alpha$ by construction, so that $\mathcal{I}_\alpha, \mathcal{J}_\alpha, \mathcal{K}_\alpha \in \mathfrak{g}$. The bracket relations for $\mathfrak{sl}(2, \mathbb{C})$ then quickly show that $[\mathcal{I}_\alpha, \mathcal{J}_\alpha] = 2\mathcal{K}_\alpha$, $[\mathcal{J}_\alpha, \mathcal{K}_\alpha] = 2\mathcal{I}_\alpha$, and $[\mathcal{K}_\alpha, \mathcal{I}_\alpha] = 2\mathcal{J}_\alpha$, so that $\mathfrak{su}(2) \cong \mathrm{span}_\mathbb{R}\{\mathcal{I}_\alpha, \mathcal{J}_\alpha, \mathcal{K}_\alpha\}$ (Exercise 4.2). For part (c), recall that $SU(2)$ is simply connected since, topologically, it is isomorphic to S^3. Thus, Theorem 4.16 provides the existence of φ_α. For part (d), observe that $d\varphi_\alpha$ is an isomorphism by definition. Thus, the kernel of φ_α is discrete and normal and therefore central by Lemma 1.21. Since the center of $SU(2)$ is $\pm I$ and since $SO(3) \cong SU(2)/\{\pm I\}$ by Lemma 1.23, the proof is complete.   □

Definition 6.21. Let G be a compact Lie group, $\mathfrak{t}$ a Cartan subalgebra of $\mathfrak{g}$, and $\alpha \in \Delta(\mathfrak{g}_\mathbb{C})$. Continuing the notation from Theorem 6.20, the set $\{E_\alpha, H_\alpha, F_\alpha\}$ is called a *standard* $\mathfrak{sl}(2, \mathbb{C})$-*triple* associated to α and the set $\{\mathcal{I}_\alpha, \mathcal{J}_\alpha, \mathcal{K}_\alpha\}$ is called a *standard* $\mathfrak{su}(2)$-*triple* associated to α.

Corollary 6.22. *Let G be a compact Lie group, $\mathfrak{t}$ a Cartan subalgebra of $\mathfrak{g}$, and $\alpha \in \Delta(\mathfrak{g}_\mathbb{C})$.*
(a) The only multiple of α in $\Delta(\mathfrak{g}_\mathbb{C})$ is $\pm\alpha$.
(b) $\dim \mathfrak{g}_\alpha = 1$.
(c) If $\beta \in \Delta(\mathfrak{g}_\mathbb{C})$, then $a(h_\beta) \in \pm\{0, 1, 2, 3\}$.
(d) If (π, V) is a representation of G and $\lambda \in \Delta(V)$, then $\lambda(h_\alpha) \in \mathbb{Z}$.

Proof. Let $\{E_\alpha, H_\alpha, F_\alpha\}$ be a standard $\mathfrak{sl}(2, \mathbb{C})$-triple associated to α and $\{\mathcal{I}_\alpha, \mathcal{J}_\alpha, \mathcal{K}_\alpha\}$ the standard $\mathfrak{su}(2)$-triple associated to α with $\varphi_\alpha : SU(2) \to G$ the corresponding embedding. Since $e^{2\pi i H} = I$, applying $\mathrm{Ad} \circ \varphi_\alpha$ shows that

$$I = \mathrm{Ad}(\varphi_\alpha e^{2\pi i H}) = \mathrm{Ad}(e^{2\pi d\varphi_\alpha i H}) = \mathrm{Ad}(e^{2\pi i H_\alpha}) = e^{2\pi i \, \mathrm{ad} \, H_\alpha}$$

on $\mathfrak{g}_\mathbb{C}$. Using the root decomposition, it follows that $\beta(H_\alpha) = \frac{2B(u_\beta, u_\alpha)}{\|u_\alpha\|^2} \in \mathbb{Z}$ where $\|\cdot\|$ is the norm corresponding to the Killing form. Now if $k\alpha \in \Delta(\mathfrak{g}_\mathbb{C})$, then $u_{k\alpha} = ku_\alpha$, so that $\frac{2}{k} = \frac{2B(u_\alpha, ku_\alpha)}{\|ku_\alpha\|^2} = \alpha(H_{k\alpha}) \in \mathbb{Z}$ and $2k = \frac{2B(ku_\alpha, u_\alpha)}{\|u_\alpha\|^2} = (k\alpha)(H_\alpha) \in \mathbb{Z}$. Thus $k \in \pm\{\frac{1}{2}, 1, 2\}$.

For part (a), it therefore suffices to show that $\alpha \in \Delta(\mathfrak{g}_\mathbb{C})$ implies $\pm 2\alpha \notin \Delta(\mathfrak{g}_\mathbb{C})$. For this, let $\mathfrak{l}_\alpha = \mathrm{span}_\mathbb{R}\{\mathcal{I}_\alpha, \mathcal{J}_\alpha, \mathcal{K}_\alpha\} \cong \mathfrak{su}(2)$, so that $(\mathfrak{l}_\alpha)_\mathbb{C} = \mathrm{span}_\mathbb{C}\{E_\alpha, H_\alpha, F_\alpha\} \cong \mathfrak{sl}(2, \mathbb{C})$. Also let $V = \mathfrak{g}_{-2\alpha} \oplus \mathfrak{g}_{-\alpha} \oplus \mathbb{C}H_\alpha \oplus \mathfrak{g}_\alpha \oplus \mathfrak{g}_{2\alpha}$, where $\mathfrak{g}_{\pm 2\alpha}$ is possibly zero in this case. By Lemma 6.19 and Theorem 6.11, V is invariant under $(\mathfrak{l}_\alpha)_\mathbb{C}$ with respect to the ad-action. In particular, V is a representation of $\mathfrak{l}_\alpha$. Of course, $(\mathfrak{l}_\alpha)_\mathbb{C} \subseteq V$ is an $\mathfrak{l}_\alpha$-invariant subspace. Thus V decomposes under the $\mathfrak{l}$-action as $V = (\mathfrak{l}_\alpha)_\mathbb{C} \oplus V'$ for some submodule V' of V. To finish parts (a) and (b), it suffices to show $V' = \{0\}$.

From the discussion in §6.1.3, we know that H_α acts on the $(n + 1)$-dimensional irreducible representation of $\mathfrak{su}(2)$ with eigenvalues $\{n, n - 2, \ldots, -n + 2, -2n\}$. In particular, if V' were nonzero, V' would certainly contain an eigenvector of H_α corresponding to an eigenvalue of either 0 or 1. Now the eigenvalues of H_α on V are contained in $\pm\{0, 2, 4\}$ by construction. Since the 0-eigenspace has multiplicity one in V and is already contained in $(\mathfrak{l}_\alpha)_\mathbb{C}$, V' must be $\{0\}$.

For part (c), let $\beta \in \Delta(\mathfrak{g}_\mathbb{C})$ and write $B(u_\beta, u_\alpha) = \|u_\beta\| \|u_\alpha\| \cos\theta$, where θ is the angle between u_β and u_β. Thus $4\cos^2\theta = \frac{2B(u_\alpha, u_\beta)}{\|u_\beta\|^2} \frac{2B(u_\beta, u_\alpha)}{\|u_\alpha\|^2} \in \mathbb{Z}$. As $\cos^2\theta \le 1$, $4\cos^2\theta = \alpha(H_\beta)\beta(H_\alpha) \in \{0, 1, 2, 3, 4\}$. To finish part (c), it only remains to rule out the possibility that $\{\alpha(H_\beta), \beta(H_\alpha)\} = \pm\{1, 4\}$. Clearly $\alpha(H_\beta)\beta(H_\alpha) = 4$ only when $\theta = 0$, i.e., when α and β are multiples of each other. By part (a), this occurs only when $\beta = \pm\alpha$ in which case $\alpha(H_\beta) = \beta(H_\alpha) = \pm 2$. In particular, $\{\alpha(H_\beta), \beta(H_\alpha)\} \ne \pm\{1, 4\}$. Thus $\alpha(H_\beta), \beta(H_\alpha) \in \pm\{0, 1, 2, 3\}$.

For part (d), simply apply $\pi \circ \varphi_\alpha$ to $e^{2\pi i H} = I$ to get $e^{2\pi i (d\pi) H_\alpha} = I$ on V. As in the first paragraph, the weight decomposition shows that $\lambda(H_\alpha) \in \mathbb{Z}$. $\quad\square$

It turns out that the above condition $\alpha(h_\beta) \in \pm\{0, 1, 2, 3\}$ is strict. In other words, there exist compact Lie groups for which each of these values are achieved.

6.2.4 Exercises

Exercise 6.15 Show that θ is a Lie algebra involution of $\mathfrak{g}_\mathbb{C}$, i.e., that θ is $\mathbb{R}$-linear, $\theta^2 = I$, and that $\theta[Z_1, Z_2] = [\theta Z_1, \theta Z_2]$ for $Z_i \in \mathfrak{g}_\mathbb{C}$.

Exercise 6.16 Let G be a connected compact Lie group and $g \in G$. Use the Maximal Torus Theorem, Lemma 6.14, and Theorem 6.9 to show that $\det \mathrm{Ad}\, g = 1$ on $\mathfrak{g}_\mathbb{C}$ and therefore on $\mathfrak{g}$ as well.

Exercise 6.17 Let G be a compact Lie group with Lie algebra $\mathfrak{g}$. Show that $\mathfrak{g}$ is simple if and only if $\mathfrak{g}_\mathbb{C}$ is an irreducible representation of $\mathfrak{g}$ under ad.

Exercise 6.18 (1) Let G be a compact Lie group with simple Lie algebra $\mathfrak{g}$. If $(\cdot, \cdot)$ is an Ad-invariant symmetric bilinear form on $\mathfrak{g}_\mathbb{C}$, show that there is a constant $c \in \mathbb{C}$ so that $(\cdot, \cdot) = cB(\cdot, \cdot)$.
(2) If $(\cdot, \cdot)$ is nonzero and $B(\cdot, \cdot)$ is replaced by $(\cdot, \cdot)$ in Definition 6.18, show that h_α is unchanged.

Exercise 6.19 Let G be a compact Lie group with a simple (c.f. Exercise 6.13) Lie algebra $\mathfrak{g} \subseteq \mathfrak{u}(n)$. Theorem 6.16 shows that there is a positive $c \in \mathbb{R}$, so that $B(X, Y) = c\,\mathrm{tr}(XY)$ for $X, Y \in \mathfrak{g}_\mathbb{C}$. In the special cases below, show that c is given as stated.
(1) $c = 2n$ for $G = SU(n), n \geq 2$
(2) $c = 2(n + 1)$ for $G = Sp(n), n \geq 1$
(3) $c = 2(n - 1)$ for $G = SO(2n), n \geq 3$
(4) $c = 2n - 1$ for $G = SO(2n + 1), n \geq 1$.

Exercise 6.20 Let G be a compact Lie group with semisimple Lie algebra $\mathfrak{g}$, $\mathfrak{t}$ a Cartan subalgebra of $\mathfrak{g}$, and $\alpha \in \Delta(\mathfrak{g}_\mathbb{C})$. If $\beta \in \Delta(\mathfrak{g}_\mathbb{C})$ with $\beta \neq \pm a$ and $B(u_\alpha, u_\alpha) \leq B(u_\beta, u_\beta)$, show that $a(h_\beta) \in \pm\{0, 1\}$.

Exercise 6.21 For each compact classical Lie group, this section lists h_α for each root α. Verify these calculations.

Exercise 6.22 Let G be a compact Lie group with semisimple Lie algebra $\mathfrak{g}$, $\mathfrak{t}$ a Cartan subalgebra of $\mathfrak{g}$, and $\alpha \in \Delta(\mathfrak{g}_\mathbb{C})$. Let V be a finite-dimensional representation of G and $\lambda \in \Delta(V)$. The α-*string through* λ is the set of all weights of the form $\lambda + n\alpha$, $n \in \mathbb{Z}$.
(1) Make use of a standard $\mathfrak{sl}(2)$-triple $\{E_\alpha, H_\alpha, F_\alpha\}$ and consider the space $\bigoplus_n V_{\lambda+n\alpha}$ to show the α-string through β is of the form $\{\lambda + n\alpha \mid -p \leq n \leq q\}$, where $p, q \in \mathbb{Z}^{\geq 0}$ with $p - q = \lambda(h_\alpha)$.
(2) If $\lambda(h_\alpha) < 0$, show show $\lambda + \alpha \in \Delta(V)$. If $\lambda(h_\alpha) > 0$, show that $\lambda - \alpha \in \Delta(V)$.
(3) Show that $d\pi(E_\alpha)^{p+q} V_{\lambda-p\alpha} \neq 0$.
(4) If $\alpha, \beta, \alpha + \beta \in \Delta(\mathfrak{g}_\mathbb{C})$, show that $[\mathfrak{g}_\alpha, \mathfrak{g}_\beta] = \mathfrak{g}_{\alpha+\beta}$.

Exercise 6.23 Show that $SL(2, \mathbb{C})$ has no nontrivial finite-dimensional unitary representations. To this end, argue by contradiction. Assume (π, V) is such a representation and compare the form $B(X, Y)$ on $\mathfrak{sl}(2, \mathbb{C})$ to the form $(X, Y)' = \mathrm{tr}\,(d\pi(X) \circ d\pi(Y))$.

6.3 Lattices

6.3.1 Definitions

Let G be a compact Lie group, $\mathfrak{t}$ a Cartan subalgebra of $\mathfrak{g}$, and $\alpha \in \Delta(\mathfrak{g}_{\mathbb{C}})$. As noted in §6.1.3, α may be viewed as an element of $(i\mathfrak{t})^*$. Use Definition 6.18 to transport the Killing form from $i\mathfrak{t}$ to $(i\mathfrak{t})^*$ by setting

$$B(\lambda_1, \lambda_2) = B(u_{\lambda_1}, u_{\lambda_2})$$

for $\lambda_1, \lambda_2 \in (i\mathfrak{t})^*$. In particular, for $\lambda \in (i\mathfrak{t})^*$,

$$\lambda(h_\alpha) = \frac{2B(\lambda, \alpha)}{B(\alpha, \alpha)}.$$

For the sake of symmetry, also note that

$$\alpha(H) = \frac{2B(H, h_\alpha)}{B(h_\alpha, h_\alpha)}$$

for $H \in i\mathfrak{t}$.

Definition 6.23. Let G be a compact Lie group and T a maximal torus of G with corresponding Cartan subalgebra $\mathfrak{t}$.
(a) The *root lattice*, $R = R(\mathfrak{t})$, is the lattice in $(i\mathfrak{t})^*$ given by

$$R = \mathrm{span}_{\mathbb{Z}}\{\alpha \mid \alpha \in \Delta(\mathfrak{g}_{\mathbb{C}})\}.$$

(b) The *weight lattice* (alternately called the set of *algebraically integral weights*), $P = P(\mathfrak{t})$, is the lattice in $(i\mathfrak{t})^*$ given by

$$P = \{\lambda \in (i\mathfrak{t})^* \mid \lambda(h_\alpha) \in \mathbb{Z} \text{ for } \alpha \in \Delta(\mathfrak{g}_{\mathbb{C}})\},$$

where $\lambda \in (i\mathfrak{t})^*$ is extended to an element of $(\mathfrak{t}_{\mathbb{C}})^*$ by $\mathbb{C}$-linearity.
(c) The set of *analytically integral weights*, $A = A(T)$, is the lattice in $(i\mathfrak{t})^*$ given by

$$A = \{\lambda \in (i\mathfrak{t})^* \mid \lambda(H) \in 2\pi i\mathbb{Z} \text{ whenever } \exp H = I \text{ for } H \in \mathfrak{t}\}.$$

To the lattices R, P, and A, there are also a number of associated dual lattices.

Definition 6.24. Let G be a compact Lie group and T a maximal torus of G with corresponding Cartan subalgebra $\mathfrak{t}$.
(a) The *dual root lattice*, $R^\vee = R^\vee(\mathfrak{t})$, is the lattice in $i\mathfrak{t}$ given by

$$R^\vee = \mathrm{span}_{\mathbb{Z}}\{h_\alpha \mid \alpha \in \Delta(\mathfrak{g}_{\mathbb{C}})\}.$$

(b) The *dual weight lattice*, $P^\vee = P^\vee(\mathfrak{t})$, is the lattice in $i\mathfrak{t}$ given by

$$P^\vee = \{H \in it \mid \alpha(H) \in \mathbb{Z} \text{ for } \alpha \in \Delta(\mathfrak{g}_\mathbb{C})\}.$$

(c) Let $\ker \mathcal{E} = \ker \mathcal{E}(T)$ be the lattice in it given by

$$\ker \mathcal{E} = \{H \in it \mid \exp(2\pi i H) = I\}.$$

(d) In general, if Λ_1 is a lattice in $(it)^*$ that spans $(it)^*$ and if Λ_2 is a lattice in it that spans it, define the *dual lattices*, Λ_1^* and Λ_2^* in it and $(it)^*$, respectively, by

$$\Lambda_1^* = \{H \in it \mid \lambda(H) \in \mathbb{Z} \text{ for } \lambda \in \Lambda_1\}$$
$$\Lambda_2^* = \{\lambda \in (it)^* \mid \lambda(H) \in \mathbb{Z} \text{ for } H \in \Lambda_2\}.$$

It is well known that Λ_1^* and Λ_2^* are lattices and that they satisfy $\Lambda_i^{**} = \Lambda_i$ (Exercise 6.24). Notice $\ker \mathcal{E}$ is a lattice by the proof of Theorem 5.2.

6.3.2 Relations

Lemma 6.25. *Let G be a compact connected Lie group with Cartan subalgebra* $\mathfrak{t}$. *For $H \in \mathfrak{t}$, $\exp H \in Z(G)$ if and only if $\alpha(H) \in 2\pi i\mathbb{Z}$ for all $\alpha \in \Delta(\mathfrak{g}_\mathbb{C})$.*

Proof. Let $g = \exp H$ and recall from Lemma 5.11 that $g \in Z(G)$ if and only if $\mathrm{Ad}(g)X = X$ for all $X \in \mathfrak{g}$. Now for $\alpha \in \Delta(\mathfrak{g}_\mathbb{C}) \cup \{0\}$ and $X \in \mathfrak{g}_\alpha$, $\mathrm{Ad}(g)X = e^{\mathrm{ad}\, H}X = e^{\alpha(H)}X$. The root decomposition finishes the proof. $\quad\square$

Definition 6.26. Let G be a compact Lie group and T a maximal torus. Write $\chi(T)$ for the *character group* on T, i.e., $\chi(T)$ is the set of all Lie homomorphisms $\xi : T \to \mathbb{C}\backslash\{0\}$.

Theorem 6.27. *Let G be a compact Lie group with a maximal torus T.*
(a) $R \subseteq A \subseteq P$.
(b) Given $\lambda \in (it)^*$, $\lambda \in A$ if and only if there exists $\xi_\lambda \in \chi(T)$ satisfying

(6.28) $$\xi_\lambda(\exp H) = e^{\lambda(H)}$$

for $H \in \mathfrak{t}$, where $\lambda \in (it)^*$ is extended to an element of $(\mathfrak{t}_\mathbb{C})^*$ by $\mathbb{C}$-linearity. The map $\lambda \to \xi_\lambda$ establishes a bijection

$$A \longleftrightarrow \chi(T).$$

(c) For semisimple $\mathfrak{g}$, $|P/R|$ is finite.

Proof. Let $\alpha \in \Delta(\mathfrak{g}_\mathbb{C})$ and suppose $H \in \mathfrak{t}$ with $\exp H = e$. Lemma 6.25 shows that $\alpha(H) \in 2\pi i\mathbb{Z}$, so that $R \subseteq A$. Next choose a standard $\mathfrak{sl}(2, \mathbb{C})$-triple $\{E_\alpha, h_\alpha, F_\alpha\}$ associated to α. As in the proof of Corollary 6.22, $\exp 2\pi i h_\alpha = I$. Thus if $\lambda \in A$, $\lambda(2\pi i h_\alpha) \in 2\pi i\mathbb{Z}$, so that $A \subseteq P$ which finishes part (a).

For part (b), start with $\lambda \in A$. Using the fact that $\exp \mathfrak{t} = T$ and using Lemma 6.25, Equation 6.28 uniquely defines a well-defined function ξ_λ on T. It is a homomorphism by Theorem 5.1. Conversely, if there is a $\xi_\lambda \in \chi(T)$ satisfying Equation 6.28, then clearly $\lambda(H) \in 2\pi i \mathbb{Z}$ whenever $\exp H = I$, so that $\lambda \in A$. Finally, to see that there is a bijection from A to $\chi(T)$, it remains to see that the map $\lambda \to \xi_\lambda$ is surjective. However, this requirement follows immediately by taking the differential of an element of $\chi(T)$ and extending via $\mathbb{C}$-linearity. Theorem 6.9 shows the differential can be viewed as an element of $(i\mathfrak{t})^*$.

Next, Theorem 6.11 shows that R spans $(i\mathfrak{t})^*$ for semisimple $\mathfrak{g}$. Part (c) therefore follows immediately from elementary lattice theory (e.g., see [3]). In fact, it is straightforward to show $|P/R|$ is equal to the determinant of the so-called *Cartan matrix* (Exercise 6.42). □

Theorem 6.29. *Let G be a compact Lie group with a semisimple Lie algebra $\mathfrak{g}$ and let T be a maximal torus of G with corresponding Cartan subalgebra $\mathfrak{t}$.*
(a) $R^* = P^\vee$.
(b) $P^* = R^\vee$.
(c) $A^* = \ker \mathcal{E}$.
(d) $P^* \subseteq A^* \subseteq R^*$, i.e., $R^\vee \subseteq \ker \mathcal{E} \subseteq P^\vee$.

Proof. The equalities $R^* = P^\vee$, $(R^\vee)^* = P$, and $(\ker \mathcal{E})^* = A$ follow immediately from the definitions. This proves parts (a), (b), and (c) (Exercise 6.24). Part (d) follows from Theorem 6.27 (Exercise 6.24). □

6.3.3 Center and Fundamental Group

The proof of part (b) of the following theorem will be given in §7.3.6. However, for the sake of comparison, part (b) is stated now.

Theorem 6.30. *Let G be a connected compact Lie group with a semisimple Lie algebra and maximal torus T.*
(a) $Z(G) \cong P^\vee / \ker \mathcal{E} \cong A/R$.
(b) $\pi_1(G) \cong \ker \mathcal{E} / R^\vee \cong P/A$.

Proof (part (a) only). By Theorem 5.1, Corollary 5.13, and Lemma 6.25, the exponential map induces an isomorphism

$$Z(G) \cong \{H \in \mathfrak{t} \mid \alpha(H) \in 2\pi i \mathbb{Z} \text{ for } \alpha \in \Delta(\mathfrak{g}_\mathbb{C})\} / \{H \in \mathfrak{t} \mid \exp H = I\}$$
$$= (2\pi i \, P^\vee) / (2\pi i \, \ker \mathcal{E}),$$

so that $Z(G) \cong P^\vee / \ker \mathcal{E}$. Basic lattice theory shows $R^*/A^* \cong A/R$ (Exercise 6.24) which finishes the proof. □

While the proof of part (b) of Theorem 6.30 is postponed until §7.3.6, in this section we at least prove the simply connected covering of a compact semisimple Lie group is still compact.

Let G be a compact connected Lie group and let $\widetilde{G}$ be the simply connected covering of G. A priori, it is not known that $\widetilde{G}$ is a *linear group* and thus our development of the theory of Lie algebras and, in particular, the exponential map is not directly applicable to $\widetilde{G}$. Indeed for more general groups, $\widetilde{G}$ may not be linear. As usual though, compact groups are nicely behaved. Instead of redoing our theory in the context of arbitrary Lie groups, we instead use the lifting property of covering spaces. Write $\exp_G : \mathfrak{g} \to G$ for the standard exponential map and let

$$\exp_{\widetilde{G}} : \mathfrak{g} \to \widetilde{G}$$

be the unique smooth lift of $\exp_G$ satisfying $\exp_{\widetilde{G}}(0) = \widetilde{e}$ and $\exp_G = \pi \circ \exp_{\widetilde{G}}$.

Lemma 6.31. *Let G be a compact connected Lie group, T a maximal torus of G, $\widetilde{G}$ the simply connected covering of G, $\pi : \widetilde{G} \to G$ the associated covering homomorphism, and $\widetilde{T} = \left[\pi^{-1}(T)\right]^0$.*
(a) Restricted to $\mathfrak{t}$, $\exp_{\widetilde{G}}$ induces an isomorphism of Lie groups $\widetilde{T} \cong \mathfrak{t}/\left(\mathfrak{t} \cap \ker \exp_{\widetilde{G}}\right)$.
(b) If $\mathfrak{g}$ is semisimple, then $\widetilde{T}$ is compact.

Proof. Elementary covering theory shows that $\widetilde{T}$ is a covering of T. From this it follows that $\widetilde{T}$ is Abelian on a neighborhood of $\widetilde{e}$ and, since $\widetilde{T}$ is connected, $\widetilde{T}$ is Abelian everywhere. Since $\pi \exp_{\widetilde{G}} \mathfrak{t} = \exp_G \mathfrak{t} = T$ and since $\exp_{\widetilde{G}} \mathfrak{t}$ is connected, $\exp_{\widetilde{G}} \mathfrak{t} \subseteq \widetilde{T}$. In particular, $\exp_{\widetilde{G}} : \mathfrak{t} \to \widetilde{T}$ is the unique lift of $\exp_G : \mathfrak{t} \to T$ satisfying $\exp_{\widetilde{G}}(0) = \widetilde{e}$. In turn, uniqueness of the lifting easily shows $\exp_{\widetilde{G}}(t_0 + t) = \exp_{\widetilde{G}}(t_0) \exp_{\widetilde{G}}(t)$. To finish part (a), it suffices to show $\exp_{\widetilde{G}} \mathfrak{t}$ contains a neighborhood $\widetilde{e}$ in $\widetilde{T}$. For this, it suffices to show the differential of $\exp_{\widetilde{G}}$ at 0 is invertible. But since π is a local diffeomorphism and since $\exp_G$ is a local diffeomorphism near 0, we are done.

For part (b), it suffices to show that $\widetilde{T}$ is a finite cover of T when $\mathfrak{g}$ is semisimple. For this, first observe that $\ker \exp_{\widetilde{G}} \subseteq \ker \exp_G = 2\pi i \ker \mathcal{E}$ since $\exp_G = \pi \circ \exp_{\widetilde{G}}$. As $T \cong \mathfrak{t}/(2\pi i \ker \mathcal{E})$, it follows that the $\ker \pi$ restricted to $\widetilde{T}$ is isomorphic to $(2\pi i \ker \mathcal{E})/(\mathfrak{t} \cap \ker \exp_{\widetilde{G}})$. By Theorems 6.27 and 6.29, it therefore suffices to show that $2\pi i \, R^\vee \subseteq \mathfrak{t} \cap \ker \exp_{\widetilde{G}}$.

Given $\alpha \in \Delta(\mathfrak{g}_{\mathbb{C}})$, let $\{\mathcal{I}_\alpha, \mathcal{J}_\alpha, \mathcal{K}_\alpha\}$ be a standard $\mathfrak{su}(2)$-triple in $\mathfrak{g}$ associated to α. Write $\varphi_\alpha : SU(2) \to G$ for the corresponding homomorphism. Since $SU(2)$ is simply connected, write $\widetilde{\varphi}_\alpha : SU(2) \to \widetilde{G}$ for the unique lift of φ_α mapping I to $\widetilde{e}$. Using the uniqueness of lifting from $\mathfrak{su}(2)$ to $\widetilde{G}$, if follows easily that $\widetilde{\varphi}_\alpha \circ \exp_{SU(2)} = \exp_{\widetilde{G}} \circ d\varphi_\alpha$. Therefore by construction,

$$\widetilde{e} = \widetilde{\varphi}_\alpha(I) = \widetilde{\varphi}_\alpha(\exp_{SU(2)} 2\pi i H) = \exp_{\widetilde{G}}(2\pi i \, d\varphi_\alpha H) = \exp_{\widetilde{G}}(2\pi i \, h_\alpha),$$

which finishes the proof.   □

Lemma 6.32. *Let G be a compact connected Lie group, T a maximal torus of G, $\widetilde{G}$ the simply connected covering of of G, $\pi : \widetilde{G} \to G$ the associated covering homomorphism, and $\widetilde{T} = \left\lfloor \pi^{-1}(T)\right\rfloor^0$.*
(a) $\widetilde{G} = \bigcup_{\widetilde{g} \in \widetilde{G}} (c_{\widetilde{g}} \widetilde{T})$.
(b) $\widetilde{G} = \exp_{\widetilde{G}}(\mathfrak{g})$.

Proof. The proof of this lemma is a straightforward generalization of the proof of the Maximal Torus theorem, Theorem 5.12 (Exercise 6.26). $\square$

Corollary 6.33. *Let G be a compact connected Lie group with semisimple Lie algebra $\mathfrak{g}$, T a maximal torus of G, $\widetilde{G}$ the simply connected covering of of G, $\pi : \widetilde{G} \to G$ the associated covering homomorphism, and $\widetilde{T} = [\pi^{-1}(T)]^0$.*
(a) $\widetilde{G}$ is compact.
(b) $\mathfrak{g}$ may be identified with the Lie algebra of $\widetilde{G}$, so that $\exp_{\widetilde{G}}$ is the corresponding exponential map.
(c) $\widetilde{T} = \pi^{-1}(T)$ and $\widetilde{T}$ is a maximal torus of $\widetilde{G}$.
(d) $\ker \pi \subseteq Z(\widetilde{G}) \subseteq \widetilde{T}$.

Proof. For part (a), observe that $\widetilde{G} = \bigcup_{\widetilde{g} \in \widetilde{G}} \left(c_{\widetilde{g}} \widetilde{T} \right)$ by Lemma 6.32. Thus $\widetilde{G}$ is the continuous image of the compact set $\widetilde{G}/Z(\widetilde{G}) \times \widetilde{T} \cong G/Z(G) \times \widetilde{T}$ (Exercise 6.26).

For part (b), recall from Corollary 4.9 that there is a one-to-one correspondence between one-parameter subgroups of $\widetilde{G}$ and the Lie algebra of $\widetilde{G}$. By the uniqueness of lifting, $\exp_{\widetilde{G}}(tX) \exp_{\widetilde{G}}(sX) = \exp_{\widetilde{G}}((t+s)X)$ for $X \in \mathfrak{g}$ and $t, s \in \mathbb{R}$, so that $t \to \exp_{\widetilde{G}}(tX)$ is a one-parameter subgroup of $\widetilde{G}$. On the other hand, if $\gamma : \mathbb{R} \to \widetilde{G}$ is a one-parameter subgroup, then so is $\pi \circ \gamma : \mathbb{R} \to G$. Thus there is a unique $X \in \mathfrak{g}$, so that $\pi(\gamma(t)) = \exp_G(tX)$. As usual, the uniqueness property of lifting from $\mathbb{R}$ to $\widetilde{G}$ shows that $\gamma(t) = \exp_{\widetilde{G}}(tX)$, which finishes part (b).

For parts (c) and (d), we already know from Lemma 6.31 that $\widetilde{T} = \exp_{\widetilde{G}}(\mathfrak{t})$. Since $\mathfrak{t}$ is a Cartan subalgebra, Theorem 5.4 shows that $\widetilde{T}$ is a maximal torus of $\widetilde{G}$. By Lemma 1.21 and Corollary 5.13, $\ker \pi \subseteq Z(\widetilde{G}) \subseteq \widetilde{T}$ so that $\pi^{-1}(T) = \widetilde{T}(\ker \pi) = \widetilde{T}$ is, in fact, connected. $\square$

6.3.4 Exercises

Exercise 6.24 Suppose Λ_i is a lattice in $(i\mathfrak{t})^*$ that spans $(i\mathfrak{t})^*$.
(1) Show that Λ_i^* is a lattice in $i\mathfrak{t}$.
(2) Show that $\Lambda_i^{**} = \Lambda_i$.
(3) If $\Lambda_1 \subseteq \Lambda_2$, show that $\Lambda_2^* \subseteq \Lambda_1^*$.
(4) If $\Lambda_1 \subseteq \Lambda_2$, show that $\Lambda_2/\Lambda_1 \cong \Lambda_1^*/\Lambda_2^*$.

Exercise 6.25 (1) Use the standard root system notation from §6.1.5. In the following table, write (θ_i) for the element $\mathrm{diag}(\theta_1, \dots, \theta_n)$ in the case of $G = SU(n)$, for the element $\mathrm{diag}(\theta_1, \dots, \theta_n, -\theta_1, \dots, -\theta_n)$ in the cases of $G = Sp(n)$ or $SO(E_{2n})$, and for the element $\mathrm{diag}(\theta_1, \dots, \theta_n, -\theta_1, \dots, -\theta_n, 0)$ in the case of $G = SO(E_{2n+1})$. Verify that the following table is correct.

G	$R^\vee$	$\ker \mathcal{E}$	$P^\vee$	$P^\vee/R^\vee$
$SU(n)$	$\{(\theta_i) \mid \theta_i \in \mathbb{Z},$ $\sum_{i=1}^n \theta_i = 0\}$	$R^\vee$	$\{(\theta_i + \frac{\theta_0}{n}) \mid \theta_i \in \mathbb{Z},$ $\sum_{i=0}^n \theta_i = 0\}$	$\mathbb{Z}_n$
$Sp(n)$	$\{(\theta_i) \mid \theta_i \in \mathbb{Z}\}$	$R^\vee$	$\{(\theta_i + \frac{\theta_0}{2}) \mid \theta_i \in \mathbb{Z}\}$	$\mathbb{Z}_2$
$SO(E_{2n})$	$\{(\theta_i) \mid \theta_i \in \mathbb{Z},$ $\sum_{i=1}^n \theta_i \in 2\mathbb{Z}\}$	$\{(\theta_i) \mid \theta_i \in \mathbb{Z}\}$	$\{(\theta_i + \frac{\theta_0}{2}) \mid \theta_i \in \mathbb{Z}\}$	$\mathbb{Z}_2 \times \mathbb{Z}_2$ n even $\mathbb{Z}_4$ n odd
$SO(E_{2n+1})$	$\{(\theta_i) \mid \theta_i \in \mathbb{Z},$ $\sum_{i=1}^n \theta_i \in 2\mathbb{Z}\}$	$P^\vee$	$\{(\theta_i) \mid \theta_i \in \mathbb{Z}\}$	$\mathbb{Z}_2$.

(2) In the following table, write (λ_i) for the element $\sum_i \lambda_i \epsilon_i$. Verify that the following table is correct.

G	R	A	P	P/R
$SU(n)$	$\{(\lambda_i) \mid \lambda_i \in \mathbb{Z},$ $\sum_{i=1}^n \lambda_i = 0\}$	P	$\{(\lambda_i + \frac{\lambda_0}{n}) \mid \lambda_i \in \mathbb{Z},$ $\sum_{i=0}^n \lambda_i = 0\}$	$\mathbb{Z}_n$
$Sp(n)$	$\{(\lambda_i) \mid \lambda_i \in \mathbb{Z},$ $\sum_{i=1}^n \lambda_i \in 2\mathbb{Z}\}$	P	$\{(\lambda_i) \mid \lambda_i \in \mathbb{Z}\}$	$\mathbb{Z}_2$
$SO(E_{2n})$	$\{(\lambda_i) \mid \lambda_i \in \mathbb{Z},$ $\sum_{i=1}^n \lambda_i \in 2\mathbb{Z}\}$	$\{(\lambda_i) \mid \lambda_i \in \mathbb{Z}\}$	$\{(\lambda_i + \frac{\lambda_0}{2}) \mid \lambda_i \in \mathbb{Z}\}$	$\mathbb{Z}_2 \times \mathbb{Z}_2$ n even $\mathbb{Z}_4$ n odd
$SO(E_{2n+1})$	$\{(\lambda_i) \mid \lambda_i \in \mathbb{Z}\}$	R	$\{(\lambda_i + \frac{\lambda_0}{2}) \mid \lambda_i \in \mathbb{Z}\}$	$\mathbb{Z}_2$.

Exercise 6.26 Let G be a compact connected Lie group, T a maximal torus of G, $\widetilde{G}$ the simply connected covering of of G, $\pi : \widetilde{G} \to G$ the associated covering homomorphism, and $\widetilde{T} = \left[\pi^{-1}(T)\right]^0$. This exercise generalizes the proof of the Maximal Torus theorem, Theorem 5.12, to show that $\widetilde{G} = \bigcup_{\widetilde{g} \in \widetilde{G}} \left(c_{\widetilde{g}} \widetilde{T}\right)$ and $\widetilde{G} = \exp_{\widetilde{G}}(\mathfrak{g})$.

(1) Make use of Lemma 5.11 and the fact that $\ker \pi$ is discrete to show that $\ker(\mathrm{Ad} \circ \pi) = Z(\widetilde{G})$.

(2) Suppose $\widetilde{\varphi} : \mathfrak{g} \to \widetilde{G}$ is lift of a map $\varphi : \mathfrak{g} \to G$. Use the fact that π is a local diffeomorphism to show that $\widetilde{\varphi}$ is a local diffeomorphism if and only if φ is a local diffeomorphism.

(3) Use the uniqueness property of lifting to show that $\exp_{\widetilde{G}} \circ \mathrm{Ad}(\pi g) = c_g \circ \exp_{\widetilde{G}}$ for $g \in \widetilde{G}$.

(4) Show that $\bigcup_{g \in \widetilde{G}} c_g \widetilde{T} = \exp_{\widetilde{G}}(\mathfrak{g})$.

(5) If $\dim \mathfrak{g} = 1$, show that $G \cong S^1$ and $\mathfrak{g} \cong \widetilde{G} \cong \mathbb{R}$ with $\exp_{\widetilde{G}}$ being the identity map. Conclude that $\widetilde{G} = \exp_{\widetilde{G}}(\mathfrak{g})$.

(6) Assume $\dim \mathfrak{g} > 1$ and use induction on $\dim \mathfrak{g}$ to show that $\widetilde{G} = \exp_{\widetilde{G}}(\mathfrak{g})$ as outlined in the remaining steps. First, in the case where $\dim \mathfrak{g}' < \dim \mathfrak{g}$, show that $G \cong \left[G' \times T^k\right]/F$, where F is a finite Abelian group. Conclude that $\widetilde{G} \cong \widetilde{G}_{ss} \times \mathbb{R}^k$. Use the fact that the exponential map from $\mathbb{R}^k$ to T^k is surjective and the inductive hypothesis to show $\widetilde{G} = \exp_{\widetilde{G}}(\mathfrak{g})$.

(7) For the remainder, assume $\mathfrak{g}$ is semisimple, so that $\widetilde{T}$ is compact. Use Lemma 1.21 to show that $\ker \pi \subseteq Z(\widetilde{G})$. Conclude that $\widetilde{G}/Z(\widetilde{G}) \cong G/Z(G)$ and use this to show that $\exp_{\widetilde{G}}(\mathfrak{g})$ is compact and therefore closed.

(8) It remains to show that $\exp_{\widetilde{G}}(\mathfrak{g})$ is open. Fix $X_0 \in \mathfrak{g}$ and write $g_0 = \exp_{\widetilde{G}}(X_0)$. Use Theorem 4.6 to show that it suffices to consider $X_0 \neq 0$.

(9) As in the proof of Theorem 5.12, let $\mathfrak{a} = \mathfrak{z}_{\mathfrak{g}}(\pi g_0)$ and $\mathfrak{b} = \mathfrak{a}^{\perp}$. Consider the map $\widetilde{\varphi} : \mathfrak{a} \oplus \mathfrak{b} \to \widetilde{G}$ given by $\widetilde{\varphi}(X, Y) = g_0^{-1} \exp_{\widetilde{G}}(Y) g_0 \exp_{\widetilde{G}}(X) \exp_{\widetilde{G}}(-Y)$. Show that $\widetilde{\varphi}$ is a local diffeomorphism near 0. Conclude that $\{\exp_{\widetilde{G}}(Y) g_0 \exp_{\widetilde{G}}(X) \exp_{\widetilde{G}}(-Y) \mid X \in \mathfrak{a}, Y \in \mathfrak{b}\}$ contains a neighborhood of g_0 in $\widetilde{G}$.

(10) Let $\widetilde{A} = (\pi^{-1}A)^0$, a covering of the compact Lie subgroup $A = Z_G(\pi g_0)^0$ of G. Show that $\exp_{\widetilde{G}}(\mathfrak{a}) \subseteq \widetilde{A}$. Conclude that $\bigcup_{g \in \widetilde{G}} g^{-1} \widetilde{A} g$ contains a neighborhood of g_0 in $\widetilde{G}$.

(11) If $\dim \mathfrak{a} < \dim \mathfrak{g}$, use the inductive hypothese to show that $\widetilde{A} = \exp_{\widetilde{G}}(\mathfrak{a})$. Conclude that $\bigcup_{g \in \widetilde{G}} g^{-1} \widetilde{A} g = \bigcup_{g \in \widetilde{G}} \exp_{\widetilde{G}}(\mathrm{Ad}(\pi g)\mathfrak{a})$, so that $\exp_{\widetilde{G}}(\mathfrak{g})$ contains a neighborhood of g_0.

(12) Finally, if $\dim \mathfrak{a} = \dim \mathfrak{g}$, show that $g_0 \in Z(\widetilde{G})$. Let $\mathfrak{t}'$ be a Cartan subalgebra containing X_0 so that $\mathfrak{g} = \bigcup_{g \in \widetilde{G}} \mathrm{Ad}(\pi g)\mathfrak{t}'$. Show that $g_0 \exp_{\widetilde{G}}(\mathfrak{g}) \subseteq \exp_{\widetilde{G}}(\mathfrak{g})$. Conclude that $\exp_{\widetilde{G}}(\mathfrak{g})$ contains a neighborhood of g_0.

6.4 Weyl Group

6.4.1 Group Picture

Definition 6.34. Let G be a compact connected Lie group with maximal torus T. Let $N = N(T)$ be the normalizer in G of T, $N = \{g \in G \mid gTg^{-1} = T\}$. The *Weyl group* of G, $W = W(G) = W(G, T)$, is defined by $W = N/T$.

If T' is another maximal torus of G, Corollary 5.10 shows that there is a $g \in G$, so $c_g T = T'$. In turn, this shows that $c_g N(T) = N(T')$, so that $W(G, T) \cong W(G, T')$. Thus, up to isomorphism, the Weyl group is independent of the choice of maximal torus.

Given $w \in N$, $H \in \mathfrak{t}$, and $\lambda \in \mathfrak{t}^*$, define an action of N on $\mathfrak{t}$ and $\mathfrak{t}^*$ by

$$(6.35) \qquad w(H) = \mathrm{Ad}(w)H$$

$$[w(\lambda)](H) = \lambda(w^{-1}(H)) = \lambda(\mathrm{Ad}(w^{-1})H).$$

As usual, extend this to an action of N on $\mathfrak{t}_{\mathbb{C}}$, $i\mathfrak{t}$, $\mathfrak{t}_{\mathbb{C}}^*$, and $(i\mathfrak{t})^*$ by $\mathbb{C}$-linearity. As $\mathrm{Ad}(T)$ acts trivially on $\mathfrak{t}$, the action of N descends to an action of $W = N/T$.

Theorem 6.36. *Let G be a compact connected Lie group with a maximal torus T.*
(a) The action of W on $i\mathfrak{t}$ and on $(i\mathfrak{t})^*$ is faithful, i.e., a Weyl group element acts trivially if and only it is the identity element.
(b) For $w \in N$ and $\alpha \in \Delta(\mathfrak{g}_{\mathbb{C}}) \cup \{0\}$, $\mathrm{Ad}(w)\mathfrak{g}_{\alpha} = \mathfrak{g}_{w\alpha}$.
(c) The action of W on $(i\mathfrak{t})^*$ preserves and acts faithfully on $\Delta(\mathfrak{g}_{\mathbb{C}})$.
(d) The action of W on $i\mathfrak{t}$ preserves and acts faithfully on $\{h_{\alpha} \mid \alpha \in \Delta(\mathfrak{g}_{\mathbb{C}})\}$. Moreover, $wh_{\alpha} = h_{w\alpha}$.
(e) W is a finite group.
(f) Given $t_i \in T$, there exists $g \in G$ so $c_g t_1 = t_2$ if and only if there exists $w \in N$, so $c_w t_1 = t_2$.

Proof. For part (a), suppose $w \in N$ acts trivially on t via Ad. Since $\exp t = T$ and since $c_w \circ \exp = \exp \circ \mathrm{Ad}(w)$, this implies that $w \in Z_G(T)$. However, Corollary 5.13 shows that $Z_G(T) = T$ so that $w \in T$, as desired.

For part (b), let $w \in N$, $H \in t_{\mathbb{C}}$, and $X_\alpha \in g_\alpha$ and calculate

$$[H, \mathrm{Ad}(w)X_\alpha] = [\mathrm{Ad}(w^{-1})H, X_\alpha] = \alpha(\mathrm{Ad}(w^{-1})H)X_\alpha = [(w\alpha)(H)]X_\alpha,$$

which shows that $\mathrm{Ad}(w)g_\alpha \subseteq g_{w\alpha}$. Since $\dim g_\alpha = 1$ and since $\mathrm{Ad}(w)$ is invertible, $\mathrm{Ad}(w)g_\alpha = g_{w\alpha}$ and, in particular, $w\alpha \in \Delta(g_{\mathbb{C}})$. Noting that W acts trivially on $\mathfrak{z}(g) \cap t$, we may reduce to the case where g is semisimple. As $\Delta(g_{\mathbb{C}})$ spans $(it)^*$, parts (b) and (c) are therefore finished.

For part (d), calculate

$$B(u_{w\alpha}, H) = B(w\alpha)(H) = \alpha(w^{-1}H) = B(u_\alpha, w^{-1}H) = B(wu_\alpha, H),$$

so that $u_{w\alpha} = wu_\alpha$. Since the action of w preserves the Killing form, it follows that $wh_\alpha = h_{w\alpha}$, which finishes part (d). As $\Delta(g_{\mathbb{C}})$ is finite and the action is faithful, part (e) is also done.

For part (f), suppose $c_g t_1 = t_2$ for $g \in G$. Consider the connected compact Lie subgroup $Z_G(t_2)^0 = \{h \in G \mid c_{t_2}h = h\}^0$ of G with Lie algebra $\mathfrak{z}_g(t_2) = \{X \in g \mid \mathrm{Ad}(t_2)X = X\}$ (Exercise 4.22). Clearly $t \subseteq \mathfrak{z}_g(t_2)$ and t is still a Cartan subalgebra of $\mathfrak{z}_g(t_2)$. Therefore $T \subseteq Z_G(t_2)$ and T is a maximal torus of $Z_G(t_2)$. On the other hand, $\mathrm{Ad}(t_2)\,\mathrm{Ad}(g)H = \mathrm{Ad}(g)\,\mathrm{Ad}(t_1)H = \mathrm{Ad}(g)H$ for $H \in t$. Thus $\mathrm{Ad}(g)t$ is also a Cartan subalgebra in $\mathfrak{z}_g(t_2)$, and so $c_g T$ is a maximal torus in $Z_G(t_2)^0$. By Corollary 5.10, there is a $z \in Z_G(t_2)$, so that $c_z (c_g T) = T$, i.e., $zg \in N(T)$. Since $c_{zg}t_1 = c_z t_2 = t_2$, the proof is finished. □

6.4.2 Classical Examples

Here we calculate the Weyl group for each of the compact classical Lie groups. The details are straightforward matrix calculations and are mostly left as an exercise (Exercise 6.27).

6.4.2.1 $U(n)$ and $SU(n)$ For $U(n)$ let $T_{U(n)} = \{\mathrm{diag}(e^{i\theta_1}, \ldots, e^{i\theta_n}) \mid \theta_i \in \mathbb{R}\}$ be a maximal torus. Write $\mathcal{S}_n$ for the set of $n \times n$ *permutation matrices*. Recall that an element of $GL(n, \mathbb{C})$ is a permutation matrix if the entries of each row and column consists of a single one and $(n-1)$ zeros. Thus $\mathcal{S}_n \cong S_n$ where S_n is the *permutation group* on n letters. Since the set of eigenvalues is invariant under conjugation, any $w \in N$ must permute, up to scalar, the standard basis of $\mathbb{R}^n$. In particular, this shows that

$$N(T_{U(n)}) = \mathcal{S}_n T_{U(n)}$$
$$W \cong S_n$$
$$|W| = n!.$$

Write (θ_i) for the element $\mathrm{diag}(\theta_1, \ldots, \theta_n) \in t$ and (λ_i) for the element $\sum_i \lambda_i \epsilon_i \in (it)^*$. It follows that W acts on $it_{U(n)} = \{(\theta_i) \mid \theta_i \in \mathbb{R}\}$ and on $(it_{U(n)})^* = \{(\lambda_i) \mid \lambda_i \in \mathbb{R}\}$ by all permutations of the coordinates.

For $SU(n)$, let $T_{SU(n)} = T_{U(n)} \cap SU(n) = \{\mathrm{diag}(e^{i\theta_1}, \dots, e^{i\theta_n}) \mid \theta_i \in \mathbb{R}, \sum_i \theta_i = 0\}$ be a maximal torus. Note that $U(n) \cong (SU(n) \times S^1)/\mathbb{Z}_n$ with S^1 central, so that $W(SU(n)) \cong W(U(n))$. In particular for the A_{n-1} root system,

$$N(T_{SU(n)}) = (S_n T_{U(n)}) \cap SU(n)$$
$$W \cong S_n$$
$$|W| = n!.$$

As before, W acts on $it_{SU(n)} = \{(\theta_i) \mid \theta_i \in \mathbb{R}, \sum_i \theta_i = 0\}$ and $(it_{SU(n)})^* = \{(\lambda_i) \mid \lambda_i \in \mathbb{R}, \sum_i \lambda_i = 0\}$ by all permutations of the coordinates.

6.4.2.2 $Sp(n)$ For $Sp(n)$ realized as $Sp(n) \cong U(2n) \cap Sp(n, \mathbb{C})$, let

$$T = \{\mathrm{diag}(e^{i\theta_1}, \dots, e^{i\theta_n}, e^{-i\theta_1}, \dots, e^{-i\theta_n}) \mid \theta_i \in \mathbb{R}\}.$$

For $1 \leq i \leq n$, write $s_{1,i}$ for the matrix realizing the linear transformation that maps e_i, the ith standard basis vector of $\mathbb{R}^{2n}$, to $-e_{i+n}$, maps e_{i+n} to e_n, and fixes the remaining standard basis vectors. In particular, $s_{1,i}$ is just the natural embedding of $\begin{pmatrix} 0 & 1 \\ -1 & 0 \end{pmatrix}$ into $Sp(n)$ in the $i \times (n+i)^{\text{th}}$ submatrix. By considering eigenvalues, it is straightforward to check that for the C_n root system,

$$N(T) = \left\{ \begin{pmatrix} s & 0 \\ 0 & s \end{pmatrix} \mid s \in S_n \right\} \left\{ \prod_i s_{1,i}^{k_i} \mid 1 \leq i \leq n, k_i \in \{0, 1\} \right\} T$$

$$W \cong S_n \ltimes (\mathbb{Z}_2)^n$$
$$|W| = 2^n n!.$$

Write (θ_i) for the element $\mathrm{diag}(\theta_1, \dots, \theta_n, -\theta_1, \dots, -\theta_n) \in it$ and (λ_i) for the element $\sum_i \lambda_i \epsilon_i \in (it)^*$. Then W acts on $it = \{(\theta_i) \mid \theta_i \in \mathbb{R}\}$ and on $(it)^* = \{(\lambda_i) \mid \lambda_i \in \mathbb{R}\}$ by all permutations and all sign changes of the coordinates.

6.4.2.3 $SO(E_{2n})$ For $G = SO(E_{2n})$, let

$$T = \{\mathrm{diag}(e^{i\theta_1}, \dots, e^{i\theta_n}, e^{-i\theta_1}, \dots, e^{-i\theta_n}) \mid \theta_i \in \mathbb{R}\}$$

be a maximal torus. For $1 \leq i \leq n$, write $s_{2,i}$ for the matrix realizing the linear transformation that maps e_i, the ith standard basis vector of $\mathbb{R}^{2n}$, to e_{i+n}, maps e_{i+n} to e_n, and fixes the remaining standard basis vectors. In particular, $s_{2,i}$ is just the natural embedding of $\begin{pmatrix} 0 & 1 \\ 1 & 0 \end{pmatrix}$ into $O(E_{2n})$ in the $i \times (n+i)^{\text{th}}$ submatrix. Then for the D_n root system,

$$N(T) = \left\{ \begin{pmatrix} s & 0 \\ 0 & s \end{pmatrix} \mid s \in S_n \right\} \left\{ \prod_i s_{2,i}^{k_i} \mid 1 \leq i \leq n, k_i \in \{0, 1\}, \sum_i k_i \in 2\mathbb{Z} \right\} T$$

$$W \cong S_n \ltimes (\mathbb{Z}_2)^{n-1}$$
$$|W| = 2^{n-1} n!.$$

Write (θ_i) for the element $\mathrm{diag}(\theta_1, \dots, \theta_n, -\theta_1, \dots, -\theta_n) \in it$ and (λ_i) for the element $\sum_i \lambda_i \epsilon_i \in (it)^*$. Then W acts on $it = \{(\theta_i) \mid \theta_i \in \mathbb{R}\}$ and on $(it)^* = \{(\lambda_i) \mid \lambda_i \in \mathbb{R}\}$ by all permutations and all even sign changes of the coordinates.

6.4.2.4 $SO(E_{2n+1})$ For $G = SO(E_{2n+1})$, let

$$T = \{\mathrm{diag}(e^{i\theta_1}, \ldots, e^{i\theta_n}, e^{-i\theta_1}, \ldots, e^{-i\theta_n}, 1) \mid \theta_i \in \mathbb{R}\}$$

be a maximal torus. For $1 \le i \le n$, write $s_{3,i}$ for the matrix realizing the linear transformation that maps e_i, the ith standard basis vector of $\mathbb{R}^{2n+1}$, to e_{i+n}, maps e_{i+n} to e_n, maps e_{2n+1} to $-e_{2n+1}$, and fixes the remaining standard basis vectors. In particular, $s_{3,i}$ is just the natural embedding of $\begin{pmatrix} 0 & 1 & 0 \\ 1 & 0 & 0 \\ 0 & 0 & -1 \end{pmatrix}$ into $SO(E_{2n+1})$ in the $i \times (n+i) \times (2n+1)^{\text{th}}$ submatrix. Then for the B_n root system,

$$N(T) = \left\{ \begin{pmatrix} s & & \\ & s & \\ & & 1 \end{pmatrix} \mid s \in \mathcal{S}_n \right\} \left\{ \prod_i s_{3,i}^{k_i} \mid 1 \le i \le n, k_i \in \{0, 1\} \right\} T$$

$$W \cong S_n \ltimes (\mathbb{Z}_2)^n$$

$$|W| = 2^n n!.$$

Write (θ_i) for the element $\mathrm{diag}(\theta_1, \ldots, \theta_n, -\theta_1, \ldots, -\theta_n, 0) \in i\mathfrak{t}$ and (λ_i) for the element $\sum_i \lambda_i \epsilon_i \in (i\mathfrak{t})^*$. Then W acts on $i\mathfrak{t} = \{(\theta_i) \mid \theta_i \in \mathbb{R}\}$ and on $(i\mathfrak{t})^* = \{(\lambda_i) \mid \lambda_i \in \mathbb{R}\}$ by all permutations and all sign changes of the coordinates.

6.4.3 Simple Roots and Weyl Chambers

Definition 6.37. Let G be compact Lie group with a Cartan subalgebra $\mathfrak{t}$. Write $\mathfrak{t}' = \mathfrak{g}' \cap \mathfrak{t}$.
(a) A *system of simple roots*, $\Pi = \Pi(\mathfrak{g}_{\mathbb{C}})$, is a subset of $\Delta(\mathfrak{g}_{\mathbb{C}})$ that is a basis of $(i\mathfrak{t}')^*$ and satisfies the property that any $\beta \in \Delta(\mathfrak{g}_{\mathbb{C}})$ may be written as

$$\beta = \sum_{\alpha \in \Pi} k_\alpha \alpha$$

with either $\{k_\alpha \mid \alpha \in \Pi\} \subseteq \mathbb{Z}_{\ge 0}$ or $\{k_\alpha \mid \alpha \in \Pi\} \subseteq \mathbb{Z}_{\le 0}$, where $\mathbb{Z}_{\ge 0} = \{k \in \mathbb{Z} \mid k \ge 0\}$ and $\mathbb{Z}_{\le 0} = \{k \in \mathbb{Z} \mid k \le 0\}$. The elements of Π are called *simple roots*.
(b) Given a system of simple roots Π, the set of *positive roots* with respect to Π is

$$\Delta^+(\mathfrak{g}_{\mathbb{C}}) = \{\beta \in \Delta(\mathfrak{g}_{\mathbb{C}}) \mid \beta = \sum_{\alpha \in \Pi} k_\alpha \alpha \text{ with } k_\alpha \in \mathbb{Z}_{\ge 0}\}$$

and the set of *negative roots* with respect to Π is

$$\Delta^-(\mathfrak{g}_{\mathbb{C}}) = \{\beta \in \Delta(\mathfrak{g}_{\mathbb{C}}) \mid \beta = \sum_{\alpha \in \Pi} k_\alpha \alpha \text{ with } k_\alpha \in \mathbb{Z}_{\le 0}\},$$

so that $\Delta(\mathfrak{g}_{\mathbb{C}}) = \Delta^+(\mathfrak{g}_{\mathbb{C}}) \coprod \Delta^-(\mathfrak{g}_{\mathbb{C}})$ and $\Delta^-(\mathfrak{g}_{\mathbb{C}}) = -\Delta^+(\mathfrak{g}_{\mathbb{C}})$.

As matters stand at the moment, we are not guaranteed that simple systems exist. In Lemma 6.42 below, this shortcoming will be rectified using the following definition.

Definition 6.38. Let G be compact Lie group with a Cartan subalgebra t.
(a) The connected components of $(it')^* \setminus \left(\cup_{\alpha \in \Delta(\mathfrak{g}_{\mathbb{C}})} \alpha^\perp \right)$ are called the (open) *Weyl chambers* of $(it)^*$. The connected components of $it' \setminus \left(\cup_{\alpha \in \Delta(\mathfrak{g}_{\mathbb{C}})} h_\alpha^\perp \right)$ are called the (open) *Weyl chambers* of it.
(b) If C is a Weyl chamber of $(it)^*$, $\alpha \in \Delta(\mathfrak{g}_{\mathbb{C}})$ is called C-*positive* if $B(C, \alpha) > 0$ and C-*negative* if $B(C, \alpha) < 0$. If α is C-positive, it is called *decomposable* with respect to C if there exist C-positive $\beta, \gamma \in \Delta(\mathfrak{g}_{\mathbb{C}})$, so that $\alpha = \beta + \gamma$. Otherwise α is called *indecomposable* with respect to C.
(c) If $C^\vee$ is a Weyl chamber of it, $\alpha \in \Delta(\mathfrak{g}_{\mathbb{C}})$ is called $C^\vee$-*positive* if $\alpha(C^\vee) > 0$ and C-*negative* if $\alpha(C^\vee) < 0$. If α is $C^\vee$-positive, it is called *decomposable* with respect to $C^\vee$ if there exist $C^\vee$-positive $\beta, \gamma \in \Delta(\mathfrak{g}_{\mathbb{C}})$, so that $\alpha = \beta + \gamma$. Otherwise α is called *indecomposable* with respect to $C^\vee$.
(d) If C is a Weyl chamber of $(it)^*$, let

$$\Pi(C) = \{\alpha \in \Delta(\mathfrak{g}_{\mathbb{C}}) \mid \alpha \text{ is } C\text{-positive and indecomposable}\}.$$

If $C^\vee$ is a Weyl chamber of it, let

$$\Pi(C^\vee) = \{\alpha \in \Delta(\mathfrak{g}_{\mathbb{C}}) \mid \alpha \text{ is } C^\vee\text{-positive and indecomposable}\}.$$

(e) If Π is a system of simple roots, the *associated Weyl chamber* of $(it)^*$ is

$$C(\Pi) = \{\lambda \in (it)^* \mid B(\lambda, \alpha) > 0 \text{ for } \alpha \in \Pi\}$$

and the *associated Weyl chamber* of it is

$$C^\vee(\Pi) = \{H \in it \mid \alpha(H) > 0 \text{ for } \alpha \in \Pi\}.$$

Each Weyl chamber is a polyhedral convex cone and its closure is called the *closed Weyl chamber*. For the sake of symmetry, note that the condition $\alpha(H) > 0$ above is equivalent to the condition $B(H, h_\alpha) > 0$. In Lemma 6.42 we will see that the mapping $C \to \Pi(C)$ establishes a one-to-one correspondence between Weyl chambers and simple systems. For the time being, we list the standard simple systems and corresponding Weyl chamber of $(it)^*$ for the classical compact groups. The details are straightforward and left to Exercise 6.30 (see §6.1.5 for the roots and notation).

In addition to a simple system and its corresponding Weyl chamber, two other pieces of data are given below. For the first, write the given simple system as $\Pi = \{\alpha_1, \ldots, \alpha_l\}$. Define the *fundamental weights* to be the basis $\{\pi_1, \ldots, \pi_l\}$ of $(it)^*$ determined by $2 \frac{B(\pi_i, \alpha_i)}{B(\alpha_i, \alpha_i)} = \delta_{i,j}$ and define $\rho = \rho(\Pi) \in (it)^*$ as

(6.39)
$$\rho = \sum_i \pi_i$$

Notice $\rho(h_{\alpha_i}) = 2 \frac{B(\rho, \alpha_i)}{B(\alpha_i, \alpha_i)} = 1$, so that $\rho \in P$ (c.f. Exercise 6.34).

The second piece of data given below is called the *Dynkin diagram* of the simple system Π. The Dynkin diagram is a graph with one vertex for each simple root, α_i, and turns out to be independent of the choice of simple system. Whenever $B(\alpha_i, \alpha_j) \neq 0$, $i \neq j$, the vertices corresponding to α_i and α_j are joined by an edge of multiplicity $m_{ij} = \alpha_i(h_{\alpha_j})\alpha_j(h_{\alpha_i})$. In this case, from the proof of Corollary 6.22 (*c.f.* Exercise 6.20), it turns out that $m_{ij} = m_{ji} = \frac{\|\alpha_i\|^2}{\|\alpha_j\|^2} \in \{1, 2, 3\}$ when $\|\alpha_i\|^2 \geq \|\alpha_j\|^2$. Furthermore, when two vertices corresponding to roots of unequal length are connected by an edge, the edge is oriented by an arrow pointing towards the vertex corresponding to the shorter root.

6.4.3.1 $SU(n)$ For $SU(n)$ with $\mathfrak{t} = \{\text{diag}(i\theta_1, \ldots, i\theta_n) \mid \theta_i \in \mathbb{R}, \sum_i \theta_i = 0\}$, i.e., the A_{n-1} root system,

$$\Pi = \{\alpha_i = \epsilon_i - \epsilon_{i+1} \mid 1 \leq i \leq n - 1\}$$
$$C = \{\text{diag}(\theta_1, \ldots, \theta_n) \mid \theta_i > \theta_{i+1}, \theta_i \in \mathbb{R}\}$$
$$\rho = \frac{1}{2}((n-1)\epsilon_1 + (n-3)\epsilon_2 + \cdots + (-n+1)\epsilon_n)$$

and the corresponding Dynkin diagram is

A_n

$\alpha_1 \quad\quad \alpha_2 \quad\quad \alpha_3 \quad \cdots \quad \alpha_{n-3} \quad\quad \alpha_{n-2} \quad\quad \alpha_{n-1}$

6.4.3.2 $Sp(n)$ For $Sp(n)$ realized as $Sp(n) \cong U(2n) \cap Sp(n, \mathbb{C})$ with

$$\mathfrak{t} = \{\text{diag}(i\theta_1, \ldots, i\theta_n, -i\theta_1, \ldots, -i\theta_n) \mid \theta_i \in \mathbb{R}\},$$

i.e., the C_n root system,

$$\Pi = \{\alpha_i = \epsilon_i - \epsilon_{i+1} \mid 1 \leq i \leq n - 1\} \cup \{\alpha_n = 2\epsilon_n\}$$
$$C = \{\text{diag}(\theta_1, \ldots, \theta_n, -\theta_1, \ldots, -\theta_n) \mid \theta_i > \theta_{i+1} > 0, \theta_i \in \mathbb{R}\}$$
$$\rho = n\epsilon_1 + (n-1)\epsilon_2 + \cdots + \epsilon_n$$

and the corresponding Dynkin diagram is

C_n

$\alpha_1 \quad\quad \alpha_2 \quad\quad \alpha_3 \quad \cdots \quad \alpha_{n-2} \quad\quad \alpha_{n-1} \quad\quad \alpha_n$

6.4.3.3 $SO(E_{2n})$ For $SO(E_{2n})$ with $\mathfrak{t} = \{\text{diag}(i\theta_1, \ldots, i\theta_n, -i\theta_1, \ldots, -i\theta_n) \mid \theta_i \in \mathbb{R}\}$, i.e., the D_n root system,

$$\Pi = \{\alpha_i = \epsilon_i - \epsilon_{i+1} \mid 1 \leq i \leq n - 1\} \cup \{\alpha_n = \epsilon_{n-1} + \epsilon_n\}$$
$$C = \{\text{diag}(\theta_1, \ldots, \theta_n, -\theta_1, \ldots, -\theta_n, 0) \mid \theta_i > \theta_{i+1}, \theta_{n-1} > |\theta_n|, \theta_i \in \mathbb{R}\}$$
$$\rho = n\epsilon_1 + (n-1)\epsilon_2 + \cdots + \epsilon_{n-1}$$

and the corresponding Dynkin diagram is

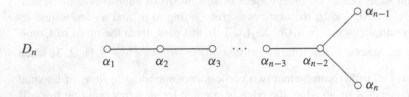

D_n

$$\alpha_1 \quad\quad \alpha_2 \quad\quad \alpha_3 \quad \cdots \quad \alpha_{n-3} \quad \alpha_{n-2} \quad\quad \begin{matrix} \alpha_{n-1} \\ \\ \alpha_n \end{matrix}$$

6.4.3.4 $SO(E_{2n+1})$ For $SO(E_{2n+1})$ with

$$\mathfrak{t} = \{\mathrm{diag}(i\theta_1, \dots, i\theta_n, -i\theta_1, \dots, -i\theta_n, 0) \mid \theta_i \in \mathbb{R}\},$$

i.e., the B_n root system,

$$\Pi = \{\alpha_i = \epsilon_i - \epsilon_{i+1} \mid 1 \le i \le n-1\} \cup \{\alpha_n = \epsilon_n\}$$
$$C = \{\mathrm{diag}(\theta_1, \dots, \theta_n, -\theta_1, \dots, -\theta_n, 0) \mid \theta_i > \theta_{i+1} > 0, \theta_i \in \mathbb{R}\}$$
$$\rho = \frac{1}{2}\left((2n-1)\epsilon_1 + (2n-3)\epsilon_2 + \cdots + \epsilon_n\right)$$

and the corresponding Dynkin diagram is

B_n

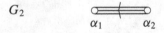

$$\alpha_1 \quad\quad \alpha_2 \quad\quad \alpha_3 \quad \cdots \quad \alpha_{n-2} \quad\quad \alpha_{n-1} \quad \alpha_n$$

It is an important fact from the theory of Lie algebras that there are only five other simple Lie algebras over $\mathbb{C}$. They are called the exceptional Lie algebras and there is a simple compact group corresponding to each one. The corresponding Dynkin diagrams are given below (see [56] or [70] for details).

G_2

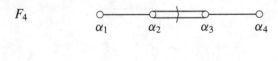

$$\alpha_1 \quad\quad \alpha_2$$

F_4

$$\alpha_1 \quad\quad \alpha_2 \quad\quad \alpha_3 \quad\quad \alpha_4$$

$$\alpha_2$$

E_6

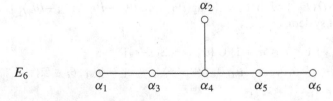

$$\alpha_1 \quad\quad \alpha_3 \quad\quad \alpha_4 \quad\quad \alpha_5 \quad\quad \alpha_6$$

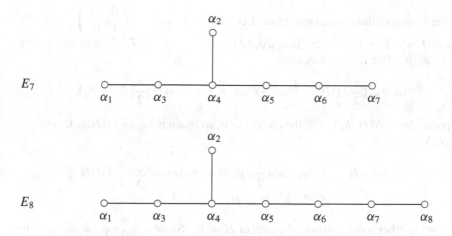

6.4.4 The Weyl Group as a Reflection Group

Definition 6.40. Let G be compact Lie group with a Cartan subalgebra $\mathfrak{t}$.
(a) For $\alpha \in \Delta(\mathfrak{g}_{\mathbb{C}})$, define $r_\alpha : (i\mathfrak{t})^* \to (i\mathfrak{t})^*$ by

$$r_\alpha(\lambda) = \lambda - 2\frac{B(\lambda, \alpha)}{B(\alpha, \alpha)}\alpha = \lambda - \lambda(h_\alpha)\alpha$$

and $r_{h_\alpha} : i\mathfrak{t} \to i\mathfrak{t}$ by

$$r_{h_\alpha}(H) = H - 2\frac{B(H, h_\alpha)}{B(h_\alpha, h_\alpha)}h_\alpha = H - \alpha(H)h_\alpha.$$

(b) Write $W(\Delta(\mathfrak{g}_{\mathbb{C}}))$ for the group generated by $\{r_\alpha \mid \alpha \in \Delta(\mathfrak{g}_{\mathbb{C}})\}$ and write $W(\Delta(\mathfrak{g}_{\mathbb{C}})^\vee)$ for the group generated by $\{r_{h_\alpha} \mid \alpha \in \Delta(\mathfrak{g}_{\mathbb{C}})\}$.

As usual, the action of $W(\Delta(\mathfrak{g}_{\mathbb{C}}))$ and $W(\Delta(\mathfrak{g}_{\mathbb{C}})^\vee)$ on $(i\mathfrak{t})^*$ and $i\mathfrak{t}$, respectively, is extended to an action on $\mathfrak{t}^*$ and $\mathfrak{t}$, respectively, by $\mathbb{C}$-linearity. Also observe that r_α acts trivially on $(\mathfrak{z}(\mathfrak{g}) \cap \mathfrak{t})^*$ and acts on $(i\mathfrak{t}')^*$ as the reflection across the hyperplane perpendicular to α. Similarly, r_{h_α} acts trivially on $\mathfrak{z}(\mathfrak{g}) \cap \mathfrak{t}$ and acts on $i\mathfrak{t}'$ as the reflection across the hyperplane perpendicular to h_α (Exercise 6.28). In particular, $r_\alpha^2 = I$ and $r_{h_\alpha}^2 = I$.

Lemma 6.41. *Let G be compact Lie group with a maximal torus T.*
(a) For $\alpha \in \Delta(\mathfrak{g}_{\mathbb{C}})$, there exists $w_\alpha \in N(T)$, so that the action of w_α on $(i\mathfrak{t})^*$ is given by r_α and the action of w_α on $i\mathfrak{t}$ is given by r_{h_α}.
(b) For $\alpha, \beta \in \Delta(\mathfrak{g}_{\mathbb{C}})$, $r_\alpha(\beta) \in \Delta(\mathfrak{g}_{\mathbb{C}})$ and $r_{h_\alpha}(h_\beta) = h_{r_\alpha(\beta)}$.

Proof. Using Theorem 6.20, choose a standard $\mathfrak{su}(2)$-triple, $\{\mathcal{I}_\alpha, \mathcal{J}_\alpha, \mathcal{K}_\alpha\}$, and a standard $\mathfrak{sl}(2, \mathbb{C})$-triple, $\{E_\alpha, H_\alpha, F_\alpha\}$, corresponding to α and let $\varphi_\alpha : SU(2) \to G$

be the corresponding homomorphism. Let $w = \exp(\frac{\pi}{2}J) = \begin{pmatrix} 0 & -1 \\ 1 & 0 \end{pmatrix} \in SU(2)$, where $J = -E + F \in \mathfrak{su}(2)$. Thus $d\varphi_\alpha(J) = \mathcal{J}_\alpha = -E_\alpha + F_\alpha$. Define $w_\alpha \in G$ by $w_\alpha = \varphi_\alpha(w)$. For $H \in i\mathfrak{t}$, calculate

$$\mathrm{ad}(d\varphi_\alpha(\frac{\pi}{2}J))H = \frac{\pi}{2}\,\mathrm{ad}(-E_\alpha + F_\alpha)H = \alpha(H)\frac{\pi}{2}[E_\alpha + F_\alpha].$$

In particular if $B(H, h_\alpha) = 0$, then $\alpha(H) = 0$, so that $\mathrm{ad}(d\varphi_\alpha(\frac{\pi}{2}J))H = 0$. Thus, if $B(H, h_\alpha) = 0$,

$$\mathrm{Ad}(w_\alpha)H = \mathrm{Ad}(\varphi_\alpha(\exp(\frac{\pi}{2}J)))H = \mathrm{Ad}(\exp(d\varphi_\alpha(\frac{\pi}{2}J)))H$$
$$= e^{\mathrm{ad}(d\varphi_\alpha(\frac{\pi}{2}J))}H = H.$$

On the other hand, consider the case of $H = h_\alpha$. Since $c_{\varphi_\alpha(w)} \circ \varphi_\alpha = \varphi_\alpha \circ c_w$, the differentials satisfy $\mathrm{Ad}(w_\alpha) \circ d\varphi_\alpha = d\varphi_\alpha \circ \mathrm{Ad}(w)$. Observing that $wHw^{-1} = -H$ in $\mathfrak{sl}(2, \mathbb{C})$, where $d\varphi_\alpha(H) = h_\alpha$, it follows that

$$\mathrm{Ad}(w_\alpha)h_\alpha = \mathrm{Ad}(w_\alpha)\varphi_\alpha(H)$$
$$= d\varphi_\alpha(\mathrm{Ad}(w)H) = -d\varphi_\alpha(H) = -h_\alpha.$$

Thus $\mathrm{Ad}(w_\alpha)$ preserves $\mathfrak{t}$ and acts on $i\mathfrak{t}$ as the reflection across the hyperplane perpendicular to h_α. In other words, $\mathrm{Ad}(w_\alpha)$ acts as r_{h_α} on $i\mathfrak{t}$. Since $T = \exp\mathfrak{t}$ and $c_{w_\alpha}(\exp H) = \exp(\mathrm{Ad}(w_\alpha)H)$, this also shows that $w_\alpha \in N(T)$. To finish part (a), calculate

$$(w_\alpha\lambda)(H) = \lambda(w_\alpha^{-1}H) = B(u_\lambda, \mathrm{Ad}(w_\alpha)^{-1}H) = B(\mathrm{Ad}(w_\alpha)u_\lambda, H)$$

for $\lambda \in (i\mathfrak{t})^*$. Thus for $\lambda = \alpha$, $\mathrm{Ad}(w_\alpha)u_\alpha = -u_\alpha$, so $w_\alpha\alpha = -\alpha$. For $\lambda \in \alpha^\perp$, $u_\lambda \in h_\alpha^\perp$, so $\mathrm{Ad}(w_\alpha)u_\lambda = u_\lambda$ and $w_\alpha\lambda = \lambda$. In particular, $\mathrm{Ad}(w_\alpha)$ acts on $(i\mathfrak{t})^*$ by r_α. Part (b) now follows from Theorem 6.36. $\square$

Lemma 6.42. *Let G be compact Lie group with a Cartan subalgebra $\mathfrak{t}$.*
(a) There is a one-to-one correspondence between

$$\{\text{systems of simple roots}\} \longleftrightarrow \{\text{Weyl chambers of } (i\mathfrak{t})^*\}.$$

The bijection maps a simple system Π to the Weyl chamber $C(\Pi)$ and maps a Weyl chamber C to the simple system $\Pi(C)$.
(b) There is a one-to-one correspondence between

$$\{\text{systems of simple roots}\} \longleftrightarrow \{\text{Weyl chambers of } i\mathfrak{t}\}.$$

The bijection maps a simple system Π to the Weyl chamber $C^\vee(\Pi)$ and maps a Weyl chamber $C^\vee$ to the simple system $\Pi(C^\vee)$.
(c) If Π is a simple system with $\alpha, \beta \in \Pi$, then $B(\alpha, \beta) \leq 0$.

Proof. Suppose $\Pi = \{\alpha_1, \ldots, \alpha_l\}$ is a simple system. From Equation 6.39, recall that $\rho \in (it)^*$ satisfies $B(\rho, \alpha_j) = \frac{\|\alpha\|^2}{2} > 0$, so that $\rho \in C(\Pi)$. For any $\lambda \in C(\Pi)$, examining the map $t \to B(t\lambda + (1 - t)\rho, \alpha)$ quickly shows that the line segment joining λ to ρ lies in $C(\Pi)$, so that $C(\Pi)$ is connected. Moreover, $B(\lambda, \alpha) > 0$ for $\alpha \in \Delta^+(\mathfrak{g}_\mathbb{C})$ and $B(\lambda, \beta) < 0$ for $\beta \in \Delta^-(\mathfrak{g}_\mathbb{C})$, so that $C(\Pi) \subseteq (it)^* \setminus (\cup_{\alpha \in \Delta(\mathfrak{g}_\mathbb{C})} \alpha^\perp)$. In particular, $C(\Pi) \subseteq C$ for some Weyl chamber C of $(it)^*$. As the sign of $B(\gamma, \alpha_j)$ is constant for $\gamma \in C$, the fact that $\rho \in C$ forces $\alpha_j(C) > 0$. Thus, $C \subseteq C(\Pi)$, so that $C(\Pi) = C$ is a Weyl chamber and the first half of part (a) is done.

Next, let C be a Weyl chamber of $(it)^*$ and fix $\lambda \in C$. If $\alpha = \beta_1 + \beta_2$ for C-positive roots α, β_i, then $B(\alpha, \lambda) > B(\beta_i, \lambda)$. Since $\{B(\lambda, \alpha) \mid \alpha \in \Delta(\mathfrak{g}_\mathbb{C})$ is C-positive$\}$ is a finite (nonempty) subset of positive real numbers, it is easy to see that $\Pi(C)$ is nonempty. It now follows from the definition of $\Pi(C)$ that any $\beta \in \Delta(\mathfrak{g}_\mathbb{C})$ may be written as $\beta = \sum_{\alpha \in \Pi} k_\alpha \alpha$ with either $\{k_\alpha \mid \alpha \in \Pi\} \subseteq \mathbb{Z}_{\geq 0}$ or $\{k_\alpha \mid \alpha \in \Pi\} \subseteq \mathbb{Z}_{\leq 0}$ depending on whether β is C-positive or C-negative. Since $\Delta(\mathfrak{g}_\mathbb{C})$ spans $(it)^*$, it only remains to see $\Pi(C)$ is an independent set.

Let $\alpha, \beta \in \Pi(C)$ be distinct and, without loss of generality, assume $B(\alpha, \alpha) \leq B(\beta, \beta)$. Positivity implies $\alpha \neq -\beta$ so that the proof of Corollary 6.22 (c.f. Exercise 6.20) shows that $\beta(h_\alpha) = 2\frac{B(\alpha, \beta)}{B(\beta, \beta)} \in \pm\{0, 1\}$. If $B(\alpha, \beta) > 0$, then $r_\alpha(\beta) = \beta - \alpha \in \Delta(\mathfrak{g}_\mathbb{C})$ by Lemma 6.41. If $\beta - \alpha$ is C-positive, then $\beta = (\beta - \alpha) + \alpha$. If $\beta - \alpha$ is C-negative, then $\alpha = -(\beta - \alpha) + \beta$. Either violates the assumption that α, β are indecomposable. Thus $\alpha, \beta \in \Pi(C)$ implies that $B(\alpha, \beta) \leq 0$ and will finish part (c) once part (a) is complete.

To see that $\Pi(C)$ is independent, suppose $\sum_{\alpha \in I_1} c_\alpha \alpha = \sum_{\beta \in I_2} c_\beta \beta$ with $c_\alpha, c_\beta \geq 0$ and $I_1 \bigsqcup I_2 = C(\Pi)$. Using the fact that $B(\alpha, \beta) \leq 0$, calculate

$$0 \leq \left\| \sum_{\alpha \in I_1} c_\alpha \alpha \right\|^2 = B(\sum_{\alpha \in I_1} c_\alpha \alpha, \sum_{\beta \in I_2} c_\beta \beta) = \sum_{\alpha \in I_1, \beta \in I_2} c_\alpha c_b B(\alpha, \beta) \leq 0.$$

Thus $0 = \sum_{\alpha \in I_1} c_\alpha \alpha$. Choosing any $\gamma \in C$, $0 = \sum_{\alpha \in I_1} c_\alpha B(\gamma, \alpha)$. Since $B(\gamma, \alpha) > 0$, $c_\alpha = 0$. Similarly $c_\beta = 0$ and part (a) is finished. As the proof of part (b) can either be done is a similar fashion or derived easily from part (a), it is omitted. $\square$

Theorem 6.43. *Let G be compact Lie group with a maximal torus T.*
(a) The action of $W(G)$ on it establishes an isomorphism $W(G) \cong W(\Delta(\mathfrak{g}_\mathbb{C})^\vee)$.
(b) The action of $W(G)$ on $(it)^*$ establishes an isomorphism $W(G) \cong W(\Delta(\mathfrak{g}_\mathbb{C}))$.
(c) $W(G)$ acts simply transitively on the set of Weyl chambers.

Proof. Using the faithful action of $W = W(G)$ on it via Ad from Theorem 6.36, identify W with the corresponding transformation group on it for the duration of this proof. Then Lemma 6.41 shows $r_{h_\alpha} \in W$ for each $\alpha \in \Delta(\mathfrak{g}_\mathbb{C})$, so that $W(\Delta(\mathfrak{g}_\mathbb{C})^\vee) \subseteq W$. It remains to show that $W \subseteq W(\Delta(\mathfrak{g}_\mathbb{C})^\vee)$ to finish the proof of part (a).

Reduce to the case where $\mathfrak{g}$ is semisimple. Fix a Weyl chamber C of it and fix $H \in C$. If $w \in W$, then W preserves $\{h_\alpha \mid \alpha \in \Delta(\mathfrak{g}_\mathbb{C})\}$. Since w leaves the Killing form invariant, w preserves $\{h_\alpha^\perp \mid \alpha \in \Delta(\mathfrak{g}_\mathbb{C})\}$, so that wC is also a Weyl chamber.

Let Ξ be the union of all intersections of hyperplanes of the form $h_\alpha^\perp$ for distinct $\alpha \in \Delta(\mathfrak{g}_\mathbb{C})$. As this is a finite union of subspaces of codimension at least 2, $(it) \setminus \Xi$

is path connected (Exercise 6.31). Thus there exists a piecewise linear path $\gamma(t)$: $[0, 1] \to i\mathfrak{t}$ from H to wH that does not intersect Ξ. Modifying $\gamma(t)$ if necessary (Exercise 6.31), there is a partition $\{s_i\}_{i=1}^N$ of $[0, 1]$, Weyl chambers C_i with $C_0 = C$ and $C_N = wC$, and roots α_i, so that $\gamma(s_{i-1}, s_i) \subseteq C_i$, $1 \le i \le N$, and $\gamma(s_i) \in h_{\alpha_i}^\perp$, $1 \le i \le N - 1$.

As $\gamma(t)$ does not intersect Ξ, there is an entire ball, B_i, around $\gamma(s_i)$ in $h_{\alpha_i}^\perp$ (of codimension 1 in $i\mathfrak{t}$) lying on the boundary of both C_{i-1} and C_i, $1 \le i \le N - 1$. For small nonzero ε, it follows that $B_i + \varepsilon h_\alpha$ lies in C_{i-1} or C_i, depending on the sign of ε. Since $r_{h_{\alpha_i}}(B_i + \varepsilon h_\alpha) = B_i - \varepsilon h_\alpha$ and since $r_{h_{\alpha_i}}$ preserves Weyl chambers, it follows that $r_{h_{\alpha_i}} C_{i-1} = C_i$. In particular, $r_{h_{\alpha_1}} \cdots r_{h_{\alpha_N}} wC = C$.

Now suppose $w_0 \in N(T)$ satisfies $\mathrm{Ad}(w_0)C = C$. To finish part (a), it suffices to show that $w_0 \in T$, so that w_0 acts by the identity operator on $i\mathfrak{t}$. Let $\Pi = \Pi(C)$ and define ρ as in Equation 6.39. By Lemma 6.42, it follows that $w\Pi = \Pi$, so that $\mathrm{Ad}(w_0)\rho = \rho$. Thus $c_{w_0} e^{itu_\rho} = e^{itu_\rho}$, $t \in \mathbb{R}$.

Choose a maximal torus S' of $Z_G(w_0)^0$ containing w_0. Note that S' is also a maximal torus of G by the Maximal Torus Theorem. Since $e^{i\mathbb{R}u_\rho}$ is in turn contained in some (other) maximal torus of $Z_G(w_0)^0$, Corollary 5.10 shows that there exists $g \in Z_G(w_0)^0$, so $c_g e^{i\mathbb{R}u_\rho} \subseteq S'$. Let S be the maximal torus of G given by $S = c_{g^{-1}}S'$. Then $w_0 \in S$ and $e^{i\mathbb{R}u_\rho} \subseteq S$ (c.f. Exercise 5.12). By definition of ρ and a simple a system, the root space decomposition of $\mathfrak{g}_\mathbb{C}$ shows $\mathfrak{z}_{\mathfrak{g}_\mathbb{C}}(u_\rho) = \mathfrak{t}$ so $\mathfrak{z}_{\mathfrak{g}}(iu_\rho) = \mathfrak{t}$. But since $\mathfrak{s} \subseteq \mathfrak{z}_{\mathfrak{g}}(iu_\rho) = \mathfrak{t}$, maximality implies $\mathfrak{s} = \mathfrak{t}$, and so $S = T$. Thus $w_0 \in T$, as desired.

Part (b) is done in a similar fashion to part (a). Part (c) is a corollary of the proof of part (a). $\square$

6.4.5 Exercises

Exercise 6.27 For each compact classical group G in §6.4.2, verify that the Weyl group and its action on $\mathfrak{t}$ and $(i\mathfrak{t})^*$ is correctly calculated.

Exercise 6.28 Let G be compact Lie group with semisimple Lie algebra and $\mathfrak{t}$ a Cartan subalgebra. For $\alpha \in \Delta(\mathfrak{g}_\mathbb{C})$, show that r_α is the reflection of $(i\mathfrak{t})^*$ across the hyperplane perpendicular to α and r_{h_α} is the reflection of $i\mathfrak{t}$ across the hyperplane perpendicular to h_α.

Exercise 6.29 Let G be a compact connected Lie group with a maximal torus T. Theorem 6.36 shows that the conjugacy classes of G are parametrized by the W-orbits in T. In fact, more is true. Show that there is a one-to-one correspondence between continuous class functions on G and continuous W-invariant functions on T (c.f. Exercise 7.10).

Exercise 6.30 For each compact classical Lie group in §6.4.3, verify that the given system of simple roots and corresponding Weyl chamber is correct.

Exercise 6.31 Suppose G is compact Lie group with semisimple Lie algebra $\mathfrak{g}$ and a Cartan subalgebra $\mathfrak{t}$.

(1) If Ξ is the union of all intersections of distinct hyperplanes of the form $h_\alpha^\perp$ for $\alpha \in \Delta(\mathfrak{g}_\mathbb{C})$, show $(it)\backslash\Xi$ is path connected.

(2) Suppose $\gamma(t) : [0, 1] \to it$ is a piecewise linear path that does not intersect Ξ with $\gamma(0)$ and $\gamma(1)$ elements of (different) Weyl chambers. Show there is a piecewise linear path $\gamma'(t) : [0, 1] \to it$ that does not intersect Ξ, satisfies $\gamma'(0) = \gamma(0)$ and $\gamma'(1) = \gamma(1)$, $\gamma(s_{i-1}, s_i) \subseteq C_i$, $1 \le i \le N$, and $\gamma(s_i) \in h_{\alpha_i}^\perp$, $1 \le i \le N - 1$, for some partition $\{s_i\}_{i=0}^N$ of $[0, 1]$, some Weyl chambers C_i, and some roots α_i.

Exercise 6.32 Let G be compact Lie group with semisimple Lie algebra $\mathfrak{g}$ and t a Cartan subalgebra of $\mathfrak{g}$. Fix a basis of $(it)^*$. With respect to this basis, the *lexicographic order* on $(it)^*$ is defined by setting $\alpha > \beta$ if the first nonzero coordinate (with respect to the given basis) of $\alpha - \beta$ is positive.

(1) Let $\Pi = \{\alpha \in \Delta(\mathfrak{g}_\mathbb{C}) \mid \alpha > 0 \text{ and } \alpha \ne \beta_1 + \beta_2 \text{ for any } \beta_i \in \Delta(\mathfrak{g}_\mathbb{C}) \text{ with } \beta_i > 0\}$. Show Π is a simple base of $\Delta(\mathfrak{g}_\mathbb{C})$ with $\Delta^+(\mathfrak{g}_\mathbb{C}) = \{\alpha \in \Delta(\mathfrak{g}_\mathbb{C}) \mid \alpha > 0\}$ and $\Delta^-(\mathfrak{g}_\mathbb{C}) = \{\alpha \in \Delta(\mathfrak{g}_\mathbb{C}) \mid \alpha < 0\}$.

(2) Show that all simple systems arise in this fashion.

(3) Show that there is a unique $\delta \in \Delta^+(\mathfrak{g}_\mathbb{C})$, so that $\delta > \beta$, $\beta \in \Delta^+(\mathfrak{g}_\mathbb{C})\backslash\{\delta\}$. The root δ is called the *highest root*. For the classical compact Lie groups, show δ is given by the following table:

G	$SU(n)$	$Sp(n)$	$SO(E_{2n})$	$SO(E_{2n+1})$
δ	$\epsilon_1 - \epsilon_n$	$2\epsilon_1$	$\epsilon_1 + \epsilon_2$	$\epsilon_1 + \epsilon_2$.

(4) Show that $B(\delta, \beta) \ge 0$ for all $\beta \in \Delta^+(\mathfrak{g}_\mathbb{C})$.

(5) For $G = SO(E_{2n+1})$, $n \ge 2$, show that there is another root besides δ satisfying the condition in part (4).

Exercise 6.33 Let G be compact Lie group with semisimple Lie algebra $\mathfrak{g}$ and t a Cartan subalgebra of $\mathfrak{g}$. Fix a simple system $\Pi = \{\alpha_i\}$ of $\Delta(\mathfrak{g}_\mathbb{C})$. For any $\beta \in \Delta^+(\mathfrak{g}_\mathbb{C})$, show that β can be written as $\beta = \alpha_{i_1} + \alpha_{i_2} + \cdots + \alpha_{i_N}$, where $\alpha_{i_1} + \alpha_{i_2} + \cdots + \alpha_{i_k} \subset \Delta^+(\mathfrak{g}_\mathbb{C})$ for $1 \le k \le N$.

Exercise 6.34 Let G be compact Lie group with a Cartan subalgebra t. Fix a simple system Π of $\Delta(\mathfrak{g}_\mathbb{C})$.

(1) For $\alpha \in \Pi$ and $\beta \in \Delta^+(\mathfrak{g}_\mathbb{C})\backslash\{\alpha\}$, write $r_\alpha\beta = \beta - 2\frac{B(\beta,\alpha)}{B(\alpha,\alpha)}\alpha$ to show that $\beta \in \Delta^+(\mathfrak{g}_\mathbb{C})$. Conclude that $r_\alpha(\Delta^+(\mathfrak{g}_\mathbb{C})\backslash\{\alpha\}) = \Delta^+(\mathfrak{g}_\mathbb{C})\backslash\{\alpha\}$.

(2) Let

$$\rho' = \frac{1}{2} \sum_{\beta \in \Delta^+(\mathfrak{g}_\mathbb{C})} \beta$$

and conclude from part (1) that $r_\alpha\rho' = \rho' - \alpha$. Use the definition of r_α to show that $\rho' = \rho$.

Exercise 6.35 Let G be compact Lie group with semisimple Lie algebra $\mathfrak{g}$ and t a Cartan subalgebra of $\mathfrak{g}$. Fix a simple system $\Pi = \{\alpha_i\}$ of $\Delta(\mathfrak{g}_\mathbb{C})$ and let $W(\Delta(\mathfrak{g}_\mathbb{C}))'$

be the subgroup of $W(\Delta(\mathfrak{g}_{\mathbb{C}}))$ generated $\{r_{\alpha_i}\}$.

(1) Given any $\beta \in \Delta(\mathfrak{g}_{\mathbb{C}})$, choose $x \in (\pm\beta)^\perp$ not lying on any other root hyperplane. For all sufficiently small $\varepsilon > 0$, show that $x + \varepsilon\beta$ lies in a Weyl chamber C' and that $\beta \in \Pi' = \Pi(C')$.

(2) Write ρ_Π for the element of $(i\mathfrak{t})^*$ satisfying $2\frac{B(\rho_\Pi, \alpha_i)}{B(\alpha_i, \alpha_i)} = 1$ from Equation 6.39 (c.f. Exercise 6.34) and choose $w \in W(\Delta(\mathfrak{g}_{\mathbb{C}}))'$ so that $B(w\rho_{\Pi'}, \rho_\Pi)$ is maximal. By examining $B(r_{\alpha_i}w\rho_{\Pi'}, \rho_\Pi)$, show that $w\rho_{\Pi'} \in C(\Pi)$. Conclude that $w\beta \in \Pi$.

(3) Show that $r_\beta = w^{-1}r_{w\beta}w$. Conclude that $W(\Delta(\mathfrak{g}_{\mathbb{C}}))' = W(\Delta(\mathfrak{g}_{\mathbb{C}}))$.

Exercise 6.36 Let G be compact Lie group with semisimple Lie algebra $\mathfrak{g}$ and $\mathfrak{t}$ a Cartan subalgebra of $\mathfrak{g}$. Fix a simple system $\Pi = \{\alpha_i\}$ of $\Delta(\mathfrak{g}_{\mathbb{C}})$ and recall Exercise 6.35. For $w \in W(\Delta(\mathfrak{g}_{\mathbb{C}}))$, let $n(w) = |\{\beta \in \Delta^+(\mathfrak{g}_{\mathbb{C}}) \mid w\beta \in \Delta^-(\mathfrak{g}_{\mathbb{C}})\}|$. For $w \neq I$, write $w = r_{\alpha_1} \cdots r_{\alpha_N}$ with N as small as possible. Then $r_{\alpha_1} \cdots r_{\alpha_N}$ is called a *reduced expression* for w. The *length* of w, with respect to Π, is defined by $l(w) = N$. For $w = I, l(I) = 0$.

(1) Use Exercise 6.34 to show

$$n(wr_{\alpha_i}) = \begin{cases} n(w) - 1 & \text{if } w\alpha_i \in \Delta^-(\mathfrak{g}_{\mathbb{C}}) \\ n(w) + 1 & \text{if } w\alpha_i \in \Delta^+(\mathfrak{g}_{\mathbb{C}}). \end{cases}$$

Conclude that $n(w) \leq l(w)$.

(2) Use Theorem 6.43 and induction on the length to show that $n(w) = l(w)$.

Exercise 6.37 (Chevalley's Lemma) Let G be compact Lie group with semisimple Lie algebra $\mathfrak{g}$ and $\mathfrak{t}$ a Cartan subalgebra of $\mathfrak{g}$. Fix $\lambda \in (i\mathfrak{t})^*$ and let $W_\lambda = \{w \in W(\Delta(\mathfrak{g}_{\mathbb{C}})) \mid w\lambda = \lambda\}$. Choose a Weyl chamber C, so that $\lambda \in \overline{C}$ and let $\Pi = \Pi(C)$.

(1) If $\beta \in \Delta(\mathfrak{g}_{\mathbb{C}})$ with $B(\lambda, \beta) > 0$, show that $\beta \in \Delta^+(\mathfrak{g}_{\mathbb{C}})$.

(2) If $\alpha \in \Pi$ and $w \in W_\lambda$ with $w\alpha \in \Delta^-(\mathfrak{g}_{\mathbb{C}})$, show $B(\lambda, \alpha) \leq 0$.

(3) *Chevalley's Lemma* states W_λ is generated by $W(\lambda) = \{r_\alpha \mid B(\lambda, \alpha) = 0\}$. Use Exercise 6.36 to prove this result. To this end, argue by contradiction and let $w \in W_\lambda \setminus \langle W(\lambda) \rangle$ be of minimal length.

(4) Show that the only reflections in $W(\Delta(\mathfrak{g}_{\mathbb{C}}))$ are of the form r_α for $\alpha \in W\Delta(\mathfrak{g}_{\mathbb{C}})$.

(5) If $W_\lambda \neq \{I\}$, show that there exists $\alpha \in \Delta(\mathfrak{g}_{\mathbb{C}})$ so $\lambda \in \alpha^\perp$.

Exercise 6.38 Let G be compact Lie group with semisimple Lie algebra $\mathfrak{g}$ and $\mathfrak{t}$ a Cartan subalgebra of $\mathfrak{g}$. Fix a simple system $\Pi = \{\alpha_i\}$ of $\Delta(\mathfrak{g}_{\mathbb{C}})$. For $w \in W(\Delta(\mathfrak{g}_{\mathbb{C}}))$, let $\text{sgn}(w) = (-1)^{l(w)}$ (c.f. Exercise 6.36). Show that $\text{sgn}(w) = \det(w)$.

Exercise 6.39 Let G be compact Lie group with semisimple Lie algebra $\mathfrak{g}$ and $\mathfrak{t}$ a Cartan subalgebra of $\mathfrak{g}$. Fix a Weyl chamber C and $H \in (i\mathfrak{t})^*$.

(1) Suppose $H \in \overline{C} \cap w\overline{C}$ for $w \in W(\Delta(\mathfrak{g}_{\mathbb{C}}))$. Show that $wH = H$.

(2) Let $H \in (i\mathfrak{t})^*$ be arbitrary. Show that $\overline{C}$ is a *fundamental chamber* for the action of $W(\Delta(\mathfrak{g}_{\mathbb{C}}))$, i.e., show that the Weyl group orbit of H intersects $\overline{C}$ in exactly one point.

Exercise 6.40 Let G be compact Lie group with semisimple Lie algebra $\mathfrak{g}$ and $\mathfrak{t}$ a Cartan subalgebra of $\mathfrak{g}$. Fix a simple system Π of $\Delta(\mathfrak{g}_{\mathbb{C}})$.

(1) Show that there is a unique $w_0 \in W(\Delta(\mathfrak{g}_{\mathbb{C}}))$, so that $w_0 \Pi = -\Pi$.

(2) Show that $w_0 = -I \in W(\Delta(\mathfrak{g}_{\mathbb{C}}))$ for G equal to $SU(2)$, $SO(E_{2n+1})$, $Sp(n)$, and $SO(E_{4n})$.

(3) Show that $w_0 \neq -I$, so $-I \notin W(\Delta(\mathfrak{g}_{\mathbb{C}}))$ for G equal to $SU(n)$ ($n \geq 3$) and $SO(E_{4n+2})$.

Exercise 6.41 Let G be compact Lie group with simple Lie algebra $\mathfrak{g}$ and $\mathfrak{t}$ a Cartan subalgebra of $\mathfrak{g}$. Fix a simple system $\Pi = \{\alpha_i\}$ of $\Delta(\mathfrak{g}_{\mathbb{C}})$.

(1) If α_i and α_j are joined by a single edge in the Dynkin diagram, show that there exists $w \in W(\Delta(\mathfrak{g}_{\mathbb{C}}))$, so that $\omega \alpha_i = \alpha_j$.

(2) If G is a classical compact Lie group, i.e., G is $SU(n)$, $Sp(n)$, $SO(E_{2n})$, or $SO(E_{2n+1})$, show the set of roots of a fixed length constitutes a single Weyl group orbit.

Exercise 6.42 Let G be compact Lie group with semisimple Lie algebra $\mathfrak{g}$ and $\mathfrak{t}$ a Cartan subalgebra of $\mathfrak{g}$. Fix a simple system $\Pi = \{\alpha_i\}$ of $\Delta(\mathfrak{g}_{\mathbb{C}})$ and let $\pi_i \in (i\mathfrak{t})^*$ be defined by the relation $2\frac{B(\pi_i, \alpha_j)}{B(\alpha_j, \alpha_j)} = \delta_{i,j}$.

(1) Show that $\{\alpha_i\}$ is a $\mathbb{Z}$-basis for the root lattice R and $\{\pi_i\}$ is a $\mathbb{Z}$-basis for the weight lattice P.

(2) Show the matrix $(B(\alpha_i, \alpha_j))$ is positive definite. Conclude $\det\left(2\frac{B(\alpha_i, \alpha_j)}{B(\alpha_j, \alpha_j)}\right) > 0$.

(3) It is well known from the study of free Abelian groups ([3]) that there exists a $\mathbb{Z}$-basis $\{\lambda_i\}$ of P and $k_i \in \mathbb{Z}$, so that $\{k_i \lambda_i\}$ is a basis for R. Thus there is a change of basis matrix from the basis $\{\lambda_i\}$ to $\{\pi_i\}$ with integral entries and determinant ± 1. Show that $|P/R| = \det\left(2\frac{B(\alpha_i, \alpha_j)}{B(\alpha_j, \alpha_j)}\right)$. The matrix $\left(2\frac{B(\alpha_i, \alpha_j)}{B(\alpha_j, \alpha_j)}\right)$ is called the *Cartan matrix* of $\mathfrak{g}_{\mathbb{C}}$.

Exercise 6.43 Let G be a compact Lie group with semisimple Lie algebra $\mathfrak{g}$ and $\mathfrak{t}$ a Cartan subalgebra of $\mathfrak{g}$. Fix a simple system $\Pi = \{\alpha_i\}$ of $\Delta(\mathfrak{g}_{\mathbb{C}})$. For each $\beta \in \Delta(\mathfrak{g}_{\mathbb{C}})$, choose a standard $\mathfrak{sl}(2, \mathbb{C})$-triple associated to β, $\{E_\beta, H_\beta, F_\beta\}$. Let $h = \sum_{\beta \in \Delta^+(\mathfrak{g}_{\mathbb{C}})} H_\beta$ and define $k_{\alpha_i} \in \mathbb{Z}_{>0}$, so $h = \sum_{\alpha_i \in \Pi} k_{\alpha_i} H_{\alpha_i}$. Set $e = \sum_i \sqrt{k_{\alpha_i}} E_{\alpha_i}$, $f = \sum_i k_{\alpha_i} F_{\alpha_i}$, and $\mathfrak{s} = \text{span}_{\mathbb{C}}\{e, h, f\}$.

(1) Show $\frac{B(h, h_{\alpha_i})}{B(h_{\alpha_i}, h_{\alpha_i})} = 1$ (c.f. Exercise 6.34).

(2) Show that $\mathfrak{s} \cong \mathfrak{sl}(2, \mathbb{C})$. The subalgebra $\mathfrak{s}$ is called the *principal three-dimensional subalgebra* of $\mathfrak{g}_{\mathbb{C}}$.

Exercise 6.44 Let G be a compact Lie group with semisimple Lie algebra $\mathfrak{g}$ and let T be a maximal torus of G. Fix a Weyl chamber C of $i\mathfrak{t}$ and let $N_G(C) = \{g \in G \mid \text{Ad}(g)C = C\}$. Show that the inclusion map of $N_G(C) \to G$ induces an isomorphism $N_G(C)/T \cong G^0/G$.

Highest Weight Theory

By studying the L^2 functions on a compact Lie group G, the Peter–Weyl Theorem gives a simultaneous construction of all irreducible representations of G. Two important problems remain. The first is to parametrize $\widehat{G}$ in a reasonable manner and the second is to individually construct each irreducible representation in a natural way. The solution to both of these problems is closely tied to the notion of *highest weights*.

7.1 Highest Weights

In this section, let G be a compact Lie group, T a maximal torus, and $\Delta^+(\mathfrak{g}_\mathbb{C})$ a system of positive roots with corresponding simple system $\Pi(\mathfrak{g}_\mathbb{C})$. Write

$$n^\pm = \bigoplus_{\alpha \in \Delta^\pm(\mathfrak{g}_\mathbb{C})} \mathfrak{g}_\alpha,$$

so that

(7.1) $$\mathfrak{g}_\mathbb{C} = n^- \oplus t_\mathbb{C} \oplus n^+$$

by the root space decomposition. Equation 7.1 is sometimes called a *triangular decomposition* of $\mathfrak{g}_\mathbb{C}$ since $n^\pm$ can be chosen to be the set of strictly upper, respectively lower, triangular matrices in the case where G is $GL(n, \mathbb{F})$. Notice that $[t_\mathbb{C} \oplus n^+, n^+] \subseteq n^+$ and $[t_\mathbb{C} \oplus n^-, n^-] \subseteq n^-$.

Definition 7.2. Let V be a representation of $\mathfrak{g}$ with weight space decomposition $V = \bigoplus_{\lambda \in \Delta(V)} V_\lambda$.
(a) A nonzero $v \in V_{\lambda_0}$ is called a *highest weight vector* of weight λ_0 with respect to $\Delta^+(\mathfrak{g}_\mathbb{C})$ if $n^+ v = 0$, i.e., if $Xv = 0$ for all $X \in n^+$. In this case, λ_0 is called a *highest weight* of V.
(b) A weight λ is said to be *dominant* if $B(\lambda, \alpha) \geq 0$ for all $\alpha \in \Pi(\mathfrak{g}_\mathbb{C})$, i.e., if λ lies in the closed Weyl chamber corresponding to $\Delta^+(\mathfrak{g}_\mathbb{C})$.

As an example, recall that the action of $\mathfrak{su}(2)_{\mathbb{C}} = \mathfrak{sl}(2, \mathbb{C})$ on $V_n(\mathbb{C}^2)$, $n \in \mathbb{Z}^{\geq 0}$, from Equation 6.7 is given by

$$E \cdot (z_1^k z_2^{n-k}) = -k \, z_1^{k-1} z_2^{n-k+1}$$
$$H \cdot (z_1^k z_2^{n-k}) = (n - 2k) \, z_1^k z_2^{n-k}$$
$$F \cdot (z_1^k z_2^{n-k}) = (k - n) \, z_1^{k+1} z_2^{n-k-1},$$

and recall that $\{V_n(\mathbb{C}^2) \mid n \in \mathbb{Z}^{\geq 0}\}$ is a complete list of irreducible representations for $SU(2)$. Taking $it = \mathrm{diag}(\theta, -\theta)$, $\theta \in \mathbb{R}$, there are two roots, $\pm \epsilon_{12}$, where $\epsilon_{12}(\mathrm{diag}(\theta, -\theta)) = 2\theta$. Choosing $\Delta^+(\mathfrak{sl}(2, \mathbb{C})) = \{\epsilon_{12}\}$, it follows that z_2^n is a highest weight vector of $V_n(\mathbb{C}^2)$ of weight $n\frac{\epsilon_{12}}{2}$. Notice that the set of dominant analytically integral weights is $\{n\frac{\epsilon_{12}}{2} \mid n \in \mathbb{Z}^{\geq 0}\}$. Thus there is a one-to-one correspondence between the set of highest weights of irreducible representations of $SU(2)$ and the set of dominant analytically integral weights. This correspondence will be established for all connected compact groups in Theorem 7.34.

Theorem 7.3. *Let G be a connected compact Lie group and V an irreducible representation of G.*
(a) V has a unique highest weight, λ_0.
(b) The highest weight λ_0 is dominant and analytically integral, i.e., $\lambda_0 \in A(T)$.
(c) Up to nonzero scalar multiplication, there is a unique highest weight vector.
(d) Any weight $\lambda \in \Delta(V)$ is of the form

$$\lambda = \lambda_0 - \sum_{\alpha_i \in \Pi(\mathfrak{g}_{\mathbb{C}})} n_i \alpha_i$$

for $n_i \in \mathbb{Z}^{\geq 0}$.
(e) For $w \in W$, $w V_\lambda = V_{w\lambda}$, so that $\dim V_\lambda = \dim V_{w\lambda}$. Here $W(G)$ is identified with $W(\Delta(\mathfrak{g}_{\mathbb{C}}))$, as in Theorem 6.43 via the Ad-action from Equation 6.35.
(f) Using the norm induced by the Killing form, $\|\lambda\| \leq \|\lambda_0\|$ with equality if and only if $\lambda = w\lambda_0$ for $w \in W(\mathfrak{g}_{\mathbb{C}})$.
(g) Up to isomorphism, V is uniquely determined by λ_0.

Proof. Existence of a highest weight λ_0 follows from the finite dimensionality of V and Theorem 6.11. Let v_0 be a highest weight vector for λ_0 and inductively define $V_n = V_{n-1} + \mathfrak{n}^- V_{n-1}$ where $V_0 = \mathbb{C}v_0$. This defines a filtration on the $(\mathfrak{n}^- \oplus \mathfrak{t}_{\mathbb{C}})$-invariant subspace $V_\infty = \cup_n V_n$ of V. If $\alpha \in \Pi(\mathfrak{g}_{\mathbb{C}})$, then $[\mathfrak{g}_\alpha, \mathfrak{n}^-] \subseteq \mathfrak{n}^- \oplus \mathfrak{t}_{\mathbb{C}}$. Since $\mathfrak{g}_\alpha V_0 = 0$, a simple inductive argument shows that $\mathfrak{g}_\alpha V_n \subseteq V_n$. In particular, this suffices to demonstrate that V_∞ is $\mathfrak{g}_{\mathbb{C}}$-invariant. Irreducibility implies $V = V_\infty$ and part (d) follows.

If λ_1 is also a highest weight, then $\lambda_1 = \lambda_0 - \sum n_i \alpha_i$ and $\lambda_0 = \lambda_1 - \sum m_i \alpha_i$ for $n_i, m_i \in \mathbb{Z}^{\geq 0}$. Eliminating λ_1 and λ_0 shows that $-\sum n_i \alpha_i = \sum m_i \alpha_i$. Thus $-n_i = m_i$, so that $n_i = m_i = 0$ and $\lambda_1 = \lambda_0$. Furthermore, the weight decomposition shows that $V_\infty \cap V_{\lambda_0} = V_0 = \mathbb{C}v_0$, so that parts (a) and (c) are complete.

The proof of part (e) is done in the same way as the proof of Theorem 6.36. For part (b), notice that $r_{\alpha_i} \lambda_0$ is a weight by part (e). Thus

$$\lambda_0 - 2\frac{B(\lambda_0, \alpha_i)}{B(\alpha_i, \alpha_i)}\alpha_i = \lambda_0 - \sum_{\alpha_j \in \Pi(\mathfrak{g}_\mathbb{C})} n_j \alpha_j$$

for $n_j \in \mathbb{Z}^{\geq 0}$. Hence $2\frac{B(\lambda_0, \alpha_i)}{B(\alpha_i, \alpha_i)} = n_i$, so that $B(\lambda_0, \alpha_i) \geq 0$ and λ_0 is dominant. Theorem 6.27 shows that λ_0 (in fact, any weight of V) is analytically integral.

For part (f), Theorem 6.43 shows that it suffices to take λ dominant by using the Weyl group action. Write $\lambda = \lambda_0 - \sum n_i \alpha_i$. Solving for λ_0 and using dominance in the second line,

$$\|\lambda_0\|^2 = \|\lambda\|^2 + 2 \sum_{\alpha_i \in \Pi(\mathfrak{g}_\mathbb{C})} n_i B(\lambda, \alpha_i) + \left\| \sum_{\alpha_i \in \Pi(\mathfrak{g}_\mathbb{C})} n_i \alpha_i \right\|^2$$

$$\geq \|\lambda\|^2 + \left\| \sum_{\alpha_i \in \Pi(\mathfrak{g}_\mathbb{C})} n_i \alpha_i \right\|^2 \geq \|\lambda\|^2 .$$

In the case of equality, it follows that $\sum_{\alpha_i \in \Pi(\mathfrak{g}_\mathbb{C})} n_i \alpha_i = 0$, so that $n_i = 0$ and $\lambda = \lambda_0$.

For part (g), suppose V' is an irreducible representation of G with highest weight λ_0 and corresponding highest weight vector v_0'. Let $W = V \oplus V'$ and define $W_n = W_{n-1} + \mathfrak{n}^- W_{n-1}$, where $W_0 = \mathbb{C}(v_0, v_0')$. As above, $W_\infty = \cup_n W_n$ is a subrepresentation of $V \oplus V'$. If U is a nonzero subrepresentation of W_∞, then U has a highest weight vector, (u_0, u_0'). In turn, this means that u_0 and u_0' are highest weight vectors of V and V', respectively. Part (a) then shows that $\mathbb{C}(u_0, u_0') = W_0$. Thus $U = W_\infty$ and W_∞ is irreducible. Projection onto each coordinate establishes the G-intertwining map $V \cong V'$. □

The above theorem shows that highest weights completely classify irreducible representations. It only remains to parametrize all possible highest weights of irreducible representations. This will be done in §7.3.5 where we will see there is a bijection between the set of dominant analytically integral weights and irreducible representations of G.

Definition 7.4. Let G be connected and let V be an irreducible representation of G with highest weight λ. As V is uniquely determined by λ, write $V(\lambda)$ for V and write χ_λ for its character.

Lemma 7.5. *Let G be connected. If $V(\lambda)$ is an irreducible representation of G, then $V(\lambda)^* \cong V(-w_0\lambda)$, where $w_0 \in W(\Delta(\mathfrak{g}_\mathbb{C}))$ is the unique element mapping the positive Weyl chamber to the negative Weyl chamber (c.f. Exercise 6.40).*

Proof. Since $V(\lambda)$ is irreducible, the character theory of Theorems 3.5 and 3.7 show that $V(\lambda)^*$ is irreducible. It therefore suffices to show that the highest weight of $V(\lambda)^*$ is $-w_0\lambda$.

Fix a G-invariant inner product, $(\,,\,)$, on $V(\lambda)$, so that $V(\lambda)^* = \{\mu_v \mid v \in V(\lambda)\}$, where $\mu_v(v') = (v', v)$ for $v' \in V(\lambda)$. By the invariance of the form, $g\mu_v = \mu_{gv}$ for $g \in G$, so that $X\mu_v = \mu_{Xv}$ for $X \in \mathfrak{g}$. Since $(\cdot, \cdot)$ is Hermitian, it follows that $Z\mu_v = \mu_{\theta(Z)v}$ for $Z \in \mathfrak{g}_\mathbb{C}$.

Let v_λ be a highest weight vector for $V(\lambda)$. Identifying $W(G)$ with $W(\Delta(\mathfrak{g}_{\mathbb{C}})^\vee)$ and $W(\Delta(\mathfrak{g}_{\mathbb{C}}))$ as in Theorem 6.43 via the Ad-action of Equation 6.35, it follows from Theorem 7.3 that $w_0 v_\lambda$ is a weight vector of weight $w_0\lambda$ (called the *lowest weight vector*). As $\theta(Y) = -Y$ for $Y \in i\mathfrak{t}$ and since weights are real valued on $i\mathfrak{t}$, it follows that $\mu_{w_0 v_\lambda}$ is a weight vector of weight $-w_0\lambda$.

It remains to see that $\mathfrak{n}^- w_0 v_\lambda = 0$ since Lemma 6.14 shows $\theta\mathfrak{n}^+ = \mathfrak{n}^-$. By construction, $w_0\Delta^+(\mathfrak{g}_{\mathbb{C}}) = \Delta^-(\mathfrak{g}_{\mathbb{C}})$ and $w_0^2 = I$, so that $\mathrm{Ad}(w_0)\mathfrak{n}^- = \mathfrak{n}^+$. Thus

$$\mathfrak{n}^- w_0 v_\lambda = w_0 \left(\mathrm{Ad}(w_0^{-1})\mathfrak{n}^- \right) v_\lambda = w_0 \mathfrak{n}^+ v_\lambda = 0$$

and the proof is complete. $\qquad\square$

7.1.1 Exercises

Exercise 7.1 Consider the representation of $SU(n)$ on $\bigwedge^p \mathbb{C}^n$. For T equal to the usual set of diagonal elements, show that a basis of weight vectors is given by vectors of the form $e_{l_1} \wedge \cdots \wedge e_{l_p}$ with weight $\sum_{i=1}^p \epsilon_{l_i}$. Verify that only $e_1 \wedge \cdots \wedge e_p$ is a highest weight to conclude that $\bigwedge^p \mathbb{C}^n$ is an irreducible representation of $SU(n)$ with highest weight $\sum_{i=1}^p \epsilon_i$.

Exercise 7.2 Recall that $V_p(\mathbb{R}^n)$, the space of complex-valued polynomials on $\mathbb{R}^n$ homogeneous of degree p, and $\mathcal{H}_p(\mathbb{R}^n)$, the harmonic polynomials, are representations of $SO(n)$. Let T be the standard maximal torus given in §5.1.2.3 and §5.1.2.4, let $h_j = E_{2j-1,2j} - E_{2j,2j-1} \in \mathfrak{t}$, $1 \le k \le m \equiv \lfloor \frac{n}{2} \rfloor$, i.e., h_j is an embedding of $\begin{pmatrix} 0 & 1 \\ -1 & 0 \end{pmatrix}$, and let $\epsilon_j \in \mathfrak{t}^*$ be defined by $\epsilon_j(h_{j'}) = -i\delta_{j,j'}$ (c.f. Exercise 6.14).
(1) Show that h_j acts on $V_p(\mathbb{R}^n)$ by the operator $-x_{2j}\partial_{x_{2j-1}} + x_{2j-1}\partial_{x_{2j}}$.
(2) For $n = 2m + 1$, conclude that a basis of weight vectors is given by polynomials of the form

$$(x_1 + ix_2)^{j_1} \cdots (x_{2m-1} + ix_{2m})^{j_m} (x_1 - ix_2)^{k_1} \cdots (x_{2m-1} - ix_{2m})^{k_m} x_{2m+1}^{l_0},$$

$l_0 + \sum_i j_i + \sum_i k_i = p$, each with weight $\sum_i (k_i - j_i)\epsilon_i$.
(3) For $n = 2m$, conclude that a basis of weight vectors is given by polynomials of the form

$$(x_1 + ix_2)^{j_1} \cdots (x_{n-1} + ix_n)^{j_m} (x_1 - ix_2)^{k_1} \cdots (x_{n-1} - ix_n)^{k_m},$$

$\sum_i j_i + \sum_i k_i = p$, each with weight $\sum_i (k_i - j_i)\epsilon_i$.
(4) Using the root system of $\mathfrak{so}(n, \mathbb{C})$ and Theorem 2.33, conclude that the weight vector $(x_1 - ix_2)^p$ of weight $p\epsilon_1$ must be the highest weight vector of $\mathcal{H}_p(\mathbb{R}^n)$ for $n \ge 3$.
(5) Using Lemma 2.27, show that a basis of highest weight vectors for $V_p(\mathbb{R}^n)$ is given by the vectors $(x_1 - ix_2)^{p-2j} \|x\|^{2j}$ of weight $(p - 2j)\epsilon_1$, $1 \le j \le m$.

Exercise 7.3 Consider the representation of $SO(n)$ on $\bigwedge^p \mathbb{C}^n$ and continue the notation from Exercise 7.2.

(1) For $n = 2m + 1$, examine the wedge product of elements of the form $e_{2j-1} \pm i e_{2j}$ as well as e_{2m+1} to find a basis of weight vectors (the weights will be of the form $\pm \epsilon_{j_1} \cdots \pm \epsilon_{j_r}$ with $1 \le j_1 < \ldots < j_r \le p$). For $p \le m$, show that only one is a highest weight vector and conclude that $\bigwedge^p \mathbb{C}^n$ is irreducible with highest weight $\sum_{i=1}^p \epsilon_i$.

(2) For $n = 2m$, examine the wedge product of elements of the form $e_{2j-1} \pm i e_{2j}$ to find a basis of weight vectors. For $p < m$, show that only one is a highest weight vector and conclude that $\bigwedge^p \mathbb{C}^n$ is irreducible with highest weight $\sum_{i=1}^p \epsilon_i$. For $p = m$, show that there are exactly two highest weights and that they are $\sum_{i=1}^{m-1} \epsilon_i \pm \epsilon_m$. In this case, conclude that $\bigwedge^m \mathbb{C}^n$ is the direct sum of two irreducible representations.

Exercise 7.4 Let G be a compact Lie group, T a maximal torus, and $\Delta^+(\mathfrak{g}_\mathbb{C})$ a system of positive roots with respect to $\mathfrak{t}_\mathbb{C}$ with corresponding simple system $\Pi(\mathfrak{g}_\mathbb{C})$.
(1) If $V(\lambda)$ and $V(\lambda')$ are irreducible representations of G, show that the weights of $V(\lambda) \otimes V(\lambda')$ are of the form $\mu + \mu'$, where μ is a weight of $V(\lambda)$ and μ' is a weight of $V(\lambda')$.
(2) By looking at highest weight vectors, show $V(\lambda + \lambda')$ appears exactly once as a summand in $V(\lambda) \otimes V(\lambda')$.
(3) Suppose $V(\nu)$ is an irreducible summand of $V(\lambda) \otimes V(\lambda')$ and write the highest weight vector of $V(\nu)$ in terms of the weights of $V(\lambda) \otimes V(\lambda')$. By considering a term in which the contribution from $V(\lambda)$ is as large as possible, show that $\nu = \lambda + \mu'$ for a weight μ' of $V(\lambda')$.

Exercise 7.5 Recall that $V_{p,q}(\mathbb{C}^n)$ from Exercise 2.35 is a representations of $SU(n)$ on the set of complex polynomials homogeneous of degree p in $z_1, \ldots, z_n$ and homogeneous of degree q in $\overline{z_1}, \ldots, \overline{z_n}$ and that $\mathcal{H}_{p,q}(\mathbb{C}^n)$ is an irreducible subrepresentation.
(1) If $H = \mathrm{diag}(t_1, \ldots, t_n)$ with $\sum_j t_j = 0$, show that H acts on $V_{p,q}(\mathbb{C}^n)$ as $\sum_j t_j(-z_j \partial_{z_j} + \overline{z_j} \partial_{\overline{z_j}})$.
(2) Conclude that $z_1^{k_1} \cdots z_n^{k_n} \overline{z_1}^{l_1} \cdots \overline{z_n}^{l_n}$, $\sum_j k_j = p$ and $\sum_j l_j = q$, is a weight vector of weight $\sum_j (l_j - k_j) \epsilon_j$.
(3) Show that $-p \epsilon_n$ is a highest weight of $V_{p,0}(\mathbb{C}^n)$.
(4) Show that $q \epsilon_1$ is a highest weight of $V_{0,q}(\mathbb{C}^n)$.
(5) Show that $q \epsilon_1 - p \epsilon_n$ is the highest weight of $\mathcal{H}_{p,q}(\mathbb{C}^n)$.

Exercise 7.6 Since $\mathrm{Spin}_n(\mathbb{R})$ is the simply connected cover of $SO(n)$, $n \ge 3$, the Lie algebra of $\mathrm{Spin}_n(\mathbb{R})$ can be identified with $\mathfrak{so}(n)$ (a maximal torus for $\mathrm{Spin}_n(\mathbb{R})$ is given in Exercise 5.5).
(1) For $n = 2m + 1$, show that the weights of the spin representation S are all weights of the form $\frac{1}{2}(\pm \epsilon_1 \cdots \pm \epsilon_m)$ and that the highest weight is $\frac{1}{2}(\epsilon_1 + \cdots + \epsilon_m)$.
(2) For $n = 2m$, show that the weights of the half-spin representation S^+ are all weights of the form $\frac{1}{2}(\pm \epsilon_1 \cdots \pm \epsilon_m)$ with an even number of minus signs and that the highest weight is $\frac{1}{2}(\epsilon_1 + \cdots + \epsilon_{m-1} + \epsilon_m)$.
(3) For $n = 2m$, show that the weights of the half-spin representation S^- are all weights of the form $\frac{1}{2}(\pm \epsilon_1 \cdots \pm \epsilon_m)$ with an odd number of minus signs and that the highest weight is $\frac{1}{2}(\epsilon_1 + \cdots + \epsilon_{m-1} - \epsilon_m)$.

7.2 Weyl Integration Formula

Let G be a compact connected Lie group, T a maximal torus, and $f \in C(G)$. We will prove the famous Weyl Integration Formula (Theorem 7.16) which says that

$$\int_G f(g)\, dg = \frac{1}{|W(G)|} \int_T d(t) \int_{G/T} f(gtg^{-1})\, dgT\, dt,$$

where $d(t) = \prod_{\alpha \in \Delta^+(\mathfrak{g}_{\mathbb{C}})} |1 - \xi_{-\alpha}(t)|^2$ for $t \in T$. Using Equation 1.42, the proof will be based on a change of variables map $\psi : G/T \times T \to G$ given by $\psi(gT, t) = gtg^{-1}$. In order to ensure all required hypothesis are met, it is necessary to first restrict our attention to a distinguished dense open subset of G called the set of *regular* elements.

7.2.1 Regular Elements

Let G be a compact Lie group with maximal torus T and $X \in \mathfrak{g}$. Recall from Definition 5.8 that X is called a *regular* element of $\mathfrak{g}$ if $\mathfrak{z}_{\mathfrak{g}}(X)$ is a Cartan subalgebra. Also recall from Theorem 6.27 the bijection between the set of analytically integral weights, $A(T)$, and the character group, $\chi(T)$, that maps $\lambda \in A(T)$ to $\xi_{\lambda} \in \chi(T)$ and satisfies

$$\xi_{\lambda}(\exp H) = e^{\lambda(H)}$$

for $H \in \mathfrak{t}$.

Definition 7.6. Let G be a compact connected Lie group with maximal torus T.
(a) An element $g \in G$ is said to be *regular* if $Z_G(g)^0$ is a maximal torus.
(b) Write $\mathfrak{g}^{\mathrm{reg}}$ for the set of regular element in $\mathfrak{g}$ and write G^{reg} for the set of regular elements in G.
(c) For $t \in T$, let

$$d(t) = \prod_{\alpha \in \Delta(\mathfrak{g}_{\mathbb{C}})} (1 - \xi_{-\alpha}(t)).$$

Theorem 7.7. *Let G be a compact connected Lie group.*
(a) *$\mathfrak{g}^{\mathrm{reg}}$ is open dense in $\mathfrak{g}$,*
(b) *G^{reg} is open dense in G,*
(c) *If T is a maximal torus and $t \in T$, $t \in T^{\mathrm{reg}}$ if and only if $d(t) \neq 0$,*
(d) *For $H \in \mathfrak{t}$, e^H is regular if and only if $H \in \Xi = \{H \in \mathfrak{t} \mid \alpha(H) \notin 2\pi i \mathbb{Z}, \alpha \in \Delta(\mathfrak{g}_{\mathbb{C}})\}$,*
(e) *$G^{\mathrm{reg}} = \bigcup_{g \in G} (gT^{\mathrm{reg}}g^{-1})$.*

Proof. Let l be the dimension of a Cartan subalgebra and $n = \dim \mathfrak{g}$. Any element $X \in \mathfrak{g}$ lies in at least one Cartan subalgebra, so that $\dim(\ker(\mathrm{ad}(X))) \geq l$. Thus

$$\det(\mathrm{ad}(X) - \lambda I) = \sum_{k=l}^{n} c_k(X) \lambda^k,$$

where $c_k(X)$ is a polynomial in X. Since $\mathrm{ad}(X)$ is diagonalizable, X is regular if and only if $\dim(\ker(\mathrm{ad}(X))) = l$. In particular, X is regular if and only if $c_l(X) \neq 0$. Thus $\mathfrak{g}^{\mathrm{reg}}$ is open in $\mathfrak{g}$. It also follows that $\mathfrak{g}^{\mathrm{reg}}$ is dense since a polynomial vanishes on a neighborhood if and only if it is zero.

For part (b), similarly observe that each $g \in G$ lies in a maximal torus so that $\dim(\ker(\mathrm{Ad}(g) - I)) \geq l$. Thus

$$\det(\mathrm{Ad}(g) - \lambda I) = \sum_{k=l}^{n} \widetilde{c}_k(g)(\lambda - 1)^k,$$

where $\widetilde{c}_k(g)$ is a smooth function of g. From Exercise 4.22, recall that the Lie algebra of $Z_G(g)$ is $\mathfrak{z}_{\mathfrak{g}}(g) = \{X \in \mathfrak{g} \mid \mathrm{Ad}(g)X = X\}$. Since $Z_G(g)^0$ is a maximal torus if and only if $\mathfrak{z}_{\mathfrak{g}}(g)$ is a Cartan subalgebra, diagonalizability implies g is regular if and only if $\widetilde{c}_l(g) \neq 0$. Thus G^{reg} is open in G.

To establish the density of G^{reg}, fix a maximal torus T of G. Since the eigenvalues of $\mathrm{Ad}(e^H)$ are of the form $e^{\alpha(H)}$ for $\alpha \in \Delta(\mathfrak{g}_{\mathbb{C}}) \cup \{0\}$, it follows that e^H is regular if and only if $H \in \Xi$. Since Ξ differs from $\mathfrak{t}$ only by a countable number of hyperplanes, Ξ is dense in $\mathfrak{t}$ by the Baire Category Theorem. Because $\exp$ is onto and continuous, T^{reg} is therefore dense in T. Since the Maximal Torus Theorem shows that $G = \cup_{g \in G} (gTg^{-1})$, counting eigenvalues of $\mathrm{Ad}(g)$ shows $G^{\mathrm{reg}} = \cup_{g \in G} (gT^{\mathrm{reg}}g^{-1})$. Density of G^{reg} in G now follows easily from the density of T^{reg} in T. $\square$

Definition 7.8. Let G be a compact connected Lie group and T a maximal torus. Define the smooth, surjective map $\psi : G/T \times T \to G$ by

$$\psi(gT, t) = gtg^{-1}.$$

Abusing notation, we also denote by ψ the smooth, surjective map $\psi : G/T \times T^{\mathrm{reg}} \to G^{\mathrm{reg}}$ defined by restriction of domain.

It will soon be necessary to understand the invertibility of the differential $d\psi : T_{gT}(G/T) \times T_t(T) \to T_{gtg^{-1}}(G)$ for $g \in G$ and $t \in T$. Calculations will be simplified by locally pulling $G/T \times T$ back to G with an appropriate cross section for G/T. Write $\pi : G \to G/T$ for the natural projection map.

Lemma 7.9. Let G be a compact connected Lie group and T a maximal torus. Then $\mathfrak{g} = \mathfrak{t} \oplus \left(\mathfrak{g} \cap \bigoplus_{\alpha \in \Delta(\mathfrak{g}_{\mathbb{C}})} \mathfrak{g}_\alpha \right)$ and there exists an open neighborhood $U_{\mathfrak{g}}$ of 0 in $\left(\mathfrak{g} \cap \bigoplus_{\alpha \in \Delta(\mathfrak{g}_{\mathbb{C}})} \mathfrak{g}_\alpha \right)$ so that, if $U_G = \exp U_{\mathfrak{g}}$ and $U_{G/T} = \pi U_G$, then:
(a) the map $U_{\mathfrak{g}} \xrightarrow{\exp} U_G \xrightarrow{\pi} U_{G/T}$ is a diffeomorphism,
(b) $U_{G/T}$ is an open neighborhood of eT in G/T,
(c) $U_G T = \{gt \mid g \in U_G, t \in T\}$ is an open neighborhood of e in G
(d) The map $\xi : U_G T \to G/T \times T$ given by $\xi(gt) = (gT, t)$ is a smooth, well-defined diffeomorphism onto $U_{G/T} \times T$.

Proof. The decomposition $\mathfrak{g} = \mathfrak{t} \oplus \left(\mathfrak{g} \cap \bigoplus_{\alpha \in \Delta(\mathfrak{g}_{\mathbb{C}})} \mathfrak{g}_\alpha \right)$ follows from Theorem 6.20. In fact, $\left(\mathfrak{g} \cap \bigoplus_{\alpha \in \Delta(\mathfrak{g}_{\mathbb{C}})} \mathfrak{g}_\alpha \right)$ is spanned by the elements $\mathcal{J}_\alpha$ and $\mathcal{K}_\alpha$ for $\alpha \in \Delta(\mathfrak{g}_{\mathbb{C}})$.

Since the map $(H, X) \to e^H e^X$, $H \in \mathfrak{t}$ and $X \in (\mathfrak{g} \cap \bigoplus_{\alpha \in \Delta(\mathfrak{g}_\mathbb{C})} \mathfrak{g}_\alpha)$, is therefore a local diffeomorphism at 0, it follows that there is an open neighborhood $U_\mathfrak{g}$ of 0 in $(\mathfrak{g} \cap \bigoplus_{\alpha \in \Delta(\mathfrak{g}_\mathbb{C})} \mathfrak{g}_\alpha)$ on which exp is a diffeomorphism onto U_G.

Recall that $T_{eT}(G/T)$ may be identified with $\mathfrak{g}/\mathfrak{t}$. Thus by construction, the differential of π restricted to $T_e(U_G)$ at e is clearly invertible, so that π is a local diffeomorphism from U_G at e. Thus, perhaps shrinking $U_\mathfrak{g}$ and U_G, we may assume that $U_{G/T}$ is an open neighborhood of eT in G/T and that the maps $U_\mathfrak{g} \overset{\exp}{\to} U_G \overset{\pi}{\to} U_{G/T}$ are diffeomorphisms. This finishes parts (a) and (b).

For part (c), $U_G T$ is a neighborhood of e since the map $(H, X) \to e^H e^X$ is a local diffeomorphism at 0. In fact, there is a subset V of T so that $U_G V$ is open. Taking the union of right translates by elements of T, it follows that $U_G T$ is open.

For part (d), suppose $gt = g't'$ with $g, g' \in U_G$ and $t, t' \in T$. Then $\pi g = \pi g'$, so that $g = g'$ and $t = t'$. Thus the map is well defined and the rest of the statement is clear. $\square$

Using Lemma 7.9, it is now possible to study the differential $d\psi : T_{gT}(G/T) \times T_t(T) \to T_{gtg^{-1}}(G)$. This will be done with the map ξ and appropriate translations to pull everything back to neighborhoods of e in G.

Lemma 7.10. *Let G be a compact connected Lie group and T a maximal torus. Choose $U_G \subseteq G$ as in Lemma 7.9. For $g \in G$ and $t \in T$, let $\phi : U_G T \to G$ be given by*

$$\phi = l_{gt^{-1}g^{-1}} \circ \psi \circ (l_{gT} \times l_t) \circ \xi,$$

where ξ is defined as in Lemma 7.9. Then the differential $d\phi : \mathfrak{g} \to \mathfrak{g}$ is given by

$$d\phi(H + X) = \text{Ad}(g) [(\text{Ad}(t^{-1}) - I) X + H]$$

for $H \in \mathfrak{t}$ and $X \in (\mathfrak{g} \cap \bigoplus_{\alpha \in \Delta(\mathfrak{g}_\mathbb{C})} \mathfrak{g}_\alpha)$ and

$$\det(d\phi) = d(t).$$

Proof. Calculate

$$d\phi(H) = \frac{d}{ds}\phi(e^{sH})|_{s=0} = \frac{d}{ds} g e^{sH} g^{-1}|_{s=0} = \text{Ad}(g)H$$

$$d\phi(X) = \frac{d}{ds}\phi(e^{sX})|_{s=0} = \frac{d}{ds} gt^{-1}e^{sX}te^{-sX}g^{-1}|_{s=0} = \text{Ad}(gt^{-1})X - \text{Ad}(g)X,$$

so that the formula for $d\phi$ is established by linearity. For the calculation of the determinant, first note that $\det \text{Ad}(g) = 1$. This follows from the three facts: (1) the determinant is not changed by complexifying, (2) each g lies in a maximal torus, and (3) the negative of a root is always a root. The problem therefore reduces to showing that the determinant of $(\text{Ad}(t^{-1}) - I)$ on $\bigoplus_{\alpha \in \Delta(\mathfrak{g}_\mathbb{C})} \mathfrak{g}_\alpha$ is $\prod_{\alpha \in \Delta(\mathfrak{g}_\mathbb{C})}(1 - e^{-\alpha(H)})$. Since $\dim \mathfrak{g}_\alpha = 1$ and $\text{Ad}(t^{-1})$ acts on $\mathfrak{g}_\alpha$ by $e^{-\alpha(\ln t)}$, where $e^{\ln t} = t$, the proof follows easily. The extra negative signs are taken care of by the even number of roots (since $\Delta(\mathfrak{g}_\mathbb{C}) = \Delta^+(\mathfrak{g}_\mathbb{C}) \amalg \Delta^-(\mathfrak{g}_\mathbb{C})$). $\square$

Theorem 7.11. *Let G be a compact connected Lie group and T a maximal torus. The map*

$$\psi : G/T \times T^{\text{reg}} \to G^{\text{reg}} \text{ given by}$$

$$\psi(gT, t) = gtg^{-1}$$

is a surjective, $|W(G)|$-to-one local diffeomorphism.

Proof. For $g \in G$ and $t \in T^{\text{reg}}$, Lemma 7.10 and Theorem 7.7 show that ψ is a surjective local diffeomorphism at (gT, t). Moreover if $w \in N(T)$, then

$$(7.12) \qquad\qquad \psi(gw^{-1}T, wtw^{-1}) = \psi(gT, t).$$

Since $gw^{-1}T = gT$ if and only if $w \in T$, it follows that $\left|\psi^{-1}(gtg^{-1})\right| \geq |W(G)|$.

To see that ψ is exactly $|W(G)|$-to-one, suppose $gtg^{-1} = hsh^{-1}$ for $h \in G$ and $s \in T^{\text{reg}}$. By Theorem 6.36, there is $w \in N(T)$, so that $s = wtw^{-1}$. Plugging this into $gtg^{-1} = hsh^{-1}$ quickly yields $w' = g^{-1}hw \in Z_G(t)$. Since t is regular, $Z_G(t)^0 = T$. Being the identity component of $Z_G(T)$, $c_{w'}$ preserves T, so that $w' \in N(T)$. Hence

$$(hT, s) = (gw'w^{-1}T, wtw^{-1}) = (gw'w^{-1}T, ww'^{-1}tw'w^{-1}).$$

Since this element was already known to be in $\psi^{-1}(gtg^{-1})$ by Equation 7.12, we see that $\left|\psi^{-1}(gtg^{-1})\right| \leq |W(G)|$, as desired. □

7.2.2 Main Theorem

Let G be a compact connected Lie group and T a maximal torus. From Theorem 1.48 we know that

$$\int_G f(g)\,dg = \int_{G/T} \left(\int_T f(gt)\,dt \right) d(tT)$$

for $f \in C(G)$. Recall that the invariant measures above are given by integration against unique (up to ± 1) normalized left-invariant volume forms $\omega_G \in \bigwedge^*_{\text{top}}(G)$ and $\omega_{G/T} \in \bigwedge^*_{\text{top}}(G/T)$. In this section we make a change of variables based on the map ψ to obtain Weyl's Integration Formula. To this end write $n = \dim G$, $l = \dim T$ (also called the *rank* of G when $\mathfrak{g}$ is semisimple), and write $\iota : T \to G$ for the inclusion map. Recall that $\pi : G \to G/T$ is the natural projection map.

Lemma 7.13. *Possibly replacing ω_T by $-\omega_T$ (which does not change integration), there exists a G-invariant form $\widetilde{\omega}_T \in \bigwedge^*_l(G)$, so that*

$$\omega_T = \iota^* \widetilde{\omega}_T$$

and

$$\omega_G = \left(\pi^* \omega_{G/T}\right) \wedge \widetilde{\omega}_T.$$

Proof. Clearly the restriction map $\iota^*|_e : \mathfrak{g}^* \to \mathfrak{t}^*$ is surjective. Choose any $(\widetilde{\omega_T})_e \in \bigwedge_l^*(G)_e$, so $\iota^*(\widetilde{\omega_T})_e = (\omega_T)_{eT}$. Using left translation, uniquely extend $(\widetilde{\omega_T})_e$ to a left-invariant form $\widetilde{\omega_T} \in \bigwedge_l^*(G)$. Since ι commutes with left multiplication by G, it follows that $\iota^*\widetilde{\omega_T} = \omega_T$. Since π also commutes with left multiplication by G, $\pi^*\omega_{G/T} \in \bigwedge_{n-l}^*(G)$ is left-invariant as well. Thus $(\pi^*\omega_{G/T}) \wedge \widetilde{\omega_T} \in \bigwedge_n^*(G)$ is left-invariant and therefore $(\pi^*\omega_{G/T}) \wedge \widetilde{\omega_T} = c\omega_G$ for some $c \in \mathbb{R}$ by uniqueness.

Write π_i for the two natural coordinate projections $\pi_1 : G/T \times T \to G/T$ and $\pi_2 : G/T \times T \to T$. Using the notation from Lemma 7.9, observe that $\pi|_{U_G T} = \pi_1 \circ \xi$, so that

$$\pi^*\omega_{G/T} = \xi^*\pi_1^*\omega_{G/T}$$

on $U_G T$. Similarly, observe that $I|_T = \pi_2 \circ \xi \circ \iota$, so that $\iota^*(\xi^*\pi_2^*\omega_T) = \omega_T$. Thus

$$\xi^*\pi_2^*\omega_T = \widetilde{\omega_T} + \omega$$

on $U_G T$ for some $\omega \in \bigwedge_l^*(U_G T)$ with $\iota^*\omega = 0$.

We claim that $(\pi^*\omega_{G/T}) \wedge \omega = 0$ on $U_G T$. Since ξ is a diffeomorphism, this is equivalent to showing $(\pi_1^*\omega_{G/T}) \wedge \omega' = 0$, where $\omega' = (\xi^{-1})^* \omega \in \bigwedge^*(U_{G/T} \times T)$ satisfies $\iota^*\xi^*\omega' = 0$. Now ω' can be written as a sum $\omega' = \sum_{j=0}^l f_j(\pi_1^*\omega_j') \wedge (\pi_2^*\omega_{l-j}'')$, where f_j is a smooth function on $G/T \times T$, $\omega_j' \in \bigwedge_j^*(U_{G/T})$, and $\omega_{l-j}'' \in \bigwedge_{l-j}^*(T)$. Without loss of generality, we may take $\pi_1^*\omega_0' = 1$. As $I|_T = \pi_2 \circ \xi \circ \iota$ and $(\pi_1 \circ \xi \circ \iota)(t) = eT$ for $t \in T$, it follows that $0 = \iota^*\xi^*\omega' = f_0\omega_l''$. Therefore $\omega' = \sum_{j=1}^l f_j(\pi_1^*\omega_j') \wedge (\pi_2^*\omega_{l-j}'')$. Since $\omega_{G/T}$ is a top degree form, $\omega_{G/T} \wedge \omega_j' = 0$, $j \geq 1$, so that $(\pi_1^*\omega_{G/T}) \wedge \omega' = 0$, as desired.

It now follows that

$$
\begin{aligned}
c\omega_G &= (\pi^*\omega_{G/T}) \wedge \widetilde{\omega_T} = (\pi^*\omega_{G/T}) \wedge (\widetilde{\omega_T} + \omega) \\
&= \xi^*[(\pi_1^*\omega_{G/T}) \wedge (\pi_2^*\omega_T)]
\end{aligned}
$$

(7.14)

on $U_G T$. Looking at local coordinates, it is clear that $(\pi_1^*\omega_{G/T}) \wedge (\pi_2^*\omega_T) \neq 0$, so $c \neq 0$. Replacing ω_T by $-\omega_T$ if necessary, we may assume $c > 0$. Choose any continuous function f supported on $U_G T$ and use the change of variables formula to calculate

$$
\begin{aligned}
c \int_{G/T} \int_T f \circ \xi^{-1}(gT, t)\, dt\, dgT &= c \int_{G/T} \int_T f(gt)\, dt\, dgT = c \int_G f(g)\, dg \\
&= \int_{U_G T} f\, c\omega_G = \int_{U_G T} f\, \xi^*[(\pi_1^*\omega_{G/T}) \wedge (\pi_2^*\omega_T)] \\
&= \int_{U_{G/T} \times T} f \circ \xi^{-1}\, (\pi_1^*\omega_{G/T}) \wedge (\pi_2^*\omega_T).
\end{aligned}
$$

Since it follows immediately from the definitions (Exercise 7.7) that

(7.15) $\displaystyle\int_{U_{G/T} \times T} f \circ \xi^{-1} \left(\pi_1^* \omega_{G/T}\right) \wedge \left(\pi_2^* \omega_T\right) = \int_{G/T} \int_T f \circ \xi^{-1}(gT, t) \, dt \, dgT,$

$c = 1$, as desired.   □

Theorem 7.16 (Weyl Integration Formula). *Let G be a compact connected Lie group, T a maximal torus, and $f \in C(G)$. Then*

$$\int_G f(g) \, dg = \frac{1}{|W(G)|} \int_T d(t) \int_{G/T} f(gtg^{-1}) \, dgT \, dt,$$

where $d(t) = \prod_{\alpha \in \Delta^+(\mathfrak{g}_\mathbb{C})} |1 - \xi_{-\alpha}(t)|^2$ for $t \in T$.

Proof. Since Theorem 7.7 shows that G^{reg} is open dense in G and T^{reg} is open dense in T, it suffices to prove that

$$\int_{G^{\mathrm{reg}}} f(g) \, dg = \frac{1}{|W(G)|} \int_{T^{\mathrm{reg}}} d(t) \int_{G/T} f(gtg^{-1}) \, dgT \, dt.$$

To this end, recall that Theorem 7.11 shows that $\psi : G/T \times T^{\mathrm{reg}} \to G^{\mathrm{reg}}$ is a surjective, $|W(G)|$-to-one local diffeomorphism. We will prove that

(7.17) $$\psi^* \omega_G = d(t) \left(\pi_1^* \omega_{G/T}\right) \wedge \left(\pi_2^* \omega_T\right),$$

where π_1 and π_2 are the projections from Lemma 7.13. Once this is done, the theorem follows immediately from Equation 1.42.

To verify Equation 7.17, first note that there is a smooth function $\delta : G/T \times T \to \mathbb{R}$, so that

$$\psi^* \omega_G|_{gtg^{-1}} = \left[\delta \left(\pi_1^* \omega_{G/T}\right) \wedge \left(\pi_2^* \omega_T\right)\right]|_{(gT, t)}$$

since the dimension of top degree form is 1 at each point. Since $U_{G/T} \times T$ is a neighborhood of (eT, e), Equation 7.14 shows $\left[\left(\pi_1^* \omega_{G/T}\right) \wedge \left(\pi_2^* \omega_T\right)\right]|_{(eT, e)} = \left(\xi^{-1}\right)^* \omega_G|_e$, so that

$$\psi^* l_{gt^{-1}g^{-1}}^* \omega_G|_e = \psi^* \omega_G|_{gtg^{-1}} = \left(l_{g^{-1}} \times l_{t^{-1}}\right)^* \left[\delta \left(\pi_1^* \omega_{G/T}\right) \wedge \left(\pi_2^* \omega_T\right)\right]|_{(eT, e)}$$
$$= \left(l_{g^{-1}} \times l_{t^{-1}}\right)^* \left(\xi^{-1}\right)^* \left[\delta \circ \left(l_g \times l_t\right) \circ \xi \, \omega_G\right]|_e.$$

Thus

$$\phi^* \omega_G|_e = \left(l_{gt^{-1}g^{-1}} \circ \psi \circ \left(l_g \times l_t\right) \circ \xi\right)^* \omega_G|_e = \left[\delta \circ \left(l_g \times l_t\right) \circ \xi \, \omega_G\right]|_e.$$

By looking at a basis of $\bigwedge_1^*(G)_e$, it follows that $\delta(gT, t) = \delta \circ \left(l_g \times l_t\right) \circ \xi|_e = \det(d\phi)$. This determinant was calculated in Lemma 7.10 and found to be

$$d(t) = \prod_{\alpha \in \Delta(\mathfrak{g}_\mathbb{C})} \left(1 - \xi_{-\alpha}(t)\right) = \prod_{\alpha \in \Delta^+(\mathfrak{g}_\mathbb{C})} |1 - \xi_{-\alpha}(t)|^2.$$ □

7.2.3 Exercises

Exercise 7.7 Verify Equation 7.15.

Exercise 7.8 Let G be a compact connected Lie group and T a maximal torus. For $H \in \mathfrak{t}$, show that

$$d(e^H) = 2^{|\Delta(\mathfrak{g}_\mathbb{C})|} \prod_{\alpha \in \Delta^+(\mathfrak{g}_\mathbb{C})} \sin^2\left(\frac{\alpha(H)}{2i}\right).$$

Note that $\alpha(H) \in i\mathbb{R}$.

Exercise 7.9 Let f be a continuous class function on $SU(2)$. Use the Weyl Integration Formula to show that

$$\int_{SU(2)} f(g) \, dg = \frac{2}{\pi} \int_0^\pi f(\operatorname{diag}(e^{i\theta}, e^{-i\theta})) \sin^2 \theta \, d\theta,$$

c.f. Exercise 3.22.

Exercise 7.10 Let G be a compact connected Lie group and T a maximal torus (c.f. Exercise 6.29).
(1) If f is an L^1-class function on G, show that

$$\int_G f(g) \, dg = \frac{1}{|W(G)|} \int_T d(t) f(t) \, dt.$$

(2) Show that the map $f \to |W(G)|^{-1} \, df|_T$ defines a norm preserving isomorphism between the L^1-class functions on G and the W-invariant L^1-functions on T.
(3) Show that the map $f \to |W(G)|^{-\frac{1}{2}} \, Df|_T$ defines a unitary isomorphism between the L^2 class functions on G to the W-invariant L^2 functions on T, where $D(e^H) = \prod_{\alpha \in \Delta^+(\mathfrak{g}_\mathbb{C})} (1 - e^{-\alpha(H)})$ for $H \in \mathfrak{t}$ (so $D\overline{D} = d$).

Exercise 7.11 For each group G below, verify $d(t)$ is correctly calculated.
(1) For $G = SU(n)$, $T = \{\operatorname{diag}(e^{i\theta_k}) \mid \sum_k \theta_k = 0\}$, and $t = \operatorname{diag}(e^{i\theta_k})$,

$$d(t) = 2^{n(n-1)} \prod_{1 \le j < k \le n} \sin^2\left(\frac{\theta_j - \theta_k}{2}\right).$$

(2) For either $G = SO(2n + 1)$, T as in §5.1.2.4, and

$$t = \operatorname{blockdiag}\left(\begin{pmatrix} \cos\theta_k & \sin\theta_k \\ -\sin\theta_k & \cos\theta_k \end{pmatrix}, 1\right)$$

or $G = SO(E_{2n+1})$, T as in Lemma 6.12, and $t = \operatorname{diag}(e^{i\theta_k}, e^{-i\theta_k}, 1)$,

$$d(t) = 2^{2n^2} \prod_{1 \le j < k \le n} \sin^2\left(\frac{\theta_j - \theta_k}{2}\right) \sin^2\left(\frac{\theta_j + \theta_k}{2}\right) \prod_{1 \le j \le n} \sin^2\left(\frac{\theta_j}{2}\right).$$

(3)   For either $G = SO(2n)$, T as in §5.1.2.3, and

$$t = \text{blockdiag}\left(\begin{pmatrix} \cos\theta_k & \sin\theta_k \\ -\sin\theta_k & \cos\theta_k \end{pmatrix}\right)$$

or $G = SO(E_{2n})$, T as in Lemma 6.12, and $t = \text{diag}(e^{i\theta_k}, e^{-i\theta_k})$,

$$d(t) = 2^{2n(n-1)} \prod_{1 \leq j < k \leq n} \sin^2\left(\frac{\theta_j - \theta_k}{2}\right) \sin^2\left(\frac{\theta_j + \theta_k}{2}\right).$$

(4)   For $G = Sp(n)$ realized as $Sp(n) \cong U(2n) \cap Sp(n, \mathbb{C})$ and $T = \{t = \text{diag}(e^{i\theta_k}, e^{-i\theta_k})\}$,

$$d(t) = 2^{2n^2} \prod_{1 \leq j < k \leq n} \sin^2\left(\frac{\theta_j - \theta_k}{2}\right) \sin^2\left(\frac{\theta_j + \theta_k}{2}\right) \prod_{1 \leq j \leq n} \sin^2(\theta_j).$$

7.3 Weyl Character Formula

Let G be a compact Lie group with maximal torus T. Recall that Theorem 3.30 shows that the set of irreducible characters $\{\chi_\lambda\}$ is an orthonormal basis for the set of L^2 class functions on G.

Assume G is connected and, for the sake of motivation, momentarily assume G is simply connected as well. In §7.3.1 we will choose a skew-W-invariant function Δ defined on T, so that $|\Delta(t)|^2 = d(t)$. It easily follows from the Weyl Integration Formula that $\{\Delta \chi_\lambda|_T\}$ is therefore an orthonormal basis for the set of L^2 skew-W-invariant functions on T with respect to the measure $|W(G)|^{-1} dt$ (c.f. Exercise 7.10).

On the other hand, it is simple to write down another basis for the set of L^2 skew-W-invariant functions on T by looking at alternating sums over the Weyl group of certain characters on T. By decomposing $\chi_\lambda|_T$ into characters on T, it will follow rapidly that these two bases are the same. In turn, this yields an explicit formula for χ_λ called the Weyl Character Formula.

7.3.1 Machinery

Let G be a compact Lie group with maximal torus T. Recall that Theorem 6.27 shows there is a bijection between the set of analytically integral weights and the character group given by mapping $\lambda \in A(T)$ to $\xi_\lambda \in \chi(T)$. The next definition sets up similar notation for more general functions on $\mathfrak{t}$.

Definition 7.18. Let G be a compact Lie group with maximal torus T.
(a) Let $f : \mathfrak{t} \to \mathbb{C}$ be a function. We say f *descends* to T if $f(H + Z) = f(H)$ for $H, Z \in \mathfrak{t}$ with $Z \in \ker(\exp)$. In that case, write $f : T \to \mathbb{C}$ for the function given by

$$f(e^H) = f(H).$$

(b) If $f : \mathfrak{t} \to \mathbb{C}$ satisfies $f(wH) = f(H)$ for $w \in W(\Delta(\mathfrak{g}_{\mathbb{C}})^{\vee})$, f is called W-*invariant.*
(c) If $F : T \to \mathbb{C}$ satisfies $F(c_w t) = F(t)$ for $w \in N(T)$, F is called W-*invariant.*
(d) If $f : \mathfrak{t} \to \mathbb{C}$ satisfies $f(wH) = \det(w) f(H)$ for $w \in W(\Delta(\mathfrak{g}_{\mathbb{C}}))$, f is called
skew-W-invariant.
(e) If $F : T \to \mathbb{C}$ satisfies $F(c_w t) = \det(\mathrm{Ad}(w)|_{\mathfrak{t}}) F(t)$ for $w \in N(T)$, F is called
skew-W-invariant.

In particular, for $\lambda \in A(T)$, the function $H \to e^{\lambda(H)}$ on $\mathfrak{t}$ descends to the function ξ_{λ} on T. Also note that $\det w \in \{\pm 1\}$ since w is a product of reflections.

Lemma 7.19. *Let G be a compact connected Lie group with maximal torus T.*
(a) If $f : \mathfrak{t} \to \mathbb{C}$ descends to T and is W-invariant, then $f : T \to \mathbb{C}$ is W-invariant.
(b) Restriction of domain establishes a bijection between the continuous class functions on G and the continuous W-invariant functions on T.

Proof. For part (a), recall that the identification of $W(G)$ with $W(\Delta(\mathfrak{g}_{\mathbb{C}})^{\vee})$ from Theorem 6.43 via the Ad-action of Equation 6.35. It follows that when f descends to T and is W-invariant, then $f(c_w t) = f(t)$ for $w \in N(T)$ and $t \in T$.

For part (b), suppose $F : T \to \mathbb{C}$ is W-invariant and fix $g_0 \in G$. By the Maximal Torus Theorem, there exists $h_0 \in G$, so $t_0 = c_{h_0} g_0 \in T$. Extend F to a class function on G by setting $F(g_0) = F(t_0)$. This is well defined by Theorem 6.36. It only remains to see that if F is continuous on T, then its extension to G is also continuous.

For this, suppose $g_n \in G$ with $g_n \to g_0$. Choose $h_n \in G$, so $t_n = c_{h_n} g_n \in T$. Since G is compact, passing to subsequences allows us to assume there is $h_0' \in G$ and $t_0' \in T$, so that $h_n \to h_0'$ and $t_n \to t_0'$. In particular, $t_0' = c_{h_0'} g_0$ so that, by Theorem 6.36, there exists $w \in N(T)$ with $w t_0 = t_0'$. Thus

$$F(g_n) = F(t_n) \to F(t_0') = F(t_0) = F(g_0).$$

Since we began with an arbitrary sequence $g_n \to g_0$, the proof is complete. $\square$

Let G be a compact Lie group, T a maximal torus, and $\Delta^+(\mathfrak{g}_{\mathbb{C}})$ a system of positive roots with corresponding simple system $\Pi(\mathfrak{g}_{\mathbb{C}}) = \{\alpha_1, \ldots, \alpha_l\}$. Recall from Equation 6.39 the unique element $\rho \in (i\mathfrak{t})^*$ satisfying $\rho(h_{\alpha_i}) = 2\frac{B(\rho, \alpha_i)}{B(\alpha_i, \alpha_i)} = 1$, $1 \leq j \leq l$.

Lemma 7.20. *Let G be a compact Lie group with a maximal torus T.*
(a) $\rho = \frac{1}{2} \sum_{\alpha \in \Delta^+(\mathfrak{g}_{\mathbb{C}})} \alpha$.
(b) For $w \in W(\Delta(\mathfrak{g}_{\mathbb{C}}))$, $w\rho - \rho \in R \subseteq A(T)$, and so the function $\xi_{w\rho - \rho}$ descends to T.

Proof. For part (a), write $\Pi(\mathfrak{g}_{\mathbb{C}}) = \{\alpha_1, \ldots \alpha_l\}$ and let $\rho' = \frac{1}{2} \sum_{\alpha \in \Delta^+(\mathfrak{g}_{\mathbb{C}})} \alpha$ (c.f. Exercise 6.34). By the definitions, it suffices to show that $r_{\alpha_j} \rho' = \rho'$. For this, it suffices to show that r_{α_j} preserves the set $\Delta^+(\mathfrak{g}_{\mathbb{C}}) \backslash \{\alpha_j\}$. If $\alpha \in \Delta^+(\mathfrak{g}_{\mathbb{C}}) \backslash \{\alpha_j\}$ is written as $\alpha = \Sigma_k n_k \alpha_k$ with $n_{k_0} > 0$, $k_0 \neq j$, then the coefficient of α_{k_0} in $r_{\alpha_j} \alpha = \alpha - \alpha(h_{\alpha_j}) \alpha_j$ is still n_{k_0}, so that $r_{\alpha_j} \alpha \in \Delta^+(\mathfrak{g}_{\mathbb{C}}) \backslash \{\alpha_j\}$.

Part (b) is straightforward. In fact, it is immediate that

$$w\rho - \rho = \sum_{\alpha \in [w \Delta^+(\mathfrak{g}_{\mathbb{C}})] \cap \Delta^-(\mathfrak{g}_{\mathbb{C}})} \alpha.   \square$$

Definition 7.21. For G a compact Lie group with a maximal torus T, let $\Delta : \mathfrak{t} \to \mathbb{C}$ be given by

$$\Delta(H) = \prod_{\alpha \in \Delta^+(\mathfrak{g}_\mathbb{C})} \left(e^{\alpha(H)/2} - e^{-\alpha(H)/2}\right)$$

for $H \in \mathfrak{t}$.

Lemma 7.22. *Let G be a compact Lie group with a maximal torus T.*
(a) The function Δ is skew-symmetric on $\mathfrak{t}$.
(b) The function Δ descends to T if and only if the function $H \to e^{-\rho(H)}$ descends to T.
(c) The function $|\Delta|^2$ always descends to T and there $|\Delta(t)|^2 = d(t)$, $t \in T$.

Proof. For part (a), it suffices to show that $\Delta \circ r_{h_\alpha} = -\Delta$ for $\alpha \in \Delta^+(\mathfrak{g}_\mathbb{C})$. This follows from three observations. The first is that composition with r_{h_α} maps $\left(e^{\alpha/2} - e^{-\alpha/2}\right)$ to $-\left(e^{\alpha/2} - e^{-\alpha/2}\right)$. The second is that if $\beta \in \Delta^+(\mathfrak{g}_\mathbb{C})$ satisfies $r_\alpha \beta = \beta$, then composition with r_{h_α} fixes $\left(e^{\beta/2} - e^{-\beta/2}\right)$. For the third, suppose $\beta \in \Delta^+(\mathfrak{g}_\mathbb{C}) \setminus \{\alpha\}$ satisfies $r_\alpha \beta \ne \beta$. Choose $\beta' \in \Delta^+(\mathfrak{g}_\mathbb{C})$, so that either $r_\alpha \beta = \beta'$ or $r_\alpha \beta = -\beta'$. Then composition with r_{h_α} fixes $\left(e^{\beta/2} - e^{-\beta/2}\right)\left(e^{\beta'/2} - e^{-\beta'/2}\right)$.

For part (b), write $\rho = \frac{1}{2}\sum_{\alpha \in \Delta^+(\mathfrak{g}_\mathbb{C})} \alpha$ to see that

(7.23) $$e^{-\rho(H)} \Delta(H) = \prod_{\alpha \in \Delta^+(\mathfrak{g}_\mathbb{C})} \left(1 - e^{-\alpha(H)}\right)$$

for $H \in \mathfrak{t}$. Since the function $H \to \prod_{\alpha \in \Delta^+(\mathfrak{g}_\mathbb{C})} \left(1 - e^{-\alpha(H)}\right)$ clearly descends to T, part (b) is complete. For part (c), calculate

$$|\Delta(H)|^2 = e^{-\rho(H)} \Delta(H) \overline{e^{-\rho(H)} \Delta(H)} = \prod_{\alpha \in \Delta^+(\mathfrak{g}_\mathbb{C})} \left|1 - e^{-\alpha(H)}\right|^2$$

to complete the proof. $\square$

Note that although $e^{-\rho}$ often descends to a function on T, it does not always descend (Exercise 7.12). Also note that the function $d(t)$ plays a prominent role in Weyl Integration Formula. In particular, we can now write the Weyl Integration Formula as

(7.24) $$\int_G f(g)\,dg = \frac{1}{|W(G)|} \int_T |\Delta(t)|^2 \int_{G/T} f(gtg^{-1})\,dg_T\,dt$$

for connected G and $f \in C(G)$.

For the next definition, recall from the proof of Theorem 7.7 that

$$\Xi = \{H \in \mathfrak{t} \mid \alpha(H) \notin 2\pi i \mathbb{Z} \text{ for all roots } \alpha\}$$

is open dense in $\mathfrak{t}$ and $\exp \Xi = T^{\text{reg}}$.

Definition 7.25. Let G be a compact Lie group with a maximal torus T. Fix an analytically integral weight $\lambda \in A(T)$. Let $\Theta_\lambda : \Xi \to \mathbb{C}$ be given by

$$\Theta_\lambda(H) = \frac{\sum_{w \in W(\Delta(\mathfrak{g}_\mathbb{C}))} \det(w) \, e^{[w(\lambda+\rho)](H)}}{\Delta(H)}$$

$$= \frac{\sum_{w \in W(\Delta(\mathfrak{g}_\mathbb{C}))} \det(w) \, e^{[w(\lambda+\rho)-\rho](H)}}{\prod_{\alpha \in \Delta^+(\mathfrak{g}_\mathbb{C})} \left(1 - e^{-\alpha(H)}\right)}$$

for $H \in \Xi$.

Lemma 7.26. *Let G be a compact connected Lie group with a maximal torus T. Fix an analytically integral weight $\lambda \in A(T)$. The function Θ_λ descends to a smooth W-invariant function on T^{reg}. In turn, this function, still denoted by Θ_λ, uniquely extends to a smooth class function on G^{reg}.*

Proof. The first expression for Θ_λ shows that it is symmetric since the numerator and denominator are skew-symmetric. The second expression for Θ_λ shows it descends to a function on T^{reg} since the numerator and denominator both descend to T and the denominator is nonzero on Ξ. The final statement follows as in Lemma 7.19. $\square$

7.3.2 Main Theorem

Let G be a compact connected Lie group with a maximal torus T. For $\lambda, \lambda' \in A(T)$, the function $\xi_\lambda : T \to \mathbb{C}$ can be viewed as a 1-dimensional irreducible representation of T. As a result, ξ_λ and $\xi_{\lambda'}$ are equivalent if and only if the are equal as functions. This happens if and only if $\lambda = \lambda'$. By the character theory of T, it follows that

$$(7.27) \qquad \int_T \xi_\lambda(t)\xi_{-\lambda'}(t) \, dt = \begin{cases} 1 & \text{if } \lambda = \lambda' \\ 0 & \text{if } \lambda \neq \lambda'. \end{cases}$$

Theorem 7.28 (Weyl Character Formula). *Let G be a compact connected Lie group with a maximal torus T. If $V(\lambda)$ is an irreducible representation of G with highest weight λ, then the character of $V(\lambda)$, χ_λ, satisfies*

$$\chi_\lambda(g) = \Theta_\lambda(g)$$

for $g \in G^{\mathrm{reg}}$.

Proof. First note it suffices to prove the theorem for $g = e^H$, $H \in \Xi$. Next for $\gamma \in A(T)$, let $D_\gamma : \mathfrak{t} \to \mathbb{C}$ be the skew-symmetric function defined by

$$D_\gamma(H) = \sum_{w \in W(\Delta(\mathfrak{g}_\mathbb{C}))} \det(w) \, e^{(w\gamma)(H)}.$$

The proof will be completed by showing that $\chi_\lambda(e^H)\Delta(H) = D_{\lambda+\rho}(H)$ for $H \in \mathfrak{t}$.

To this end, by considering the weight decomposition of $V(\lambda)$, write $\chi_\lambda = \sum_{\gamma_j \in A(T)} n_j \xi_{\gamma_j}$ as a finite sum on T for $n_j \in \mathbb{Z}^{\geq 0}$. Thus

$$\chi_\lambda(e^H)\Delta(H) = e^{\rho(H)} \prod_{\alpha \in \Delta^+(\mathfrak{g}_\mathbb{C})} \left(1 - e^{-\alpha(H)}\right) \sum_{\gamma_j \in A(T)} n_j e^{\gamma_j(H)}$$

$$= \sum_{\gamma_j \in A(T)} m_j e^{(\gamma_j + \rho)(H)}$$

for some $m_j \in \mathbb{Z}$. Since χ_λ is symmetric and Δ is skew-symmetric, $\chi_\lambda(e^H)\Delta(H)$ is skew-symmetric as well. Noting that the set of functions $\{e^{\gamma_j + \rho} \mid \gamma_j \in A(T)\}$ is independent, the action of r_α coupled with skew-symmetry shows that $m_j = 0$ if $\gamma_j + \rho$ is on a Weyl chamber wall. Recalling that the Weyl group acts simply transitively on the open Weyl chambers (Theorem 6.43), examination of the the Weyl group orbits of $A(T) + \rho$ and skew-symmetry imply that

$$\chi_\lambda(e^H)\Delta(H) = \sum_{\gamma_j \in A(T),\, \gamma_j + \rho \text{ strictly dominant}} m_j D_{\gamma_j + \rho}(H),$$

where *strictly dominant* means $B(\gamma_j + \rho, \alpha_i) > 0$ for $\alpha_i \in \Pi(\mathfrak{g}_\mathbb{C})$, i.e., $\gamma_j + \rho$ lies in the open positive Weyl chamber.

Next, character theory shows that $\int_G |\chi_\lambda|^2 \, dg = 1$. Thus the Weyl Integration Formula gives

(7.29)
$$1 = \frac{1}{|W(G)|} \int_T |\Delta|^2 |\chi_\lambda|^2 \, dt$$

$$= \frac{1}{|W(G)|} \int_T \left| \sum_{\gamma_j \in A(T),\, \gamma_j + \rho \text{ str. dom.}} m_j D_{\gamma_j + \rho} \right|^2 \, dt.$$

Here $\left| \sum_{\gamma_j \in A(T),\, \gamma_j + \rho \text{ str. dom.}} m_j D_{\gamma_j + \rho} \right|^2$ descends to T since $|\Delta|^2 |\chi_\lambda|^2$ descends to T. In fact, the function $H \to e^{-\rho(H)} D_{\gamma_j + \rho}(H)$ descends to T since $e^{w(\gamma_j + \rho) - \rho}$ does. Therefore $D_{\gamma_j + \rho} \overline{D_{\gamma_{j'} + \rho}} = \left(e^{-\rho} D_{\gamma_j + \rho}\right) \overline{\left(e^{-\rho} D_{\gamma_{j'} + \rho}\right)}$ descends to T and

$$\frac{1}{|W(G)|} \int_T D_{\gamma_j + \rho} \overline{D_{\gamma_{j'} + \rho}} \, dt = \frac{1}{|W(G)|} \sum_{w, w' \in W(\Delta(\mathfrak{g}_\mathbb{C}))} \det(w w') \int_T \xi_{w(\gamma_j + \rho)} \xi_{-w'(\gamma_{j'} + \rho)} \, dt.$$

Since $\gamma_j + \rho$ and $\gamma_{j'} + \rho$ are in the open Weyl chamber, $w(\gamma_j + \rho) = w'(\gamma_{j'} + \rho)$ if and only if $w = w'$ and $j = j'$. Thus

$$\frac{1}{|W(G)|} \int_T D_{\gamma_j + \rho} \overline{D_{\gamma_{j'} + \rho}} \, dt = \begin{cases} 1 & \text{if } j = j' \\ 0 & \text{if } j \neq j'. \end{cases}$$

In particular, this simplifies Equation 7.29 to

$$1 - \sum_{\gamma_j \in A(T),\, \gamma_j + \rho \text{ str. dom.}} m_j^2$$

Finally, since $m_j \in \mathbb{Z}$, all but one are zero. Thus there is a $\gamma \in A(T)$ with $\gamma + \rho$ strictly dominant so that $\chi_\lambda(e^H)\Delta(H) = \pm D_{\gamma + \rho}(H)$. To determine γ and the $\pm$

sign, notice that the weight decomposition shows that $\chi_\lambda(e^H) = e^{\lambda(H)} + \dots$ where the ellipses denote weights strictly lower than λ. Writing

$$\chi_\lambda(e^H)\Delta(H) = e^{\rho(H)}\chi_\lambda(e^H)e^{-\rho(H)}\Delta(H) = \left(e^{(\lambda+\rho)(H)} + \dots\right) \prod_{\alpha \in \Delta^+(\mathfrak{g}_\mathbb{C})} \left(1 - e^{-\alpha(H)}\right),$$

we see $\chi_\lambda(e^H)\Delta(H) = e^{(\lambda+\rho)(H)} + \dots$. In particular, expanding the function $H \to \chi_\lambda(e^H)\Delta(H)$ in terms of $\{e^{\gamma_j+\rho} \mid \gamma_j \in A(T)\}$, it follows that $e^{\lambda+\rho}$ appears with coefficient 1. On the other hand, similarly expanding $\pm D_{\gamma+\rho}$, we see that the only term of the form $e^{\gamma_j+\rho}$ appearing for which $\gamma_j + \rho$ is dominant is $\pm e^{\gamma+\rho}$. Therefore $\lambda = \gamma$, the undetermined $\pm$ sign is a $+$.   □

7.3.3 Weyl Denominator Formula

Theorem 7.30 (Weyl Denominator Formula). *Let G be a compact connected Lie group with a maximal torus T. Then*

$$\Delta(H) = \sum_{w \in W(\Delta(\mathfrak{g}_\mathbb{C}))} \det(w)\, e^{(w\rho)(H)}$$

for $H \in \mathfrak{t}$.

Proof. Simply take the trivial representation $V(0) = \mathbb{C}$ with $\chi_0(g) = 1$ and apply the Weyl Character Formula to $g = e^H$ for $H \in \Xi$. The formula extends to all $\mathfrak{t}$ by continuity.   □

Note the Weyl Denominator Formula allows the Weyl Character Formula to be rewritten in the form

$$(7.31) \qquad \chi_\lambda(e^H) = \frac{\sum_{w \in W(\Delta(\mathfrak{g}_\mathbb{C}))} \det(w)\, e^{[w(\lambda+\rho)](H)}}{\sum_{w \in W(\Delta(\mathfrak{g}_\mathbb{C}))} \det(w)\, e^{(w\rho)(H)}}$$

for $H \in \mathfrak{t}$ with $e^H \in T^{\text{reg}}$, *i.e.*, $H \in \Xi$.

7.3.4 Weyl Dimension Formula

Theorem 7.32 (Weyl Dimension Formula). *Let G be a compact connected Lie group with a maximal torus T. If $V(\lambda)$ is the irreducible representation of G with highest weight λ, then*

$$\dim V(\lambda) = \prod_{\alpha \in \Delta^+(\mathfrak{g}_\mathbb{C})} \frac{B(\lambda + \rho, \alpha)}{B(\rho, \alpha)}.$$

Proof. Since $\dim V(\lambda) = \chi_\lambda(e)$, we ought to evaluate Equation 7.31 at $H = 0$. Unfortunately, Equation 7.31 is not defined at $H = 0$, so we take a limit. Let $u_\rho \in i\mathfrak{t}$, so that $\rho(H) = B(H, u_\rho)$ for $H \in \mathfrak{t}$. Then it is easy to see that $itu_\rho \in \Xi$ for small positive t (Exercise 7.13), so that

$$\dim V(\lambda) = \lim_{t \to 0} \Theta_\lambda(itu_\rho)$$

$$(7.33) \qquad = \lim_{t \to 0} \frac{\sum_{w \in W(\Delta(\mathfrak{g}_\mathbb{C}))} \det(w)\, e^{[w(\lambda+\rho)](itu_\rho)}}{\sum_{w \in W(\Delta(\mathfrak{g}_\mathbb{C}))} \det(w)\, e^{(w\rho)(itu_\rho)}}.$$

Now observe that

$$(w(\lambda+\rho))(itu_\rho) = it(\lambda+\rho)(w^{-1}u_\rho) = it B(u_{\lambda+\rho}, w^{-1}u_\rho)$$
$$= it B(wu_{\lambda+\rho}, u_\rho) = it\rho(wu_{\lambda+\rho}) = (w^{-1}\rho)(itu_{\lambda+\rho}).$$

Since $\det w = \det(w^{-1})$, the Weyl Denominator Formula rewrites the numerator in Equation 7.33 as

$$\sum_{w \in W(\Delta(\mathfrak{g}_\mathbb{C}))} \det(w)\, e^{[w(\lambda+\rho)](itu_\rho)} = \sum_{w \in W(\Delta(\mathfrak{g}_\mathbb{C}))} \det(w)\, e^{(w\rho)(itu_{\lambda+\rho})} = \Delta(itu_{\lambda+\rho})$$

$$= \prod_{\alpha \in \Delta^+(\mathfrak{g}_\mathbb{C})} \left(e^{\alpha(itu_{\lambda+\rho})/2} - e^{-\alpha(itu_{\lambda+\rho})/2} \right)$$

$$= \prod_{\alpha \in \Delta^+(\mathfrak{g}_\mathbb{C})} \left(it\alpha(u_{\lambda+\rho}) + \cdots \right)$$

$$= (it)^{|\Delta^+(\mathfrak{g}_\mathbb{C})|} \prod_{\alpha \in \Delta^+(\mathfrak{g}_\mathbb{C})} B(\alpha, \lambda+\rho) + \cdots$$

where the ellipses denote higher powers of t. Similarly, the Weyl Denominator Formula rewrites denominator in Equation 7.33 as

$$\sum_{w \in W(\Delta(\mathfrak{g}_\mathbb{C}))} \det(w)\, e^{(w\rho)(itu_\rho)} = (it)^{|\Delta^+(\mathfrak{g}_\mathbb{C})|} \prod_{\alpha \in \Delta^+(\mathfrak{g}_\mathbb{C})} B(\alpha, \rho) + \cdots$$

which finishes the proof. $\qquad\qquad\qquad\qquad\qquad\qquad\qquad\qquad\qquad\qquad\square$

7.3.5 Highest Weight Classification

Theorem 7.34 (Highest Weight Classification). *For a connected compact Lie group G with maximal torus T, there is a one-to-one correspondence between irreducible representations and dominant analytically integral weights given by mapping $V(\lambda) \to \lambda$ for dominant $\lambda \in A(T)$.*

Proof. We saw in Theorem 7.3 that the map $V(\lambda) \to \lambda$ is well defined and injective. It remains to see it is surjective. For any $\lambda \in A(T)$, Lemma 7.26 shows the function Θ_λ descends to a smooth class function on G^{reg}. The Weyl Integral Formula to calculates

$$\int_G |\Theta_\lambda|^2 \, dg = \frac{1}{|W(G)|} \int_{T^{\text{reg}}} |\Delta(t)\Theta_\lambda|^2 \, dt$$

$$= \frac{1}{|W(G)|} \int_T \left| \sum_{w \in W(\Delta(\mathfrak{g}_\mathbb{C}))} \det(w)\, \xi_{w(\lambda+\rho)} \right|^2 dt$$

$$= \frac{1}{|W(G)|} \sum_{w, w' \in W(\Delta(\mathfrak{g}_\mathbb{C}))} \det(ww') \int_T \xi_{w(\lambda+\rho)} \bar{\xi}_{-w'(\lambda+\rho)} \, dt.$$

When λ is also dominant, $\lambda + \rho$ is strictly dominant so that, as in the proof of the Weyl Character Formula, Equation 7.27 shows that

$$\int_T \xi_{w(\lambda+\rho)} \xi_{-w'(\lambda+\rho)} \, dt = \delta_{w,w'}.$$

As a result, $\int_G |\Theta_\lambda|^2 \, dg = 1$ for any dominant $\lambda \in A(T)$. In particular, Θ_λ is a nonzero L^2 class function on G.

Now choose any irreducible representation $V(\mu)$ of G and note that the function Θ_μ extends to the character χ_μ. By the now typical calculation,

$$\begin{aligned}
\int_G \chi_\mu \overline{\Theta_\lambda} \, dg &= \frac{1}{|W(G)|} \int_{T^{\text{reg}}} |\Delta(t)|^2 \, \Theta_\mu \overline{\Theta_\lambda} \, dt \\
&= \frac{1}{|W(G)|} \sum_{w,w' \in W(\Delta(\mathfrak{g}_{\mathbb{C}}))} \det(ww') \int_{T^{\text{reg}}} \xi_{w(\mu+\rho)} \xi_{w'(\lambda+\rho)} \, dt \\
&= \begin{cases} 1 & \text{if } \mu = \lambda \\ 0 & \text{if } \mu \neq \lambda. \end{cases}
\end{aligned}$$

Since Theorems 7.3 and 3.30 imply that $\{\chi_\mu \mid$ there exists an irreducible representation with highest weight $\mu\}$ is an orthonormal basis for the set of L^2 class functions on G, the value of $\int_G \chi_\mu \overline{\Theta_\lambda} \, dg$ cannot be zero for every such μ. In particular, this means that there is an irreducible representation with highest weight λ. □

7.3.6 Fundamental Group

Here we finish the proof of Theorem 6.30. This is especially important in light of the Highest Weight Classification. Of special note, it shows that when G is a simply connected compact Lie group with semisimple Lie algebra, then the irreducible representations are parametrized by the set of dominant algebraic weights, P. In turn, this also classifies the irreducible representations of $\mathfrak{g}$ (Theorem 4.16). At the opposite end of the spectrum, Theorem 6.30 shows that the irreducible representations of $\text{Ad}(G) \cong G/Z(G)$ (Lemma 5.11) are parametrized by the dominant elements of the root lattice, R. The most general group lies between these two extremes.

Lemma 7.35. *Let G be a compact connected Lie group with maximal torus T. Let $G^{\text{sing}} = G \backslash G^{\text{reg}}$. Then G^{sing} is a closed subset with* codim $G^{\text{sing}} \geq 3$ *in G.*

Proof. It follows from Theorem 7.7 that G^{sing} is closed and the map $\psi : G/T \times T^{\text{sing}} \to G^{\text{sing}}$ is surjective. Moreover $t \in T^{\text{sing}}$ if and only if there exists $\alpha \in \Delta^+(\mathfrak{g}_{\mathbb{C}})$, so $\xi_\alpha(t) = 1$ so that $T^{\text{sing}} = \cup_{\in \Delta^+(\mathfrak{g}_{\mathbb{C}})} \ker \xi_\alpha$. As a Lie subgroup of T, $\ker \xi_\alpha$ is a closed subgroup of codimension 1. Let $U_\alpha = \{gtg^{-1} \mid g \in G \text{ and } t \in \ker \xi_\alpha\}$, so that $G^{\text{sing}} = \cup_{\in \Delta^+(\mathfrak{g}_{\mathbb{C}})} U_\alpha$.

Recall that $\mathfrak{z}_{\mathfrak{g}}(t) = \{X \in \mathfrak{g} \mid \text{Ad}(t)X = X\}$ (Exercise 4.22). Since $\text{Ad}(t)$ acts on $\mathfrak{g}_\alpha$ as $\xi_\alpha(t)$, it follows that $\mathfrak{g} \cap (\mathfrak{g}_{-\alpha} \oplus \mathfrak{g}_\alpha) \subseteq \mathfrak{z}_{\mathfrak{g}}(t)$ when $t \in \ker \xi_\alpha$. Now choose a standard embedding $\varphi_\alpha : SU(2) \to G$ corresponding to α and let V_α be the compact manifold $V_\alpha = G/(\varphi_\alpha(SU(2))T) \times \ker \xi_\alpha$. Observe that dim $V_\alpha = \dim G - 3$ and that ψ maps V_α onto U_α. Therefore the precise version of this lemma is that G^{sing} is a finite union of closed images of compact manifolds each of which has codimension 3 with respect to G. □

Thinking of a homotopy of loops as a two-dimensional surface, Lemma 7.35 coupled with standard transversality theorems ([42]), show that loops in G with a base point in G^{reg} can be homotoped to loops in G^{reg}. As a corollary, it is straightforward to see that

$$\pi_1(G) \cong \pi_1(G^{\text{reg}}).$$

Let G be a compact Lie group with maximal torus T. Recall from Theorem 7.7 that $e^H \in T^{\text{reg}}$ if and only if $H \in \{H \in \mathfrak{t} \mid \alpha(H) \notin 2\pi i \mathbb{Z}$ for all roots $\alpha\}$. The connected regions of $\{H \in \mathfrak{t} \mid \alpha(H) \notin 2\pi i \mathbb{Z}$ for all roots $\alpha\}$ are convex and are given a special name.

Definition 7.36. Let G be a compact Lie group with maximal torus T. The connected components of $\{H \in \mathfrak{t} \mid \alpha(H) \notin 2\pi i \mathbb{Z}$ for all roots $\alpha\}$ are called *alcoves*.

Lemma 7.37. *Let G be a compact connected Lie group with maximal torus T and fix a base $t_0 = e^{H_0} \in T^{\text{reg}}$ with $H_0 \in \mathfrak{t}$.*
(a) Any continuous loop $\gamma : [0, 1] \to G^{\text{reg}}$ with $\gamma(0) = t_0$ can be written as

$$\gamma(s) = c_{g_s} e^{H(s)}$$

with $g_0 = e$, $H(0) = H_0$, and the maps $s \to g_s T \in G/T$ and $s \to H(s) \in \mathfrak{t}^{\text{reg}}$ continuous. In that case, $g_1 \in N(T)$ and

$$H(1) = \text{Ad}(g_1)^{-1} H_0 + X_\gamma$$

for some $X_\gamma \in 2\pi i \ker \mathcal{E}$. The element X_γ is independent of the homotopy class of γ.
(b) Write A_0 for the alcove containing H_0. Keeping the same base t_0, the map

$$\pi_1(G^{\text{reg}}) \to A_0 \cap \{w H_0 + Z \mid w \in W(\Delta(\mathfrak{g}_\mathbb{C})^\vee)$ and $Z \in 2\pi i A(T)^*\}$$

induced by $\gamma \to X_\gamma$ is well defined and bijective.

Proof. Using the Maximal Torus Theorem, write $\gamma(s) = c_{g_s}\tau(s)$ with $\tau(s) \in T^{\text{reg}}$, $\tau(0) = t_0$, and $g_0 = e$. In fact, since $\psi : G/T \times T^{\text{reg}} \to G^{\text{reg}}$ is a covering, the lifts $s \to \tau(s) \in T^{\text{reg}}$ and $s \to g_s T \in G/T$ are uniquely determined by these conditions and continuity. Since exp : $\mathfrak{t}^{\text{reg}} \to T^{\text{reg}}$ is also a local diffeomorphism (Theorem 5.14), there exists a unique continuous lift $s \to H(s) \in \mathfrak{t}^{\text{reg}}$ of τ, so $H(0) = H_0$ and $\gamma(s) = c_{g_s} e^{H(s)}$.

As γ is a loop, $\gamma(0) = \gamma(1)$, so $e^{H_0} = c_{g_1} e^{H(1)}$. Because e^{H_0} and $e^{H(1)}$ are regular, $T = Z_G(e^{H_0})^0 = c_{g_1} Z_G(e^{H(1)})^0 = c_{g_1} T$, so that $g_1 \in N(T)$ is a Weyl group element. Writing $w = \text{Ad}(g_1)$, it follows that $H_0 \equiv w H(1)$ modulo $2\pi i \ker \mathcal{E}$, the kernel of exp : $\mathfrak{t} \to T$. Therefore write $H(1) = w^{-1}H_0 + X_\gamma$ for some $X_\gamma \in 2\pi i \ker \mathcal{E}$.

To see that X_γ is independent of the homotopy class of γ, suppose $\gamma' : [0, 1] \to G^{\mathrm{reg}}$ with $\gamma'(0) = t_0$ is another loop and that $\gamma(s, t)$ is a homotopy between γ and γ'. Thus $\gamma(s, 0) = \gamma(s)$, $\gamma(s, 1) = \gamma'(s)$, and $\gamma(0, t) = \gamma(1, t) = t_0$. Using the same arguments as above and similar notational conventions, write $\gamma'(s) = c_{g'_s} e^{H'(s)}$ and $H'(1) = w'^{-1}H_0 + X'_\gamma$. Similarly, write $\gamma(s, t) = c_{g_{s,t}} e^{H(s,t)}$ and $H(1, s) = w_s^{-1}H_0 + X_\gamma(s)$. Notice that $w_0 = w$, $w_1 = w'$, $X_\gamma(0) = X_\gamma$, and $X_\gamma(1) = X'_\gamma$. Since w_s and $X_\gamma(s)$ vary continuously with s and since $W(T)$ and $2\pi i \ker \mathcal{E}$ are discrete, w_s and $X_\gamma(s)$ are constant. This finishes part (a).

For part (b), first note that continuity of $H(s)$ implies that $H(1)$ is still in A_0, so that the map is well defined. To see surjectivity, fix $H' \in A_0 \cap \{wH_0 + Z \mid w \in W(\Delta(\mathfrak{g}_{\mathbb{C}})^\vee)$ and $Z \in 2\pi i A(T)^*\}$ and write $H' = w'^{-1}H_0 + Z'$ for $w' \in W(\Delta(\mathfrak{g}_{\mathbb{C}})^\vee)$ and $Z' \in 2\pi i \ker \mathcal{E}$. Choose a continuous path $s \to g'_s \in G$, so that $g'_0 = e$ and $\mathrm{Ad}(g'_1) = w'$. Let $H'(s) = H_0 + s(H' - H_0) \in A_0$ and consider the curve $\gamma'(s) = c_{g_s} e^{H'(s)}$. Since $\gamma'(0) = t_0$ and $\gamma'(1) = e^{w'H'} = e^{H_0 + Z'} = t_0$, γ' is a loop with base point t_0. By construction, $X_{\gamma'} = H'$, as desired. To see injectivity, observe that if $X_\gamma = X_{\gamma''}$ with $\gamma(s) = c_{g_s} e^{H'(s)}$ and $\gamma''(s) = c_{g_s} e^{H''(s)}$, then $\gamma(s, t) = c_{g_s} e^{(1-t)H'(s)+tH''(s)}$ is a homotopy between the two.   □

Lemma 7.38. *Let G be a compact connected Lie group with maximal torus T.*
(a) Each homotopy class in G with base e can be represented by a loop of the form

$$\gamma(s) = e^{sX_\gamma}$$

for some $X_\gamma \in 2\pi i \ker \mathcal{E}$, i.e., for some X_γ in the kernel of exp $: \mathfrak{t} \to T$. The surjective map from $2\pi i \ker \mathcal{E}$ to $\pi_1(G)$ induced by $X_\gamma \to \gamma$ is a homomorphism.
(b) Fix an alcove A_0 and $H_0 \in A_0$. The above map restricts to a bijection on $\{Z \in 2\pi i \ker \mathcal{E} \mid wH_0 + Z \in A_0 \text{ for some } w \in W(\Delta(\mathfrak{g}_{\mathbb{C}})^\vee)\}$.

Proof. Lemma 7.37 shows that each homotopy class in G with base t_0 can be represented by a curve of the form $\gamma(s) = c_{g_s} e^{H(s)}$ with $H(1) = \mathrm{Ad}(g_1)^{-1}H_0 + X_\gamma$ for some $X_\gamma \in 2\pi i \ker \mathcal{E}$. Using the homotopy $\gamma(s, t) = c_{g_s} e^{(1-t)H(s)+t[H_0+s(H(1)-H_0)]}$, we may assume $H(s)$ is of the form $H(s) = H_0 + s\left(\mathrm{Ad}(g_1)^{-1}H_0 + X_\gamma - H_0\right)$.

Translating back to the identity, it follows that each homotopy class in G with base e can be represented by a curve of the form

$$\gamma(s) = e^{-H_0} c_{g_s} e^{H_0+s\left(\mathrm{Ad}(g_1)^{-1}H_0+X_\gamma-H_0\right)}.$$

Using the homotopy $\gamma(s, t) = e^{-tH_0} c_{g_s} e^{tH_0+s\left(t\,\mathrm{Ad}(g_1)^{-1}H_0+X_\gamma-tH_0\right)}$, we may assume $\gamma(s) = c_{g_s} e^{sX_\gamma}$. Finally, using the homotopy $\gamma(s, t) = c_{g_{st}} e^{sX_\gamma}$, we may assume $\gamma(s) = e^{sX_\gamma}$. Verifying that the map $\gamma \to X_\gamma$ is a homomorphism is straightforward and left as an exercise (Exercise 7.24). Part (b) follows from Lemma 7.37.   □

Note that a corollary of Lemma 7.38 shows that the inclusion map $T \to G$ induces a surjection $\pi_1(T) \to \pi_1(G)$.

Definition 7.39. Let G be a compact connected Lie group with maximal torus T. The *affine Weyl group* is the group generated by the transformations of t of the form $H \to wH + Z$ for $w \in W(\Delta(\mathfrak{g}_\mathbb{C})^\vee)$ and $Z \in 2\pi i R^\vee$.

Lemma 7.40. *Let G be a compact connected Lie group with maximal torus T.*
(a) The affine Weyl group is generated by the reflections across the hyperplanes $\alpha^{-1}(2\pi i n)$ for $\alpha \in \Delta(\mathfrak{g}_\mathbb{C})$ and $n \in \mathbb{Z}$.
(b) The affine Weyl group acts simply transitively on the set of alcoves.

Proof. Recall that $h_\alpha \in R^\vee$ and notice the reflection across the hyperplane $\alpha^{-1}(2\pi i n)$ is given by $r_{h_\alpha,n}(H) = r_{h_\alpha}H + 2\pi i h_\alpha$ (Exercise 7.25). Since the Weyl group is generated by the reflections r_{h_α}, part (a) is finished. The proof of part (b) is very similar to Theorem 6.43 and the details are left as an exercise (Exercise 7.26). $\square$

Theorem 7.41. *Let G be a connected compact Lie group with semisimple Lie algebra and maximal torus T. Then $\pi_1(G) \cong \ker \mathcal{E}/R^\vee \cong P/A(T)$.*

Proof. By Lemma 7.38, it suffices to show that the loop $\gamma(s) = e^{sX_\gamma}$, $X_\gamma \in 2\pi i \ker \mathcal{E}$, is trivial if and only if $X_\gamma \in 2\pi i R^\vee$. For this, first consider the standard $\mathfrak{su}(2)$-triple corresponding to $\alpha \in \Delta(\mathfrak{g}_\mathbb{C})$ and let $\varphi_\alpha : SU(2) \to G$ be the corresponding embedding. The loop $\gamma_\alpha(s) = e^{2\pi i s h_\alpha}$ is the image under φ_α of the loop $s \to \mathrm{diag}(e^{2\pi i s}, e^{-2\pi i s})$ in $SU(2)$. As $SU(2)$ is simply connected, γ_α is trivial. Thus there is a well-defined surjective map $2\pi i \ker \mathcal{E}/2\pi i R^\vee \to \pi_1(G)$.

It remains to see that it is injective. Fix an alcove A_0 and $H_0 \in A_0$. Since $2\pi i \ker \mathcal{E} \subseteq 2\pi i P^\vee$, $A_0 - X_\gamma$ is another alcove. By Lemma 7.40, there is a $w \in W(\Delta(\mathfrak{g}_\mathbb{C})^\vee)$ and $H \in 2\pi i R^\vee$, so that $wH_0 + H \in A_0 - X_\gamma$. Thus $wH_0 + (X_\gamma + H) \in A_0$. Because the loop $s \to e^{sH}$ is trivial, we may use a homotopy on γ and assume $H = 0$, so that $wH_0 + X_\gamma \in A_0$. But as $H_0 + 0 \in A_0$, Lemma 7.38 shows that γ must be homotopic to the trivial loop $s \to e^{s0}$. $\square$

7.3.7 Exercises

Exercise 7.12 Show that the function e^ρ descends to the maximal torus for $SU(n)$, $SO(2n)$, and $Sp(2n)$, but not for $SO(2n + 1)$.

Exercise 7.13 Let G be a compact Lie group with a maximal torus T. Let $u_\rho \in i\mathfrak{t}$, so that $\rho(H) = B(H, u_\rho)$ for $H \in \mathfrak{t}$. Show that $itu_\rho \in \Xi$ for small positive t.

Exercise 7.14 Show that the dominant analytically integral weights of $SU(3)$ are all expressions of the form $\lambda = n\pi_1 + m\pi_2$ for $n, m \in \mathbb{Z}^{\geq 0}$ where π_1, π_2 are the fundamental weights $\pi_1 = \frac{2}{3}\epsilon_{1,2} + \frac{1}{3}\epsilon_{2,3}$ and $\pi_2 = \frac{1}{3}\epsilon_{1,2} + \frac{2}{3}\epsilon_{2,3}$. Conclude that

$$\dim V(\lambda) = \frac{(n+1)(m+1)(n+m+2)}{2}.$$

Exercise 7.15 Let G be a compact Lie group with semisimple $\mathfrak{g}$ and a maximal torus T. The set of dominant weight vectors are of the form $\lambda = \sum_i n_i \pi_i$ where $\{\pi_i\}$ are the fundamental weights and $n_i \in \mathbb{Z}^{\geq 0}$. Verify the following calculations.

(1) For $G = SU(n)$,

$$\dim V(\lambda) = \prod_{1 \leq i < j \leq n} \left(1 + \frac{n_i + \cdots + n_{j-1}}{j - i} \right).$$

(2) For $G = Sp(n)$,

$$\dim V(\lambda) = \prod_{1 \leq i < j \leq m} \left(1 + \frac{n_i + \cdots + n_{j-1}}{j - i} \right)$$
$$\cdot \prod_{1 \leq i < j \leq m} \left(1 + \frac{n_i + \cdots + n_{j-1} + 2\left(n_j + \cdots + n_{m-1} \right)}{2n + 2 - i - j} \right)$$
$$\cdot \prod_{1 \leq i \leq m} \left(1 + \frac{n_i + \cdots + n_{m-1} + n_m}{n + 1 - i} \right).$$

(3) For $G = \mathrm{Spin}_{2m+1}(\mathbb{R})$,

$$\dim V(\lambda) = \prod_{1 \leq i < j \leq m} \left(1 + \frac{n_i + \cdots + n_{j-1}}{j - i} \right)$$
$$\cdot \prod_{1 \leq i < j \leq m} \left(1 + \frac{n_i + \cdots + n_{j-1} + 2\left(n_j + \cdots n_{m-1} \right) + n_m}{2m + 1 - i - j} \right)$$
$$\cdot \prod_{1 \leq i \leq m} \left(1 + \frac{2\left(n_i + \cdots + n_{m-1} \right) + n_m}{2n + 1 - 2i} \right).$$

(4) For $G = \mathrm{Spin}_{2m}(\mathbb{R})$,

$$\dim V(\lambda) = \prod_{1 \leq i < j \leq m} \left(1 + \frac{n_i + n_{j-1}}{j - i} \right)$$
$$\cdot \prod_{1 \leq i < j \leq m} \left(1 + \frac{n_i + \cdots + n_{j-1} + 2\left(n_j + \cdots + n_{m-1} \right) + n_m}{2m - i - j} \right).$$

Exercise 7.16 For each group G below, show that the listed representation(s) V of G has minimal dimension among nontrivial irreducible representations.
(1) For $G = SU(n)$, V is the standard representation on $\mathbb{C}^n$ or its dual.
(2) For $G = Sp(n)$, V is the standard representation on $\mathbb{C}^{2n}$.
(3) For $G = \mathrm{Spin}_{2m+1}(\mathbb{R})$ with $m \geq 2$, $V = \mathbb{C}^{2m+1}$ and the action comes from the covering $\mathrm{Spin}_{2m+1}(\mathbb{R}) \to SO(2m + 1)$.
(4) For $G = \mathrm{Spin}_{2m}(\mathbb{R})$ with $m > 4$, $V = \mathbb{C}^{2m}$ and the action comes from the covering $\mathrm{Spin}_{2m}(\mathbb{R}) \to SO(2m)$.

Exercise 7.17 Let G be a compact Lie group with a maximal torus T. Suppose V is a representation of G that possesses a highest weight of weight λ. If $\dim V = \dim V(\lambda)$, show that $V \cong V(\lambda)$ and, in particular, irreducible.

Exercise 7.18 Use Exercise 7.17 and the Weyl Dimension Formula to show that the following representation V of G is irreducible:
(1) $G = SU(n)$ with $V = \bigwedge^p \mathbb{C}^n$ (c.f. Exercise 7.1).
(2) $G = SO(n)$ with $V = \mathcal{H}_m(\mathbb{R}^n)$ (c.f. Exercise 7.2).
(3) $G = SO(2n + 1)$ with $V = \bigwedge^p \mathbb{C}^{2n+1}$, $1 \le p \le n$ (c.f. Exercise 7.3).
(4) $G = SO(2n)$ with $V = \bigwedge^p \mathbb{C}^{2n}$, $1 \le p < n$ (c.f. Exercise 7.3).
(5) $G = SU(n)$ with $V = V_{p,0}(\mathbb{C}^n)$ (c.f. Exercise 7.5).
(6) $G = SU(n)$ with $V = V_{0,q}(\mathbb{C}^n)$ (c.f. Exercise 7.5).
(7) $G = SU(n)$ with $V = \mathcal{H}_{p,q}(\mathbb{C}^n)$ (c.f. Exercise 7.5).
(8) $G = \mathrm{Spin}_{2m+1}(\mathbb{R})$ with $V = S$ (c.f. Exercise 7.6).
(9) $G = \mathrm{Spin}_{2m}(\mathbb{R})$ with $V = S^{\pm}$ (c.f. Exercise 7.6).

Exercise 7.19 Let λ be a dominant analytically integral weight of $U(n)$ and write $\lambda = \lambda_1\epsilon_1 + \cdots + \lambda_n\epsilon_n$, $\lambda_j \in \mathbb{Z}$ with $\lambda_1 \ge \cdots \ge \lambda_n$. For $H = \mathrm{diag}(H_1, \ldots, H_n) \in \mathfrak{t}$, show that the Weyl Character Formula can be written as

$$\chi_\lambda(e^H) = \frac{\det\left(e^{(\lambda_j + j - 1)H_k}\right)}{\det\left(e^{(j-1)H_k}\right)}.$$

Exercise 7.20 Let G be a compact connected Lie group with maximal torus T.
(1) If G is not Abelian, show that the dimensions of the irreducible representations of G are unbounded.
(2) If $\mathfrak{g}$ is semisimple, show that there are at most a finite number of irreducible representations of any given dimension.

Exercise 7.21 Let G be a compact connected Lie group with maximal torus T. For $\lambda \in (i\mathfrak{t})^*$, the *Kostant partition function* evaluated at λ, $\mathcal{P}(\lambda)$, is the number of ways of writing $\lambda = \sum_{\alpha \in \Delta^+(\mathfrak{g}_{\mathbb{C}})} m_\alpha \alpha$ with $m_\alpha \in \mathbb{Z}^{\ge 0}$.
(1) As a formal sum of functions on $\mathfrak{t}$, show that

$$\prod_{\alpha \in \Delta^+(\mathfrak{g}_{\mathbb{C}})} \left(1 + e^{-\alpha} + e^{-2\alpha} + \cdots\right) = \sum_\lambda \mathcal{P}(\lambda)e^{-\lambda}$$

to conclude that

$$1 = \left(\sum_\lambda \mathcal{P}(\lambda)e^{-\lambda}\right) \prod_{\alpha \in \Delta^+(\mathfrak{g}_{\mathbb{C}})} \left(1 - e^{-\alpha}\right).$$

For what values of $H \in \mathfrak{t}$ can this expression be evaluated?
(2) The *multiplicity*, m_μ, of μ in $V(\lambda)$ is the dimension of the μ-weight space in $V(\lambda)$. Thus $\chi_\lambda = \sum_\mu m_\mu \xi_\mu$. Use the Weyl Character Formula, part (1), and gather terms to show that m_μ is given by the expression

$$m_\mu = \sum_{w \in W(\Delta(\mathfrak{g}_\mathbb{C}))} \det(w) \, \mathcal{P} \left(w(\lambda + \rho) - (\mu + \rho) \right).$$

This formula is called the *Kostant Multiplicity Formula*.

(3) For $G = SU(3)$, calculate the weight multiplicities for $V(\epsilon_{1,2} + 3\epsilon_{2,3})$.

Exercise 7.22 Let G be a compact connected Lie group with maximal torus T. The *multiplicity*, m_μ, of $V(\mu)$ in $V(\lambda) \otimes V(\lambda')$ is the number of times $V(\mu)$ appears as a summand in $V(\lambda) \otimes V(\lambda')$. Thus $\chi_\lambda \chi_{\lambda'} = \sum_\mu m_\mu \chi_\mu$. Use part (1) of Exercise 7.21 and compare dominant terms to show m_μ is given by the expression

$$m_\mu = \sum_{w,w' \in W(\Delta(\mathfrak{g}_\mathbb{C}))} \det(ww') \, \mathcal{P} \left(w(\lambda + \rho) + w'(\lambda' + \rho) - (\mu + 2\rho) \right).$$

This formula is called *Steinberg's Formula*.

Exercise 7.23 Let G be a compact connected Lie group with maximal torus T and $\alpha \in \Delta(\mathfrak{g}_\mathbb{C})$. Show that $\ker \xi_\alpha$ in T may be disconnected.

Exercise 7.24 Show that the map $\gamma \to X_\gamma$ from Lemma 7.38 is a homomorphism.

Exercise 7.25 Let G be a compact connected Lie group with maximal torus T. Show that the reflection across the hyperplane $\alpha^{-1}(2\pi i n)$ is given by the formula $r_{h_\alpha,n}(H) = r_{h_\alpha} H + 2\pi i n h_\alpha$ for $H \in \mathfrak{t}$.

Exercise 7.26 Let G be a compact connected Lie group with maximal torus T. Show that the affine Weyl group acts simply transitively on the set of alcoves.

7.4 Borel–Weil Theorem

The Highest Weight Classification gives a parametrization of the irreducible representations of a compact Lie group. Lacking is an explicit realization of these representations. The Borel–Weil Theorem repairs this gap.

7.4.1 Induced Representations

Definition 7.42. (a) A complex *vector bundle* $\mathcal{V}$ of rank n on a manifold M is a manifold $\mathcal{V}$ and a smooth surjective map $\pi : \mathcal{V} \to M$ called the *projection*, so that: (i) for each $x \in M$, the *fiber* over x, $\mathcal{V}_x = \pi^{-1}(x)$, is a vector space of dimension n and (ii) for each $x \in M$, there is a neighborhood U of x in M and a diffeomorphism $\varphi : \pi^{-1}(U) \to U \times \mathbb{C}^n$, so that $\varphi(\mathcal{V}_y) = (y, \mathbb{C}^n)$ for $y \in U$.

(b) The set of smooth (continuous) *sections* of $\mathcal{V}$ are denoted by $\Gamma(M, \mathcal{V})$ and consists of all smooth (continuous) maps $s : M \to \mathcal{V}$, so that $\pi \circ s = I$.

(c) An action of a Lie group G on $\mathcal{V}$ is said to *preserve fibers* if for each $g \in G$ and $x \in M$, there exists $x' \in M$, so that $g\mathcal{V}_x \subseteq \mathcal{V}_{x'}$. In this case, the action of G on $\mathcal{V}$ naturally descends to an action of G on M.

(d) $\mathcal{V}$ is a *homogeneous* vector bundle over M for the Lie group G if **(i)** the action of G on $\mathcal{V}$ preserves fibers; **(ii)** the resulting action of G on M is transitive; and **(iii)** each $g \in G$ maps $\mathcal{V}_x$ to $\mathcal{V}_{gx}$ linearly for $x \in M$.

(e) If $\mathcal{V}$ is a homogeneous vector bundle over M, the vector space $\Gamma(M, \mathcal{V})$ carries an action of G given by

$$(gs)(x) = g(s(g^{-1}x))$$

for $s \in \Gamma(M, \mathcal{V})$.

(f) Two homogeneous vector bundles $\mathcal{V}$ and $\mathcal{V}'$ over M for G are *equivalent* if there is a diffeomorphism $\varphi : \mathcal{V} \to \mathcal{V}'$, so that $\pi' \circ \varphi = \varphi \circ \pi$.

Note it suffices to study manifolds of the form $M = G/H$, H a closed subgroup of G, when studying homogenous vector bundles.

Definition 7.43. Let G be a Lie group and H a closed subgroup of G. Given a representation V of H, define the homogeneous vector bundle $G \times_H V$ over G/H by

$$G \times_H V = (G \times V)/^\sim,$$

where $\sim$ is the equivalence relation given by

$$(gh, v) \sim (g, hv)$$

for $g \in G$, $h \in H$, and $v \in V$. The projection map $\pi : G \times_H V \to G/H$ is given by $\pi(g, v) = gH$ and the G-action is given by $g'(g, v) = (g'g, v)$ for $g' \in G$.

It is necessary to verify that $G \times_H V$ is indeed a homogeneous vector bundle over G/H. Since H is a regular submanifold, this is a straightforward argument and left as an exercise (Exercise 7.27).

Theorem 7.44. *Let G be a Lie group and H a closed subgroup of G. There is a bijection between equivalence classes of homogenous vector bundles $\mathcal{V}$ on G/H and representations of H.*

Proof. The correspondence maps $\mathcal{V}$ to $\mathcal{V}_{eH}$. By definition $\mathcal{V}_{eH}$ is a representation of H. Conversely, given a representation V of H, the vector bundle $G \times_H V$ inverts the correspondence. □

Definition 7.45. Let G be a Lie group and H a closed subgroup of G. Given a representation (π, V) of H, define the smooth (continuous) *induced representation* of G by

$$\text{Ind}_H^G(V) = \text{Ind}_H^G(\pi) = \{\text{smooth (continuous) } f : G \to V \mid f(gh) = h^{-1}f(g)\}$$

with action $(g_1 f)(g_2) = f(g_1^{-1}g_2)$ for $g_i \in G$.

Theorem 7.46. *Let G be a Lie group, H a closed subgroup of G, and V a representation of H. There is a linear G-intertwining bijection between $\Gamma(G/H, G \times_H V)$ and $\text{Ind}_H^G(V)$.*

Proof. Identify $(G \times_H V)_{eH}$ with V by mapping $(h, v) \in (G \times_H V)_{eH}$ to $h^{-1}v \in V$. Given $s \in \Gamma(G/H, G \times_H V)$, let $f_s \in \mathrm{Ind}_H^G(V)$ be defined by $f_s(g) = g^{-1}(s(gH))$. Conversely, given $f \in \mathrm{Ind}_H^G(V)$, let $s_f \in \Gamma(G/H, G \times_H V)$ be defined by $s_f(gH) = (g, f(g))$. It is easy to use the definitions to see these maps are well defined, inverses, and G-intertwining (Exercise 7.28). $\qquad \square$

Theorem 7.47 (Frobenius Reciprocity). *Let G be a Lie group and H a closed subgroup of G. If V is a representation of H and a W is a representation of G, then as vector spaces*

$$\mathrm{Hom}_G(W, \mathrm{Ind}_H^G(V)) \cong \mathrm{Hom}_H(W|_H, V).$$

Proof. Map $T \in \mathrm{Hom}_G(W, \mathrm{Ind}_H^G(V))$ to $S_T \in \mathrm{Hom}_H(W|_H, V)$ by $S_T(w) = (T(w))(e)$ for $w \in W$ and map $S \in \mathrm{Hom}_H(W|_H, V)$ to $T_S \in \mathrm{Hom}_G(W, \mathrm{Ind}_H^G(V))$ by $(T_S(w))(g) = S(g^{-1}w)$. Verifying these maps are well defined and inverses is straightforward (Exercise 7.28). $\qquad \square$

In the special case of $H = \{e\}$ and $V = \mathbb{C}$, the continuous version gives $\Gamma(G/H, G \times_H V) \cong \mathrm{Ind}_H^G(V) = C(G)$. In this setting, Frobenius Reciprocity already appeared in Lemma 3.23.

7.4.2 Complex Structure on G/T

Definition 7.48. Let G be a compact connected Lie group with maximal torus T.
(a) Choosing a faithful representation, assume $G \subseteq U(n)$ for some n. By Theorem 4.14 there exists a unique connected Lie subgroup of $GL(n, \mathbb{C})$ with Lie algebra $\mathfrak{g}_\mathbb{C}$. Write $G_\mathbb{C}$ for this subgroup and call it the *complexification* of G.
(b) Fix $\Delta^+(\mathfrak{g}_\mathbb{C})$ a system of positive roots and recall $\mathfrak{n}^+ = \bigoplus_{\alpha \in \Delta^+(\mathfrak{g}_\mathbb{C})} \mathfrak{g}_\alpha$. The corresponding *Borel subalgebra* is $\mathfrak{b} = \mathfrak{t}_\mathbb{C} \oplus \mathfrak{n}^+$.
(c) Let N, B, and A be the unique connected Lie subgroups of $GL(n, \mathbb{C})$ with Lie algebras $\mathfrak{n}^+$, $\mathfrak{b}$, and $\mathfrak{a} = i\mathfrak{t}$, respectively.

For example, if $G = U(n)$ with the usual positive root system, $G_\mathbb{C} = GL(n, \mathbb{C})$, N is the subgroup of upper triangular matrices with 1's on the diagonal, B is the subgroup of all upper triangular matrices, and A is the subgroup of diagonal matrices with entries in $\mathbb{R}^{>0}$. Although not obvious from Definition 7.48, $G_\mathbb{C}$ is in fact unique up to isomorphism when G is compact. More generally for certain types of noncompact groups, complexifications may not be unique or even exist (e.g., [61], VII §1). In any case, what is important for the following theory is that $G_\mathbb{C}$ is a complex manifold.

Lemma 7.49. *Let G be a compact connected Lie group with maximal torus T.*
(a) The map $\exp : \mathfrak{n}^+ \to N$ is a bijection.
(b) The map $\exp : \mathfrak{a} \to A$ is a bijection.
(c) N, B, A, and AN are closed subgroups.
(d) The map from $T \times \mathfrak{a} \times \mathfrak{n}^+$ to B sending $(t, X, H) \to te^X e^H$ is a diffeomorphism.

Proof. Since T consists of commuting unitary matrices, we may assume T is contained in the set of diagonal matrices of $GL(n, \mathbb{C})$. By using the Weyl group of $GL(n, \mathbb{C})$, we may further assume $u_\rho = \text{diag}(c_1, \dots, c_n)$ with $c_i \geq c_{i+1}$. Therefore if $X \in \mathfrak{g}_\alpha$, $\alpha \in \Delta^+(\mathfrak{g}_\mathbb{C})$, with $X = \sum_{i,j} k_{i,j} E_{i,j}$, then

$$\sum_{i,j} (c_i - c_j) k_{i,j} E_{i,j} = [u_\rho, X] = \alpha(u_\rho) X = \sum_{i,j} B(\alpha, \rho) k_{i,j} E_{i,j}.$$

Since $B(\alpha, \rho) > 0$, it follows that $k_{i,j} = 0$ whenever $c_i - c_j \leq 0$. In turn, this shows that X is strictly upper triangular.

It is well known and easy to see that the set of nilpotent matrices are in bijection with the set of unipotent matrices by the polynomial map $M \to e^M$ with polynomial inverse $M \to \ln(I + (M - I)) = \sum_k \frac{(-1)^{k+1}}{k} (M - I)^k$. In particular if $X, Y \in \mathfrak{n}^+$, there is a unique strictly upper triangular $Z \in \mathfrak{gl}(n, \mathbb{C})$, so that $e^X e^Y = e^Z$.

Dynkin's formula is usually only applicable to small X and Y. However, $\Delta^+(\mathfrak{g}_\mathbb{C})$ is finite, so $[X_n^{(i_n)}, \dots, X_1^{(i_1)}]$ is 0 for sufficiently large i_j for $X_j \in \mathfrak{n}^+$. Thus all the sums in the proof of Dynkin's formula are finite and the formula for Z is a polynomial in X and Y. Coupled with the already mentioned polynomial formula for Z, Dynkin's Formula therefore actually holds for all $X, Y \in \mathfrak{n}^+$. As a consequence, $Z \in \mathfrak{n}^+$ and $\exp \mathfrak{n}^+$ is a subgroup. Since N is generated by $\exp \mathfrak{n}^+$, part (a) is finished. The group N is closed since $\exp : \mathfrak{n}^+ \to N$ is a bijection and the exponential map restricted to the strictly upper triangular matrices has a continuous inverse.

Part (b) and the fact that A is closed in $G_\mathbb{C}$ follows from the fact that $\mathfrak{a}$ is Abelian and real valued. Next note that AN is a subgroup. This follows from the two observations that $(an)(a'n') = (aa')((c_{a'^{-1}}n)n')$, $a, a' \in A$ and $n, n' \in N$, and that $c_{e^H} e^X = \exp(e^{\text{ad}(H)} X)$, $H \in \mathfrak{a}$ and $X \in \mathfrak{n}^+$. Since the map from $\mathfrak{b} = \mathfrak{t} \oplus \mathfrak{a} \oplus \mathfrak{n}^+ \to G_\mathbb{C}$ given by $(H_1, H_2, X) \to e^{H_1} e^{H_2} e^X$ is a local diffeomorphism near 0, products of the form tan, $t \in T$, $a \in A$, and $n \in N$, generate B. Just as with AN, TAN is a subgroup, so that $B = T e^{\mathfrak{a}} e^{\mathfrak{n}^+}$. It is an elementary fact from linear algebra that this decomposition is unique and the proof is complete.   □

The point of the next theorem is that G/T has a G-invariant complex structure inherited from the fact that $G_\mathbb{C}/B$ is a complex manifold. This will allow us to study holomorphic sections on G/T.

Theorem 7.50. *Let G be a compact connected Lie group with maximal torus T. The inclusion $G \hookrightarrow G_\mathbb{C}$ induces a diffeomorphism*

$$G/T \cong G_\mathbb{C}/B.$$

Proof. Recall that $\mathfrak{g} = \{X + \theta X \mid X \in \mathfrak{g}_\mathbb{C}\}$, so that $\mathfrak{g}/\mathfrak{t}$ and $\mathfrak{g}_\mathbb{C}/\mathfrak{b}$ are both spanned by the projections of $\{X_\alpha + \theta X_\alpha \mid X_\alpha \in \mathfrak{g}_\alpha, \alpha \in \Delta^+(\mathfrak{g}_\mathbb{C})\}$. In particular, the differential of the map $G \to G_\mathbb{C}/B$ is surjective. Thus the image of G contains a neighborhood of eB in $G_\mathbb{C}/B$. As left multiplication by g and g^{-1}, $g \in G$, is continuous, the image of G is open in $G_\mathbb{C}/B$. Compactness of G shows that the image is closed so that connectedness shows the map $G \to G_\mathbb{C}/B$ is surjective.

It remains to see that $G \cap B = T$. Let $g \in G \cap B$. Clearly $\mathrm{Ad}(g)$ preserves $\mathfrak{g} \cap \mathfrak{b} = \mathfrak{t}$, so that $g \in N(T)$. Writing w for the corresponding element of $W(\Delta(\mathfrak{g}_{\mathbb{C}}))$, the fact that $g \in B$ implies that w preserves $\Delta^+(\mathfrak{g}_{\mathbb{C}})$. In turn, this means w preserves the corresponding Weyl chamber. Since Theorem 6.43 shows that $W(\Delta(\mathfrak{g}_{\mathbb{C}}))$ acts simply transitively on the Weyl chambers, $w = I$ and $g \in T$. $\square$

7.4.3 Holomorphic Functions

Definition 7.51. Let G be a compact Lie group with maximal torus T. For $\lambda \in A(T)$, write $\mathbb{C}_\lambda$ for the one-dimensional representation of T given by ξ_λ and write L_λ for the *line bundle*

$$L_\lambda = G \times_T \mathbb{C}_\lambda.$$

By Frobenius Reciprocity, $\Gamma(G/T, L_\lambda)$ is a huge representation of G. However by restricting our attention to holomorphic sections, we will obtain a representation of manageable size.

Definition 7.52. Let G be a compact connected Lie group with maximal torus T and $\lambda \in A(T)$.
(a) Extend $\xi_\lambda : T \to \mathbb{C}$ to a homomorphism $\xi_\lambda^{\mathbb{C}} : B \to \mathbb{C}$ by

$$\xi_\lambda^{\mathbb{C}}(te^{iH}e^X) = \xi_\lambda(t)e^{i\lambda(H)}$$

for $t \in T$, $H \in \mathfrak{t}$, and $X \in \mathfrak{n}^+$.
(b) Let $L_\lambda^{\mathbb{C}} = G_{\mathbb{C}} \times_B \mathbb{C}_\lambda$ where $\mathbb{C}_\lambda$ is the one-dimensional representation of B given by $\xi_\lambda^{\mathbb{C}}$.

Lemma 7.53. *Let G be a compact connected Lie group with maximal torus T and $\lambda \in A(T)$. Then $\Gamma(G/T, L_\lambda) \cong \Gamma(G_{\mathbb{C}}/B, L_\lambda^{\mathbb{C}})$ and $\mathrm{Ind}_T^G(\xi_\lambda) \cong \mathrm{Ind}_B^{G_{\mathbb{C}}}(\xi_\lambda^{\mathbb{C}})$ as G-representations.*

Proof. Since the map $G \to G_{\mathbb{C}}/B$ induces an isomorphism $G/T \cong G_{\mathbb{C}}/B$, any $h \in G_{\mathbb{C}}$ can be written as $h = gb$ for $g \in G$ and $b \in B$. Moreover, if $h = g'b'$, $g' \in G$ and $b' \in B$, then there is $t \in T$ so $g' = gt$ and $b' = t^{-1}b$.

On the level of induced representations, map $f \in \mathrm{Ind}_T^G(\xi_\lambda)$ to $F_f \in \mathrm{Ind}_B^{G_{\mathbb{C}}}(\xi_\lambda^{\mathbb{C}})$ by $F_f(gb) = f(g)\xi_{-\lambda}^{\mathbb{C}}(b)$ for $g \in G$ and $b \in B$ and map $F \in \mathrm{Ind}_B^{G_{\mathbb{C}}}(\xi_\lambda^{\mathbb{C}})$ to $f_F \in \mathrm{Ind}_T^G(\xi_\lambda)$ by $f_F(g) = F(g)$. It is straightforward to verify that these maps are well defined, G-intertwining, and inverse to each other (Exercise 7.31). $\square$

Definition 7.54. Let G be a compact connected Lie group with maximal torus T and $\lambda \in A(T)$.
(a) A section $s \in \Gamma(G/T, L_\lambda)$ is said to be *holomorphic* if the corresponding function $F \in \mathrm{Ind}_B^{G_{\mathbb{C}}}(\xi_\lambda^{\mathbb{C}})$, c.f. Theorem 7.46 and Lemma 7.53, is a holomorphic function on $G_{\mathbb{C}}$, i.e., if

$$dF(iX) = idF(X)$$

at each $g \in G_{\mathbb{C}}$ and for all $X \in T_g(G_{\mathbb{C}})$ where $dF(X) = X(\mathrm{Re}\, F) + iX(\mathrm{Im}\, F)$.
(b) Write $\Gamma_{\mathrm{hol}}(G/T, L_\lambda)$ for the set of all holomorphic sections.

Since the differential dF is always $\mathbb{R}$-linear, the condition of being holomorphic is equivalent to saying that dF is $\mathbb{C}$-linear. Written in local coordinates, this condition gives rise to the standard Cauchy–Riemann equations (Exercise 7.32).

Definition 7.55. Let G be a connected (linear) Lie group with maximal torus T. Write $C^\infty(G_\mathbb{C})$ for the set of smooth functions on $G_\mathbb{C}$ and use similar notation for G.
(a) For $Z \in \mathfrak{g}_\mathbb{C}$ and $F \in C^\infty(G_\mathbb{C})$, let

$$[dr(Z)F](h) = \frac{d}{dt}F(he^{tZ})|_{t=0}$$

for $h \in G_\mathbb{C}$. For $X \in \mathfrak{g}$ and $f \in C^\infty(G)$, let

$$[dr(X)f](g) = \frac{d}{dt}f(ge^{tX})|_{t=0}$$

for $g \in G$.
(b) For $Z = X + iY$ with $X, Y \in \mathfrak{g}$, let

$$dr_\mathbb{C}(Z) = dr(X) + idr(Y).$$

Note that $dr_\mathbb{C}(Z)$ is a well-defined operator on $C^\infty(G)$ but that $dr(Z)$ is not (except when $Z \in \mathfrak{g}$).

Lemma 7.56. *Let G be a compact connected Lie group with maximal torus T, $\lambda \in A(T)$, $F \in \mathrm{Ind}_B^{G_\mathbb{C}}(\xi_\lambda^\mathbb{C})$, and $f = F|_G$ the corresponding function in $\mathrm{Ind}_T^G(\xi_\lambda)$.
(a) Then F is holomorphic if and only if*

$$dr_\mathbb{C}(Z)F = 0$$

*for $Z \in \mathfrak{n}^+$.
(b) Equivalently, F is holomorphic if and only if*

$$dr_\mathbb{C}(Z)f = 0$$

for $Z \in \mathfrak{n}^+$.

Proof. Since $dl_g : T_e(G_\mathbb{C}) \to T_g(G_\mathbb{C})$ is an isomorphism, F is holomorphic if and only if

(7.57) $$dF(dl_g(iZ)) = idF(dl_g Z)$$

for all $g \in G_\mathbb{C}$ and $X \in \mathfrak{g}_\mathbb{C}$ where, by definition,

$$dF(dl_g Z) = \frac{d}{dt}F(ge^{tZ})|_{t=0} = [dr(Z)F](g).$$

If $Z \in \mathfrak{n}^+$, then $e^{tZ} \in N$, so that $F(ge^{tZ}) = F(g)$. Thus for $Z \in \mathfrak{n}^+$, Equation 7.57 is automatic since both sides are 0. If $Z \in \mathfrak{t}_\mathbb{C}$, $F(ge^{tZ}) = F(g)e^{-t\lambda(Z)}$. Thus for $Z \in \mathfrak{t}_\mathbb{C}$, Equation 7.57 also holds since both sides are $-i\lambda(Z)F(g)$.

Since $\mathfrak{g}_{\mathbb{C}} = \mathfrak{n}^- \oplus \mathfrak{t}_{\mathbb{C}} \oplus \mathfrak{n}^+$, part (a) will be finished by showing Equation 7.57 holds for $Z \in \mathfrak{n}^-$. However, Equation 7.57 is equivalent to requiring $dr(iZ)F = idr(Z)F$ which in turn is equivalent to requiring $dr(Z)F = dr_{\mathbb{C}}(Z)F$. If $Z \in \mathfrak{n}^-$, then $\theta Z \in \mathfrak{n}^+$ and $Z + \theta Z \in \mathfrak{g}$. Thus

$$dr(Z)F = dr(Z + \theta Z)F - dr(\theta Z)F = dr_{\mathbb{C}}(Z + \theta Z)F = dr_{\mathbb{C}}(Z) + dr_{\mathbb{C}}(\theta Z),$$

so that $dr(Z)F = dr_{\mathbb{C}}(Z)F$ if and only if $dr_{\mathbb{C}}(\theta Z) = 0$, as desired.

For part (b), first, assume F is holomorphic. Since $f = F|_G$, it follows that $dr_{\mathbb{C}}(\mathfrak{n}^+)f = 0$. Conversely, suppose $dr_{\mathbb{C}}(\mathfrak{n}^+)f = 0$. Restricting the above arguments from $G_{\mathbb{C}}$ to G shows $dr(Z)F|_g = dr_{\mathbb{C}}(Z)F|_g$ for $g \in G$ and $Z \in \mathfrak{g}_{\mathbb{C}}$. Hence if $X \in \mathfrak{g}$,

$$(dr(X)F)(gb) = \frac{d}{dt}F(gbe^{tX})|_{t=0} = \frac{d}{dt}F(ge^{t\,\mathrm{Ad}(b)X}b)|_{t=0}$$

$$= \xi_{-\lambda}(b)\frac{d}{dt}F(ge^{t\,\mathrm{Ad}(b)X})|_{t=0}$$

$$= \xi_{-\lambda}(b)(dr(\mathrm{Ad}(b)X)F)(g) = \xi_{-\lambda}(b)(dr_{\mathbb{C}}(\mathrm{Ad}(b)X)F)(g)$$

for $g \in G$ and $b \in B$. Thus if $Z = X + iY \in \mathfrak{n}^+$ with $X, Y \in \mathfrak{g}$, note $\mathrm{Ad}(b)Z \in \mathfrak{n}^+$ and calculate

$$(dr_{\mathbb{C}}(Z)F)(gb) = (dr(X)F)(gb) + i(dr(Y)F)(gb)$$

$$= \xi_{-\lambda}(b)[(dr_{\mathbb{C}}(\mathrm{Ad}(b)X)F)(g) + (dr_{\mathbb{C}}(i\,\mathrm{Ad}(b)Y)F)(g)]$$

$$= \xi_{-\lambda}(b)(dr_{\mathbb{C}}(\mathrm{Ad}(b)Z)F)(g) = 0,$$

as desired.   □

7.4.4 Main Theorem

The next theorem gives an explicit realization for each irreducible representation.

Theorem 7.58 (Borel–Weil). *Let G be a compact connected Lie group and $\lambda \in A(T)$.*

$$\Gamma_{\mathrm{hol}}(G/T, L_\lambda) \cong \begin{cases} V(w_0\lambda) & \text{for } -\lambda \text{ dominant} \\ \{0\} & \text{else}, \end{cases}$$

where $w_0 \in W(\Delta(\mathfrak{g}_{\mathbb{C}}))$ is the unique Weyl group element mapping the positive Weyl chamber to the negative Weyl chamber (c.f. Exercise 6.40).

Proof. The elements of $\Gamma_{\mathrm{hol}}(G/T, L_\lambda)$ correspond to holomorphic functions in $\mathrm{Ind}_T^G(\xi_\lambda)$. It follows that the elements of $\Gamma_{\mathrm{hol}}(G/T, L_\lambda)$ correspond to the set of smooth functions f on G, satisfying

(7.59) $$f(gt) = \xi_{-\lambda}(t)f(g)$$

for $g \in G$ and $t \in T$ and

(7.60)
$$d r_{\mathbb{C}}(Z) f = 0$$

for $Z \in \mathfrak{n}^+$.

Using the C^∞-topology on $C^\infty(G)$, Corollary 3.47 shows that $C^\infty(G)_{G\text{-fin}} = C(G)_{G\text{-fin}}$ so that, by Theorem 3.24 and the Highest Weight Theorem,

$$C^\infty(G)_{G\text{-fin}} \cong \bigoplus_{\text{dom. } \gamma \in A(T)} V(\gamma)^* \otimes V(\gamma)$$

as a $G \times G$-module with respect to the $r \times l$-action. In this decomposition, tracing through the identifications (Exercise 7.33) **??** shows that the action of G on $\Gamma_{\text{hol}}(G/T, L_\lambda)$ intertwines with the trivial action on $V(\gamma)^*$ and the standard action on $V(\gamma)$. Recalling that Lemma 7.5, write φ for the intertwining operator

$$\varphi : \bigoplus_{\text{dom. } \gamma \in A(T)} V(-w_0 \gamma) \otimes V(\gamma) \xrightarrow{\sim} C^\infty(G)_{G\text{-fin}}.$$

Given $f \in C^\infty(G)$, use Theorem 3.46 to write $f = \sum_{\text{dom. } \gamma \in A(T)} f_\gamma$ with respect to convergence in the C^∞-topology, where $f_\gamma = \varphi(x_\gamma)$ with $x_\gamma \in V(-w_0 \gamma) \otimes V(\gamma)$.

Equation 7.60 is then satisfied by f if and only if it is satisfied by each f_γ. Tracing through the identifications, the action of $dr_{\mathbb{C}}(Z)$ corresponds to the standard (complexified) action of Z on $V(-w_0 \gamma)$ and the trivial action on $V(\gamma)$. In particular, Theorem 7.3 shows that x_γ can be written as $x_\gamma = v_{-w_0 \gamma} \otimes y_\gamma$ where $v_{-w_0 \gamma}$ is a highest weight vector of $V(-w_0 \gamma)$ and $y_\gamma \in V(\gamma)$.

Tracing through the identifications again, Equation 7.59 is then satisfied if and only if $t v_{-w_0 \gamma} = \xi_{-\lambda}(t) v_{-w_0 \gamma}$. But since $t v_{-w_0 \gamma} = \xi_{-w_0 \gamma}(t) v_{-w_0 \gamma}$, it follows that Equation 7.59 is satisfied if and only if $w_0 \gamma = \lambda$ and the proof is complete.   □

As an example, consider $G = SU(2)$ with T the usual subgroup of diagonal matrices. Realizing $\Gamma_{\text{hol}}(G/T, \xi_{-n\frac{c_{12}}{2}})$ as the holomorphic functions in $\text{Ind}_B^{G_{\mathbb{C}}}(\xi_{-n\frac{c_{12}}{2}}^{\mathbb{C}})$,

$$\Gamma_{\text{hol}}(G/T, \xi_{-n\frac{c_{12}}{2}}) \cong$$

$$\{\text{holomorphic } f : SL(2, \mathbb{C}) \to \mathbb{C} \mid f(g \begin{pmatrix} a & b \\ 0 & a^{-1} \end{pmatrix}) = a^n f(g), g \in SL(2, \mathbb{C})\}.$$

Since $\begin{pmatrix} z_1 & z_3 \\ z_2 & z_4 \end{pmatrix} \begin{pmatrix} 1 & b \\ 0 & 1 \end{pmatrix} = \begin{pmatrix} z_1 & bz_1 + z_3 \\ z_2 & bz_2 + z_4 \end{pmatrix}$, the induced condition in the case of $a = 1$ shows $f \in \text{Ind}_B^{G_{\mathbb{C}}}(\xi_{-n\frac{c_{12}}{2}}^{\mathbb{C}})$ is determined by its restriction to the first column of $SL(2, \mathbb{C})$. Since $\begin{pmatrix} z_1 & z_3 \\ z_2 & z_4 \end{pmatrix} \begin{pmatrix} a & 0 \\ 0 & a^{-1} \end{pmatrix} = \begin{pmatrix} az_1 & a^{-1}z_3 \\ az_2 & a^{-1}z_4 \end{pmatrix}$, the induced condition for the case of $b = 0$ shows that f is homogeneous of degree n as a function on the first column of $SL(2, \mathbb{C})$. Finally, the holomorphic condition shows $\Gamma_{\text{hol}}(G/T, \xi_{-n\frac{c_{12}}{2}})$ can be identified with the set of homogeneous polynomials of degree n on the first column of $SL(2, \mathbb{C})$. In other words, $\Gamma_{\text{hol}}(G/T, \xi_{-n\frac{c_{12}}{2}}) \cong V_n(\mathbb{C}^2)$ as expected.

As a final remark, there is a (dualized) generalization of the Borel–Weil Theorem to the Dolbeault cohomology setting called the Bott–Borel–Weil Theorem. Although we only state the result here, it is fairly straightforward to reduce the calculation to a fact from Lie algebra cohomology ([97]). In turn this is computed by a theorem of Kostant ([64]), an efficient proof of which can be found in [86].

Given a complex manifold M, write $A^p(M) = \bigwedge_p^* T^{0,1}(M)$ for the smooth differential forms of type $(0, p)$ ([93]). The $\overline{\partial}_M$ operator maps $A^p(M)$ to $A^{p+1}(M)$ and is given by

$$\left(\overline{\partial}_M \omega\right)(X_0, \dots, X_p) = \sum_{k=0}^{p} (-1)^k X_k \omega(X_0, \dots, \widehat{X_k}, \dots, X_p)$$
$$+ \sum_{i<j} (-1)^{i+j} \omega([X_i, X_j], X_0, \dots, \widehat{X_i}, \dots, \widehat{X_j}, \dots, X_p)$$

for antiholomorphic vector fields X_j. If $\mathcal{V}$ is a holomorphic vector bundle over M, the sections of $\mathcal{V} \otimes A^p(M)$ are the $\mathcal{V}$-valued differential forms of type $(0, p)$ and the set of such is denoted $A^p(M, \mathcal{V})$. The operator $\overline{\partial} : A^p(M, \mathcal{V}) \to A^{p+1}(M, \mathcal{V})$ is given by $\overline{\partial} = 1 \otimes \overline{\partial}_M$ and satisfies $\overline{\partial}^2 = 0$. The *Dolbeault cohomology* spaces are defined as

$$H^p(M, \mathcal{V}) = \ker \overline{\partial} / \operatorname{Im} \overline{\partial}.$$

Theorem 7.61 (Bott–Borel–Weil Theorem). *Let G be a compact connected Lie group and $\lambda \in A(T)$. If $\lambda + \rho$ lies on a Weyl chamber wall, then $H^p(G/T, L_\lambda) = \{0\}$ for all p. Otherwise,*

$$H^p(G/T, L_\lambda) \cong \begin{cases} V(w(\lambda + \rho) - \rho) \text{ for } p = \left|\{\alpha \in \Delta^+(\mathfrak{g}_{\mathbb{C}}) \mid B(\lambda + \rho, \alpha) < 0\}\right| \\ \{0\} \qquad\qquad\qquad\qquad\qquad else, \end{cases}$$

where $w \in W(\Delta(\mathfrak{g}_{\mathbb{C}}))$ is the unique Weyl group element making $w(\lambda + \rho)$ dominant.

7.4.5 Exercises

Exercise 7.27 Let G be a Lie group and H a closed subgroup of G. Given a representation V of H, verify $G \times_H V$ is a homogeneous vector bundle over G/H.

Exercise 7.28 Verify the details of Theorems 7.46 and 7.47.

Exercise 7.29 Let G be a compact connected Lie group with maximal torus T and $\lambda \in A(T)$.
(1) Show that $\xi_\lambda^{\mathbb{C}}$ is a homomorphism.
(2) Show that $\xi_\lambda^{\mathbb{C}}$ is the unique extension of ξ_λ from T to B as a homomorphism of complex Lie groups.

Exercise 7.30 Let G be a compact connected Lie group with maximal torus T and $\lambda \in A(T)$. If V is an irreducible representation, show that $V \cong V(\lambda)$ if and only if there is a nonzero $v \in V$ satisfying $bv = \xi_\lambda^{\mathbb{C}}(b)v$ for $b \in B$. In this case, show that v is unique up to nonzero scalar multiplication and is a highest weight vector.

Exercise 7.31 Verify the details of Lemma 7.53.

Exercise 7.32 Let $G_{\mathbb{C}}$ be a *complex* (linear) connected Lie group with maximal torus T. Recall that a complex-valued function F on $G_{\mathbb{C}}$ is *holomorphic* if $dF(dl_g(iX)) = idF(dl_gX)$ for all $g \in G_{\mathbb{C}}$ and $X \in \mathfrak{g}_{\mathbb{C}}$, where $dF(dl_gX) = \frac{d}{dt}F(ge^{tX})|_{t=0}$. Note that dF is $\mathbb{R}$-linear.
(1) In the special case of $G_{\mathbb{C}} = \mathbb{C}\backslash\{0\} \cong GL(1, \mathbb{C})$, $z \in G_{\mathbb{C}}$, and $X = 1$, show that $dF(dl_z(iX)) = \frac{\partial}{\partial y}F|_z$ and $idF(dl_zX) = i\frac{\partial}{\partial x}F|_z$, where $z = x + iy$. Conclude that dF is not $\mathbb{C}$-linear for general F and that, in this case, F is holomorphic if and only if $u_x = v_y$ and $u_y = -v_x$, where $F = u + iv$.
(2) Let $\{X_j\}_{j=1}^n$ be a basis over $\mathbb{C}$ for $\mathfrak{g}_{\mathbb{C}}$. For $g \in G_{\mathbb{C}}$, show that the map $\varphi : \mathbb{R}^{2n} \to G_{\mathbb{C}}$ given by

$$\varphi(x_1, \ldots, x_n, y_1, \ldots, y_n) = ge^{x_1X_1}\cdots e^{x_nX_n}e^{iy_1X_1}\cdots e^{iy_nX_n}$$

is a local diffeomorphism near 0, c.f. Exercise 4.12.
(3) Identifying $\mathfrak{g}_{\mathbb{C}}$ with $T_e(G_{\mathbb{C}})$, show $d\varphi(\partial_{x_j}|_0) = dl_gX_j$ and $d\varphi(\partial_{y_j}|_0) = dl_g(iX_j)$.
(4) Given a smooth function F on $G_{\mathbb{C}}$, write F in local coordinates near g as $f = \varphi^*F$. Show that F is holomorphic if and only if for each $g \in G_{\mathbb{C}}$, $u_{x_j} = v_{y_j}$ and $u_{y_j} = -v_{x_j}$ where $f = u + iv$. In other words, F is holomorphic if and only if it satisfies the Cauchy–Riemann equations in local coordinates.

Exercise 7.33 In the proof of the Borel–Weil theorem, trace through the various identifications to verify that the claimed actions are correct.

Exercise 7.34 Let B be the subgroup of upper triangular matrices in $GL(n, \mathbb{C})$. Let $\lambda = \lambda_1\epsilon_1 + \cdots + \lambda_n\epsilon_n$ be a dominant integral weight of $U(n)$, i.e., $\lambda_k \in \mathbb{Z}$ and $\lambda_1 \geq \ldots \lambda_n$.
(1) Let $f : GL(n, \mathbb{C}) \to \mathbb{C}$ be smooth. For $i < j$, show that $dr(iE_{j,k})f|_g = idr(E_{j,k})f|_g$ if and only if

$$0 = \sum_{l=1}^n \overline{z_{l,j}}\frac{\partial f}{\partial \overline{z_{l,k}}}|_g,$$

where $g = (z_{j,k}) \in GL(n, \mathbb{C})$ and $\frac{\partial}{\partial \overline{z_{j,k}}} = \frac{1}{2}\left(\frac{\partial}{\partial x_{j,k}} + i\frac{\partial}{\partial y_{j,k}}\right)$ with $z_{j,k} = x_{j,k} + iy_{j,k}$. Conclude that $dr(iE_{j,k})f = idr(E_{j,k})f$ if and only if $\frac{\partial f}{\partial \overline{z_{l,k}}} = 0$.
(2) Show that the irreducible representation of $U(n)$ with highest weight λ is realized by

$$V_\lambda = \{\text{holomorphic } F : GL(n, \mathbb{C}) \to \mathbb{C} \mid F(gb) = \xi^{\mathbb{C}}_{-\lambda_n\epsilon_1\cdots-\lambda_1\epsilon_n}(b)F(g),$$
$$g \in GL(n, \mathbb{C}), b \in B\}$$

with action given by left translation of functions, i.e., $(g_1F)(g_2) = F(g_1^{-1}g_2)$.
(3) Let $F_{w_0\lambda} : GL(n, \mathbb{C}) \to \mathbb{C}$ be given by

$$F_{w_0\lambda}(g) = (\det_1 g)^{\lambda_{n-1}-\lambda_n}\cdots(\det_{n-1} g)^{\lambda_1-\lambda_2}(\det_n g)^{-\lambda_1},$$

where $\det_k(g_{i,j}) = \det_{i,j \le k}(g_{i,j})$. Show that $F_{w_0\lambda}$ is holomorphic, invariant under right translation by N, and invariant under left translation by N^t.

(4) Show that $F_{w_0\lambda} \in V_\lambda$ and show $F_{w_0\lambda}$ has weight $\lambda_n \epsilon_1 + \cdots + \lambda_1 \epsilon_n$. Conclude that $F_{w_0\lambda}$ is the lowest weight vector of V_λ, i.e., that $F_{w_0\lambda}$ is the highest weight vector for the positive system corresponding to the opposite Weyl chamber.

(5) Let $F_\lambda(g) = F_{w_0\lambda}(w_0 g)$, where $w_0 = E_{1,n} + E_{2,n-2} + \ldots, E_{n,1}$. Write F_λ in terms of determinants of submatrices and show F_λ is a highest weight for V_λ.

Exercise 7.35 Let G be a compact Lie group. Show G is *algebraic* by proving the following:

(1) Suppose G acts on a vector space V and $\mathcal{O}$ and $\mathcal{O}'$ are two distinct orbits. Show there is a continuous function f on V that is 1 on $\mathcal{O}$ and -1 on $\mathcal{O}'$.

(2) Show there is a polynomial p on V, so that $|p(x) - f(x)| < 1$ for $x \in \mathcal{O} \cup \mathcal{O}'$. Conclude that $p(x) > 0$ when $x \in \mathcal{O}$ and $p(x) < 0$ when $x \in \mathcal{O}'$.

(3) Let $\mathcal{P}$ be the convex set of all polynomials p on V satisfying $p(x) > 0$ when $x \in \mathcal{O}$ and $p(x) < 0$ when $x \in \mathcal{O}'$. With respect to the usual action, $(g \cdot p)(x) = p(g^{-1}x)$ for $g \in G$, use integration to show that there exists $p \in \mathcal{P}$ that is G-invariant.

(4) Show that G-invariant polynomials on V are constant on G-orbits.

(5) Let $\mathcal{I}$ be the ideal of all G-invariant polynomials on V that vanish on $\mathcal{O}$. Show that there is $p \in \mathcal{I}$, so that p is nonzero on $\mathcal{O}'$. Conclude that the set of zeros of $\mathcal{I}$ is exactly $\mathcal{O}$.

(6) By choosing a faithful representation, assume $G \subseteq GL(n, \mathbb{C})$ and consider the special case of $V = M_{n,n}(\mathbb{C})$ with G-action given by left multiplication of matrices. Show that G is itself an orbit in V and is therefore algebraic.

References

1. Adams, J., *Lectures on Lie Groups*, W. A. Benjamin, New York, 1969.
2. Artin, E., *Geometric Algebra*, Interscience, New York, 1957.
3. Artin, M., *Algebra*, Prentice-Hall, Englewood Cliffs, N.J., 1991.
4. Authur, J., Characters, harmonic analysis, and an L^2-Lefschetz formula, *The Mathematical Heritage of Hermann Weyl* (Durham, NC, 1987), 167-179, *Proc. Sympos. Pure Math.*, 48, *Amer. Math. Soc.*, Providence, RI, 1988.
5. Baker, B., Alternants and continuous groups, *Proc. London Math. Soc.*, 3:24-47, 1905.
6. Bekka, M. and de la Harpe, P., Irreducibility of unitary group representations and reproducing kernels Hilbert spaces; Appendix by the authors in collaboration with R. Grigorchuk, *Expo. Math.* 21(2):115-149, 2003.
7. Berstein, I., Gelfand, I., and Gelfand, S., Structure of representations generated by highest weight vectors, *Funktsionalnyi Analiz i Ego Prilozheniya 5*, 1:1-9, 1971.
8. Boothby, W., *An Introduction to Differentiable Manifolds and Riemannian Geometry*, second edition, Pure and Applied Mathematics, 120, Academic Press, Inc., Orlando, FL, 1986.
9. Borel, A., Hermann Weyl and Lie groups, in *Hermann Weyl, 1885-1985*, Centenary lectures (Zürich 1986), 53-82, ed. K. Chandrasekharan, Springer Verlag, Berlin-Heidelberg-New York, 1986.
10. Borel, A., *Linear Algebraic Groups*, W. A. Benjamin, New York, 1987.
11. Borel, A., Sous groupes commutatifs et torsion des groupes de Lie compacts connexes, *Tohoku Math. J.*, 13:216-240, 1961.
12. Borel, A., and Serre, J-P., Sur certains sous-groupes des groupes de Lie compacts, *Comment. Math. Helv.*, 27:128-139, 1953.
13. Bott, R., Homogeneous vector bundles, *Annals of Math.*, 66:203-248, 1957.
14. Bourbaki, N., *Eléments de Mathématique, Groupes et Algèbres de Lie: Chapitre 1*, Actualités scientifiques et industreielles 1285, Hermann, Paris, 1960.
15. Bourbaki, N., *Eléments de Mathématique, Groupes et Algèbres de Lie: Chapitres 4, 5, et 6*, Actualités scientifiques et industreielles 1337, Hermann, Paris, 1968.
16. Bourbaki, N., *Eléments de Mathématique, Groupes et Algèbres de Lie: Chapitre II et III*, Hermann, Paris, 1972.
17. Bredon, G., *Introduction to Compact Transformation Groups*, Academic Press, New York London, 1972.
18. Bremmer, M., Moody, R., and Patera, J., *Tables of Cominant Weight Multiplicities for Representations of Simple Lie Algebras*, Marcel Dekker, New York, 1985.

19. Bröcker, T., and tom Dieck, T., *Representations of Compact Lie Groups*, translated from the German manuscript, corrected reprint of the 1985 translation, Graduate Texts in Mathematics, 98, Springer-Verlag, New York, 1995.

20. Burnside, W., On the representation of a group of finite order as an irreduible group of linear substitutions and the direct establishment of the relations between the group characters, *Proc. London Math. Soc.*, 1:117-123, 1904.

21. Campbell, J., On the law of combination of operators bearin on the theory of continuous transformation groups, *Proc. London Math. Soc.*, 28: 381-190, 1897.

22. Cartan, É., La géométries des groupes simples, *Ann. Math. Pura Appl.*, 4:209-256, 1927.

23. Cartan, É., Les groupes projectifs qui ne laissent invariante aucune multiplicité plane, *Bull. Soc. Math. France*, 41:53-96, 1913.

24. Cartan, É., Les groupes réels simples finis et continus, *Annales Sci. École Norm. Sup.*, 31:263-255, 1914.

25. Cartan, É., Les tenseurs irréductibles et les groupes linéaires simples et semis-simples, *Bull. des Sci. Math.*, 49:130-152, 1925.

26. Cartan, É., Sur la structure des groupes de transformations finis et continus, *Thèse*, Nony, Paris, 1894. Second edition: Vuibert, 1933.

27. Cartan, É., Sur certaines formes riemanniennes remarquables des géométries à groupe fondamental simple, *Annales Sci. École Norm. Sup.*, 44:345-467, 1927.

28. Chevalley, C., Sur la classification des algèbres de Lie simples et de leur représentations, *C. R. Acad. Sci. Paris*, 227, 1136-1138, 1948.

29. Chevalley, C., *Theory of Lie Groups I*, Princeton University Press, Princeton, 1946.

30. Conway, J., *A Course in Functional Analysis*, second edition, Graduate Texts in Mathematics, 96, Springer-Verlag, New York, 1990.

31. Coxeter, H., Discrete groups generated by reflections, *Annals of Math.*, 35:588-621, 1934.

32. Curtis, C., and Reiner, I., *Methods of Representation Theory*, Volumes I and II, John Wiley & Sons, New York Chichester Brisbane Toronto, 1981/87.

33. Dixmier, J., *Algèbres Enveloppantes*, Gauthier-Villars Editeur, Paris, 1974.

34. Duistermaat, J., and Kolk, J, *Lie Groups*, Universitext. Springer-Verlag, Berlin, 2000.

35. Dynkin, E., Calculation of the coefficients of the Campbell-Hausdorff formula, *Dokl. Akad. Nauk.*, 57:323-326, 1947.

36. Dynkin, E., Classification of the simple Lie groups, *Mat. Sbornik*, 18(60):347-352, 1946.

37. Folland, G., *Real Analysis. Modern Techniques and Their Applications*, second edition, Pure and Applied Mathematics (New York), A Wiley-Interscience Publication, John Wiley & Sons, Inc., New York, 1999.

38. Freudenthal, H., and de Vries, H., *Linear Lie Groups*, Academic Press, New York London, 1969.

39. Frobenius, F., Über lineare Substitutionen und bilineare Formen, *J. reine angew. Math.*, 84:1-63, 1878.

40. Fulton, W., and Harris, J., *Representation Theory. A first course*, Graduate Texts in Mathematics, 129, Readings in Mathematics, Springer-Verlag, New York, 1991.

41. Goodman, R., and Wallach, N., *Representations and Invariants of the Classical Groups*, Encyclopedia of Mathematics and its Applications, 68, Cambridge University Press, Cambridge, 1998.

42. Guillemin and Pollack, *Differential Topology*, Prentice-Hall, Englewood Cliffs, 1974.

43. Haar, A., Der Massbegriff in der Theorie der kontinuierlichen Gruppen, *Annals of Math.*, 34:147-169, 1933.

44. Hall, B., *Lie Groups, Lie Algebras, and Representations: An Elementary Introduction*, Graduate Texts in Mathematics, 222, Springer-Verlag, New York, 2003.

45. Harish-Chandra, Characters of semi-simple Lie groups, *Symposia on Theoretical Physics*, 4:137-142, 1967.

46. Harish-Chandra, Harmonic analysis on real reductive groups I, *J. Func. Anal.*, 19:104-204, 1975.

47. Harish-Chandra, On a lemma of F. Bruhat, *J. Math Pures Appl.*, 35(9):203-210, 1956.

48. Harish-Chandra, On some appliations of the universal enveloping algebra of a semi-simple Lie algebra, *Trans. Amer. Math. Soc.*, 70:28-95, 1951.

49. Hausdorff, F., Die symbolische exponentialformel in der gruppentheorie, *Leibzier Ber.*, 58:19-48, 1906.

50. Helgason, S., *Differential Geometry, Lie Groups, and Symmetric Spaces*, corrected reprint of the 1978 original, Graduate Studies in Mathematics, 34, American Mathematical Society, Providence, RI, 2001.

51. Hocking, J., and Young, G., *Topology*, second edition, Dover Publications, Inc., New York, 1988.

52. Hofmann, K., and Morris, S., *The Structure of Compact Groups. A Primer for the Student—a Handbook for the Expert*, de Gruyter Studies in Mathematics, 25, Walter de Gruyter & Co., Berlin, 1998.

53. Hopf, H., Über den Rang geschlossener Liescher Gruppen, *Comment. Math. Helv.*, 13:119-143, 1940-41.

54. Hopf, H., Maximale Toroide unde singuläre Elemente in geschlossenen Lieschen Gruppen, *Comment. Math. Helv.*, 15:59-70, 1942-43.

55. Howe, R., and Tan, E., *Non-abelian Harmonic Analysis. Applications of $SL(2, \mathbb{R})$*, Universitext, Springer-Verlag, New York, 1992.

56. Humphreys, J., *Introduction to Lie Algebras and Representation Theory*, second printing, revised, Graduate Texts in Mathematics, 9, Springer-Verlag, New York-Berlin, 1978.

57. Humphreys, J., *Reflection Groups and Coxeter Groups*, Cambridge University Press, Cambridge, 1990.

58. Hurwitz, A., Über die Erzeugung der Invarianten durch Intergration, *Nachrichten von der Königlichen Gesellschaft der Wissenschaften zu Göttingen*, 71-90, 1897.

59. Jacobson, N., *Lie Algebras*, Interscience Publishers, New York London Sydney, 1962.

60. Killing, W., Die Zusammensetzung der stetigen endlichen Transformationsgruppen, I, II, III, IV, *Math. Ann.*, 31:252-290, 1888; 33:1-48, 1889; 34:57-122, 1889; 36:161-189, 1890.

61. Knapp, A., *Lie Groups Beyond an Introduction*, second edition, Progress in Mathematics, 140, Birkhäuser Boston, Inc., Boston, MA, 2002.

62. Knapp, A., *Representation Theory of Semisimple Groups. An Overview Based on Examples*, reprint of the 1986 original, Princeton Landmarks in Mathematics, Princeton University Press, Princeton, NJ, 2001.

63. Kostant, B., A formula for the multiplicity of a weight, *Trans. Amer. Math. Soc.*, 93:53-73, 1959.

64. Kostant, B., Lie algebra cohomology and the generalized Borel-Weil theorem, *Annal of Math.*, 74:329-387, 1961.

65. Loos, O., *Symmetric Spaces, II: Compact Spaces and Classification*, W. A. Benjamin, New York Amsterdam, 1969.

66. Macdonald, I., The volume of a compact Lie group, *Invent. Math.*, 56:93-95, 1980.

67. McKay, W., and Patera, J., *Tables of Dimensions, Indices, and Branching Rules for Representations of Simple Lie Algebras*, Marcell Dekker, New York, 1981.

68. Montgomery, D., and Zippin, L., *Topological Transformation Groups*, Interscience, New York, 1955.

69. Munkres, J., *Topology: a First Course*, Prentice-Hall, Inc., Englewood Cliffs, N.J., 1975.

70. Onishchik, A., and Vinberg, E., *Lie Groups and Algebraic Groups*, translated from the Russian and with a preface by D. A. Leites, Springer Series in Soviet Mathematics, Springer-Verlag, Berlin, 1990.

71. Peter, F., and Weyl, H., Die Vollständigkeit der primitiven Darstellungen einer geschlossenen kontinuierlichen Gruppe, *Math. Ann.*, 97:737-755, 1927.

72. Rossmann, W., *Lie Groups. An Introduction Through Linear Groups*, Oxford Graduate Texts in Mathematics, 5, Oxford University Press, Oxford, 2002.

73. Rudin, W, *Real and complex analysis*, third edition, McGraw-Hill Book Co., New York, 1987.

74. Rudin, W, *Functional Analysis*, Second edition. International Series in Pure and Applied Mathematics. McGraw-Hill, Inc., New York, 1991.

75. Schur, I., Neue Anwendung der Integralrechnung and Probleme der Invariententheorie, *Sitzungsberichte der Königlich Preussichen Akademie der Wissenschaften*, 406-423, 1905.

76. Serre, J.P., Représentations linéaires et espaces homogènes Kählérians des groupes de Lie compacts, Exposé 100, *Séminaire Bourbaki, 6ᵉ année 1953/54*, Inst. Henri Poincaré, Paris 1954.

77. Serre, J.P., Tores maximaux des groupes de Lie compacts, Exposé 23, Séminaire "Sophus Lie," *Théorie des algèbres de Lie topologie des groupes de Lie*, 1954-55, École Normale Supérieure, Paris, 1955.

78. de Siebenthal, J., Sur les groupes de Lie compacts non-connexes, *Comment. Math. Helv.*, 31:41-89, 1956.

79. Springer, T., *Linear Algebraic Groups*, Birkhäuser, Boston, 1981.

80. Springer, T., and Steinberg, R., Conjugacy classes, *Seminar on Algebraic Groups and Related Finite Groups*, Lecture Notes in Math., 131:167-266, 1970.

81. Steinberg, R., *Endomorphisms of Linear Algebraic Groups*, Mem. Amer. Math. Soc., 80, Amer. Math. Soc., Providence., 1968.

82. Steinberg, R., A general Clebsch-Gordan theorem, *Bull. Amer. Math. Soc.*, 67:406-407, 1961.

83. Tits, J., Sur certaines classes d'espaces homogènes de groupes de Lie, Acad. Roy. Belg. Cl. Sci. Mém. Coll. 29, No. 3, 1955.

84. Varadarajan, V., *An Introduction to Harmonic Analysis on Semisimple Lie Groups*, Cambridge University Press, Cambridge, 1989.

85. Varadarajan, V., *Lie Groups, Lie Algebras, and Their Representations*, reprint of the 1974 edition, Graduate Texts in Mathematics, 102, Springer-Verlag, New York, 1984.

86. Vogan, D., *Representations of Real Reductive Lie Groups*, Birkhäuser, Boston-Basel-Stuttgart, 1981.

87. Vogan, D., *Unitary Representations of Reductive Lie Groups*, Princeton University Press, Princeton, 1987.

88. Warner, F., *Foundations of Differentiable Manifolds and Lie Groups*, corrected reprint of the 1971 edition, Graduate Texts in Mathematics, 94, Springer-Verlag, New York-Berlin, 1983.

89. Warner, G., *Harmonic Analysis on Semi-simple Lie Groups. I*, Die Grundlehren der mathematischen Wissenschaften, Band 188, Springer-Verlag, New York-Heidelberg, 1972.

90. Warner, G., *Harmonic Analysis on Semi-simple Lie Groups. II*, Die Grundlehren der mathematischen Wissenschaften, Band 189. Springer-Verlag, New York-Heidelberg, 1972.

91. Weil, A., Démonstration topologique d'un théorème fondamental de Cartan, *C. R. Acad. Sci. Paris*, 200:518-520, 1935.

92. Weil, A., *L'intégration dans les groupes topologiques et ses applications*, Actualités scientifiques et industrielles 869, Hermann, Paris, 1940.

93. Wells, R., *Differential Analysis on Complex Manifolds*, second edition, Graduate Texts in Mathematics, 65, Springer-Verlag, New York-Berlin, 1980.

94. Weyl, H., *The Classical Groups, Their Invariants and Representations*, 2nd ed., Princeton University Press, Princeton, 1946.

95. Weyl, H., Theorie der Darstellung kontinuierlicher halbeinfacher Gruppen durch lineare Transformationen I, II, II, and Nachtrag, *Math. Zeitschrift* 23 (1925), 271-309; 24 (1926), 238-376; 24 (1926), 277-395; 24 (1926), 789-791.

96. Zelobenko, D., *Compact Lie Groups and Their Representations*, Translations of Mathematical Monographs, Vol. 40, American Mathematical Society, Providence, R.I., 1973.

97. Zierau, R., Representations in Dolbeault Cohomology in *Representation Theory of Lie Groups*, Park City Math Institute, vol. 8, AMS, Providence RI, 2000.

Index